Microwave Engineering of Materials

Microwave Engineering of Materials

From Mesoscale to Nanoscale

edited by

Erwann Guénin

Pan Stanford Publishing

Published by

Pan Stanford Publishing Pte. Ltd.
Penthouse Level, Suntec Tower 3
8 Temasek Boulevard
Singapore 038988

Email: editorial@panstanford.com
Web: www.panstanford.com

British Library Cataloguing-in-Publication Data
A catalogue record for this book is available from the British Library.

Microwave Engineering of Materials: From Mesoscale to Nanoscale

ISBN 978-981-4669-42-9 (Hardcover)
ISBN 978-981-4669-43-6 (eBook)

Printed in the USA

Contents

Foreword xiii

1 Introduction to Microwave Chemistry **1**
Dariusz Bogdal and Mateusz Galica
1.1 Nature of Microwave Irradiation 1
1.2 Microwave Thermal Effect vs. Nonthermal Effect 8
1.2.1 Nonthermal Effects 10
1.2.2 Thermal Effects 14
1.2.3 Thermal Gradients 15
1.3 Conclusion 25

2 General Features of MW Interaction with Materials **31**
S. Takayama, M. Sato, and J. Fukushima
2.1 Introduction to General Features of MW Interaction with Materials: Microwave, Heat, and Materials 31
2.1.1 Maxwell's Equations 32
2.1.2 Waves in Materials and the Dispersion Relation 34
2.1.3 Energy Transfers from Microwaves to Materials 36
2.1.4 Effect on Magnetization 37
2.1.5 Internal Heating, Volumetric Heating, and Rapid Heating 38
2.1.6 Selective Heating 39
2.2 Nonthermal Microwave Effects 39
2.2.1 Decrystallization and Formation of Nanodomain Structures 39
2.2.2 Enhancement of Nonthermal Reduction of CuO and TiO_{2-x} during Microwave Heating 43
2.2.3 Sintering Behavior of Metal Powders Involving Microwave-Enhanced Chemical Reaction 48

2.3 Energy Transfer Mechanism on Microwave Processing 55
2.3.1 Introduction 55
2.3.2 Chemical Reactions and Phase Transitions under Microwave Irradiation 57
2.3.3 Excitation of Ultrasound Waves by External Microwaves 58
2.3.4 Collisionless Energy Transfers from Electrokinetic Waves to Materials 61
2.3.5 Rate Equations with Perturbations by Microwave in Velocity Space 63
2.3.6 Energy Balance Estimations 65
2.3.7 Magnetic Effects 69
2.4 Conclusion 70

3 Microwave-Assisted Synthesis and Modification of Polymers 73
Martin Fimberger and Frank Wiesbrock
3.1 Introduction 73
3.2 Controlled Radical Polymerization 74
3.2.1 Reversible Addition–Fragmentation Chain Transfer 75
3.2.2 Atom Transfer Radical Polymerization 78
3.2.3 Emulsion Polymerization 79
3.3 Step-Growth Polymerizations 81
3.4 Ring-Opening Polymerizations 84
3.4.1 Poly(ε-Caprolactone)s, Poly(Glycolic Acid), and Polylactides 84
3.4.1.1 Homopolymers of ε-caprolactone 85
3.4.1.2 Copolymers of ε-caprolactone 87
3.4.1.3 Copolymers of poly(glycolic acid) 88
3.4.1.4 Homopolymers of lactic acid 89
3.4.2 Poly(2-oxazoline)s and Poly(2-oxazine)s 91
3.5 Highlights 98
3.5.1 Ionic Liquids and Solvent-Free Polymerizations 98
3.5.2 Click Chemistry 102
3.6 Scale-Up 104
3.7 Conclusion 105
3.7.1 Safety Rules 106

4 Microwave Processing of Ceramics and Glasses **115**
D. C. Folz, D. E. Clark, and R. L. Schulz
4.1 Introduction 116
4.1.1 The Ceramics Industry 116
4.1.2 Brief Review of Microwave–Ceramic Interactions 117
4.1.2.1 Methods of heating 121
4.1.2.2 Benefits of microwave processing 123
4.2 Low-Temperature Microwave Applications: Drying 124
4.3 Moderate-Temperature Applications: Glass Nucleation and Crystallization 127
4.4 High-Temperature Applications: Sintering 130
4.5 Summary 135

5 Microwave Processing of Composites, Glass-Ceramic Coatings and Metallic Alloys: An Overview **141**
Sumana Ghosh
5.1 Introduction 141
5.2 Microwave Processing of Composites 142
5.2.1 Microwave Processing of Glass-Ceramic Coatings 149
5.3 Microwave Processing of Metallic Alloys 151
5.4 Conclusion 156

6 Microwave Engineering for Synthesizing Clays and Modifying Properties in Zeolites **163**
Olga Bergadà, Fiseha B. Gebretsadik, M. Dolores González, Judith Granados-Reyes, Elena Pérez, Tatiana Sánchez, Isabel Vicente, Pilar Salagre, and Yolanda Cesteros
6.1 Introduction 164
6.2 Synthesis of Clays Using Microwaves 167
6.2.1 Cationic Clays: Hectorites and Saponites 168
6.2.1.1 Hectorites 168
6.2.1.2 Saponites 171
6.2.2 Anionic Clays: Hydrotalcites and Hydrocalumites 174
6.3 Microwave-Assisted Modification of Zeolites 175

6.3.1 Effect of Microwaves on the Acidic Properties of Ni-Exchanged Mordenites 176
6.3.2 Modification of Surface and Acidic Properties during Dealumination of Zeolites by Acid Treatment With Microwaves 179
6.3.3 Microwave-Assisted Sulfonic Acid Functionalization of Zeolites by Postsynthesis 182
6.4 Future Prospects 186

7 Microwave Engineering of Carbon Nanotubes 195
Madhulika Sinha and Yong-Chien Ling
7.1 Introduction 195
7.2 Microwave Chemistry 197
7.3 General Synthesis of CNTs 199
7.3.1 Properties of CNTs 199
7.3.2 Arc Discharge Method 200
7.3.3 Laser Ablation Method 201
7.3.4 Chemical Vapor Deposition Method 202
7.4 Microwave-Assisted Synthesis of CNTs 202
7.4.1 CNT Applications 202
7.4.2 Interaction Mechanisms between CNTs and MW 203
7.4.3 Effect of Modifiers 204
7.4.4 MW-Assisted Purification of CNTs 205
7.5 Covalent and Noncovalent Functionalization of CNTs 210
7.5.1 MW-Assisted Synthesis of CNTs-Based Nanocomposites 210
7.5.2 MW-Assisted Synthesis of CNTs-Based Hybrids 212
7.6 Summary and Future Scope of Microwave-Assisted Synthesis of CNTs 217

8 Microwave Synthesis Porous Zeolitic Metal–Organic Framework Materials 229
Hossein Kazemian and Sohrab Rohani
8.1 Introduction 229
8.2 Metal–Organic Frameworks 230
8.3 Applications of MOFs 236
8.4 Microwaves for Chemical Synthesis 239

8.5 MOF Synthetic Approaches 244
8.6 MOF Synthesis Using the Microwave and Ultrasonic Hybrid Technique 249
8.7 Conclusions 257

9 Microwave-Assisted Synthesis of Metallic Nanoparticles 263
Stephany García, Graham W. Piburn, and Simon M. Humphrey
9.1 Introduction 263
9.2 Equipment 267
9.3 Common Reagents and Their Compatibility with Microwave Heating 270
9.4 Direct Microwave-Assisted Synthesis of Unsupported MNPs 273
9.4.1 Monometallic MNPs 273
9.4.2 Heterobimetallic MNPs 278
9.4.2.1 Core–shell MNPs 279
9.4.2.2 Alloy MNPs 281
9.4.3 Ex situ Supported MNP Catalysts 285
9.4.4 In situ Microwave-Assisted Synthesis of MNPs Inside Support Media 286

10 Microwave-Assisted Synthesis of Semiconductor Nanomaterials for Energy Conversion 305
Xinjuan Liu, Likun Pan, and Chang Q. Sun
10.1 Introduction 305
10.2 Microwave-Assisted Synthesis of Semiconductor Nanomaterials for Photocatalysis 307
10.2.1 Photocatalysts 309
10.2.1.1 Semiconductor oxide 309
10.2.1.2 Metal tungstate 316
10.2.1.3 Metal vanadates 320
10.2.1.4 Metal molybdates 324
10.2.1.5 Perovskite-type materials 325
10.2.1.6 Hydroxy salts 328
10.2.1.7 Bismuth oxyhalides 331
10.2.1.8 Other semiconductors 334
10.2.2 Summary 336

10.3 Microwave-Assisted Synthesis of Semiconductor Nanomaterials for Sensitized Solar Cells 337
10.3.1 QD-Sensitized Solar Cells 337
10.3.1.1 Semiconductor sulfides 339
10.3.1.2 Semiconductor selenide 340
10.3.2 Dye-Sensitized Solar Cells 342
10.3.2.1 Semiconductor oxide 342
10.3.3 Summary 347
10.4 Summary and Outlook 347

11 Microwave Nanosurface Engineering **369**
S. Richard, R. Aufaure, E. Nehlig, L. Motte, and E. Guénin
11.1 Introduction 369
11.2 Microwave Fundamentals 371
11.3 Microwave Surface Modification of Nanomaterials 372
11.3.1 MW Modification of Porous Silicon-Based Nanomaterials 373
11.3.1.1 MW functionalization of porous silicon surfaces 374
11.3.1.2 MW functionalization of nanosilica surfaces 377
11.3.2 MW Functionalization of Carbon-Based Nanomaterials 382
11.3.2.1 MW functionalization of single-walled and multiwalled carbon nanotubes 382
11.3.2.2 MW synthesis of composites with CNT and formation of inorganic/organic nanohybrids 387
11.3.2.3 MW functionalization of other carbon nanomaterials 388
11.3.3 MW Functionalization of Inorganic Nanoparticles 389
11.3.3.1 MW formation of SiO_2 or TiO_2 shell on inorganic nanoparticles 389
11.3.3.2 MW formation of a metallic shell on inorganic nanoparticles 393

11.3.3.3 MW formation of an organic polymer shell on inorganic nanoparticles 396
11.3.3.4 MW functionalization of small molecules at the inorganic nanoparticle surface 398
11.3.4 MW Functionalization of the Plasmonic Nanostructure 402
11.4 Conclusion 403

Index 421

Foreword

I first came upon reports of microwave-assisted reactions while I was doing my PhD in the 1990s. At that time microwaves were not developed as they are now in chemistry laboratories. I remember well the reaction of most of the senior researchers to my questioning on microwave synthesis. I am not ashamed to say that at that time in France scientists using this methodology were not considered as pioneers but more as charlatans. Hopefully and in great part due to the priceless work of people such as André Loupy, the scientific community vision evolved rapidly. Nevertheless there are still unbelievers, often among the older researchers, and I remember that a few years ago, while defending my project "Habilitation à Diriger les Recherches" (a French diploma needed to supervise my PhD), I had an animated discussion over microwaves. I was told by one member of my jury that studying reactions done under microwave irradiation was not a proper research project as microwaves were only a different heating methodology. My response was simple: "Yes, it is a different source of heating methodology as horse riding and high-speed trains are two different transportation methodologies, but to travel from one point of the planet to another would you really ride a horse? And was the development of high-speed trains not due to scientific research?" Note that having unbelievers among the scientific community is important because it obliges us to be more rigorous to describe phenomena and improve the quality of our research.

Since I am a senior scientist working first in organic chemistry and more recently in surface science, I have tried to test microwave solutions for my projects, whenever it was possible, and it was often conclusive. It rapidly became for me an essential tool, and I even managed to convert some colleagues to microwaves. More recently, when I started to work in nanosciences, I discovered that microwaves were was used in various areas of materials science, not only for the synthesis of materials and nanomaterials, but also for their structure and surface modifications. This is what prompted me to edit this book. The idea was to describe the use of microwave irradiation for the engineering of materials and nanomaterials. This covers an area of science at the crossroads of many disciplines: organic chemistry, polymer chemistry, materials science, surface science, nanoscience, etc. That is why I tried to gather the best scientists in their areas for them to explain the improvement brought by microwaves in their disciplines. The first two chapters consist of an introduction to microwaves from two different points of view. In Chapter 1, Bogdal and Galica introduce microwave chemistry from the organic chemistry point of view. Then Takayama, Sato, and Fukushima, in Chapter 2, give the general features of microwave interaction with materials. In the following chapters the authors show the application of microwaves for engineering of various materials and nanomaterials. In Chapter 3, Fimberger and Wiesbrock review the proceedings of microwave-supported polymerizations and modifications of polymers. In Chapter 4, Folz, Clark, and Schulz present microwave processing of ceramics and glasses. Ghosh then gives further insight into the microwave processing of composites, glass-ceramic coatings, and metallic alloys in Chapter 5. Continuing with the processing of materials Salagre and Cesteros's group reviews microwave-assisted synthesis of clay and the microwave modifications of zeolites. In Chapter 7 Sinha and Ling propose a literature survey of microwave-based methods for synthesis of carbon nanotubes. In Chapter 8 Rohani and Kazemian focus on another nanomaterial—metallic organic frameworks—and describe the use of microwaves for their synthesis and functionalization. The next two chapters describe the use of microwaves for nanoparticle synthesis. García, Piburn, and Humphrey treat the synthesis of metallic nanoparticles in Chapter 9. Liu, Pan, and Sun present the synthesis of semiconductor nanomaterials in Chapter 10. Finally in the last chapter Motte, our students, and I describe the use of microwaves for the surface modification of various types of nanomaterials.

I hope that this book will be of some help for young chemists and other scientists wanting to learn more about microwave engineering of materials and nanomaterials. And I finally hope that some of the unbelievers will be converted after reading this cooperative work.

Erwann Guénin

Chapter 1

Introduction to Microwave Chemistry

Dariusz Bogdal and Mateusz Galica
Faculty of Chemical Engineering and Technology, Cracow University of Technology, Warszawska 24, Krakow 31-155, Poland
pcbogdal@cyf-kr.edu.pl

Microwave irradiation is a rapid means of material heating for domestic, industrial, and medical purposes. Microwaves offer a number of advantages over conventional heating, such as noncontact heating (reduction of overheating of material surfaces), energy transfer instead of heat transfer (penetrative radiation), material-selective and volumetric heating, fast start-up and stopping, and, last but not least, the reverse thermal effect. Moreover, the reduced time of processing under microwave conditions found for a great number of chemical reactions was the main reason that microwave techniques became so attractive for chemists, who, in the last three decades, have begun to apply this technique as a routine in their everyday practice [1].

1.1 Nature of Microwave Irradiation

Microwaves are the part of the electromagnetic spectrum with frequencies of 300 GHz to 300 MHz that corresponds to wavelengths

Microwave Engineering of Materials: From Mesoscale to Nanoscale
Edited by Erwann Guénin

ISBN 978-981-4669-42-9 (Hardcover), 978-981-4669-43-6 (eBook)
www.panstanford.com

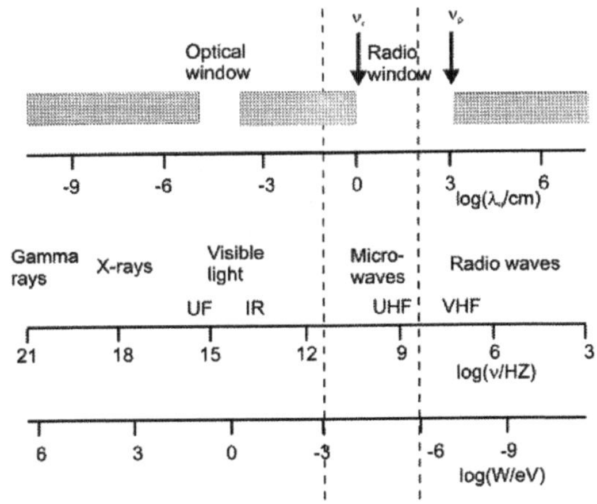

Figure 1.1 Spectrum of electromagnetic radiation: ν_r is the lowest resonance frequency in the rotational spectrum of water, and ν_p is the plasma frequency of the ionosphere.

of 1 mm to 1 m, respectively (Fig. 1.1). Depending on whether they are used for transmission of information (telecommunication) or transmission of energy, the major applications fall into these two categories. In fact, the application of microwaves in telecommunication (mobile phones, radar, and radar line transmissions) caused that only specially assigned frequencies are allocated for energy transmission, that is, for industrial, scientific, or medical applications. For example, the most common microwave applicators (i.e., domestic microwave ovens) utilize the frequency of 2.45 GHz, which corresponds to the wavelength of 12.2 cm. For this reason, commercially available microwave reactors devoted for chemical use operate at the same frequency; however, some other frequencies are also available for heating [2].

To apply microwaves to carry out chemical processes, it is most important to find at least one reagent component that is polarizable and whose dipoles can reorient (couple) rapidly in response to the changing electric field of microwave radiation. In fact, the electric field component of microwave radiation is responsible for dielectric heating mechanisms since it can cause molecular motion either by migration of ionic species (conduction mechanism) or by rotation of dipolar species (dipolar polarization mechanism). At 2.45 GHz the

Table 1.1 Dielectric constants (ε_r) of some common materials at 20°C

Material	Dielectric constant (ε_r)	Material	Dielectric constant (ε_r)
Vacuum	1	Titanium dioxide	100
Air (1 atm)	1.00059	Water	80
Air (100 atm)	1.0548	Acetonitrile	38
Glass	5–10	Liquid ammonia (−78°C)	5
Quartz glass	5	Ethyl alcohol	25
Porcelain	5–6	Benzene	2
Mica	3–6	Carbon tetrachloride	2
Rubber	2–4	Hexane	2
Nylon	3–22	Plexiglass	3
Paper	1–3	Polyvinyl chloride	3
Paraffin	2–3	Polyethylene	2
Soil (dry)	2.5–3	Teflon	2
Wood (dry)	1–3	Polystyrene (foam)	1.05

field oscillates 4.9×10^9 times per second, and the strong agitation, provided by cyclic reorientation of molecules, can result in intense internal heating, which can lead to heating rates in excess of 10°C/s when microwave radiation of a kilowatt-capacity source is used [3].

Fortunately, a number of organic compounds and solvents fulfill these requirements and are the best candidates for microwave applications. These materials are characterized by dielectric properties, among which the dielectric constant ε_r, sometimes called electric permeability, is of the greatest importance. The dielectric constant (ε_r) is defined as the ratio of the electric permeability of the material to the electric permeability of free space (i.e., vacuum). Dielectric constants for some common materials are given in Table 1.1.

Nonpolar organic solvents (i.e., benzene, carbon tetrachloride, *n*-hexane) have a low value of ε_r and, in fact, show negligible heating effects under microwave irradiation. Polar organic solvents (i.e., water, acetonitrile, alcohols) are characterized by relatively high values of ε_r and, in turn, can be heated by the dielectric heating mechanism under microwave irradiation. Most plastics range in the low values of ε_r (i.e., between 2 and 3); therefore some of these materials besides glass and quartz glass are used to manufacture reaction vessels for microwave application due to their good chemical as well as temperature resistance (e.g.,

Teflon, polyether ether ketone [PEEK]). Thus, heating materials in microwave ovens is based upon the ability of some liquids as well as solids to absorb and transform electromagnetic energy into heat. However, in the case of highly oscillating electric fields, the dielectric constant (ε_r) is turned into complex permeability $\varepsilon^* = \varepsilon' - j\varepsilon''$. The real permeability (ε') component characterizes the ability of the material to be polarized by the electric field and thus the ability of microwaves to propagate into the material. The imaginary part of the complex electric permeability (ε'') is usually called the loss factor and indicates the ability of the material to dissipate the energy, that is, the efficiency of conversion of electromagnetic radiation into heat. Loss tangent ($\tan\delta$), which is calculated as the ratio of ε'' and ε' ($\tan\delta = \varepsilon''/\varepsilon'$), is another commonly used parameters that characterizes heating of materials under microwave irradiation and even better describes the ability of a material to absorb microwave energy.

The loss tangent of water as a function of frequency is presented in Fig. 1.2, together with ε' and ε'', while $\tan\delta$ of some other common materials is presented in Table 1.2. It is apparent that appreciable values of $\tan\delta$ exist over a wide frequency range. For instance, for water, the most effective heating, as measured by $\tan\delta$, reaches its maximum at ca. 80 GHz, while most microwave ovens operate at a much lower frequency, that is, 2.45 GHz (Fig. 1.2). The practical

Table 1.2 Loss tangents ($\tan\delta$) of different materials at 25°C (2.45 GHz)

Solvent	Loss tangent ($\tan\delta$)	Material	Loss tangent ($\tan\delta$)
Water	0.12	Fused quartz	0.00006
Ice	0.0009	Porcelain	0.001
Ethanol	0.94	Borosilicate glass	0.001
Butanol	0.57	Phosphate glass	0.0046
Acetic acid	0.17	Teflon	0.00015
DMF	0.16	Polyethylene	0.0003
Acetonitrile	0.062	Polystyrene	0.0003
Acetone	0.054	Polycarbonate	0.0006
THF	0.047	Plexiglass	0.005
Methylene chloride	0.042	Polyamide	0.005
Hexane	0.021	Polyvinylchloride	0.005
Vaseline	~0.0007	ABS (plastics)	0.006–0.019

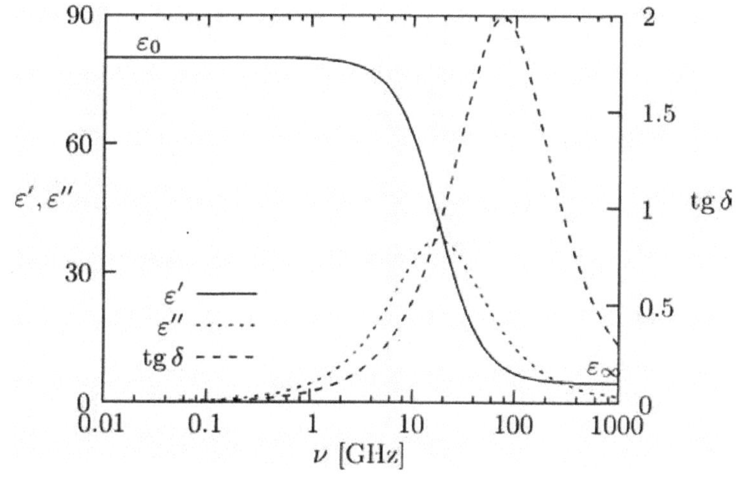

Figure 1.2 The loss tangent of water as a function of frequency, together with ε' and ε''.

reason for the application of the lower frequency is to heat the material throughout its interior so that the radiation is not totally absorbed by the first layer of the material that it encounters and may penetrate further into the material, heating it more evenly. In the other words, if the frequency is optimal for the maximum heating rate, the microwaves are absorbed in the outer region of the material and penetrate only a short distance (skin effect) [4].

The penetration depth is defined as the distance from the sample surface where the absorbed power is $1/e$ of the absorbed power at the surface. Beyond this depth, volumetric heating due to absorption of microwave energy is negligible. The penetration depth D_p is proportional to the wavelength of the radiation and depends on the dielectric properties of the material. For lossy dielectrics ($\varepsilon''/\varepsilon' << 1$) the penetration depth (D_p) can be calculated as follows:

$$D_p = \frac{\lambda_0}{2\pi}\frac{(\varepsilon')^{1/2}}{\varepsilon''}$$

where λ_0 is the length of electromagnetic wave (nm).

Materials with relatively high values of the loss factor (ε'') are characterized by low values of the penetration depth (D_p), and therefore, microwaves can be totally absorbed within the outer layers of these materials. For example, the penetration depth of

Table 1.3 Microwave (2.45 GHz) penetration depths (D_p) of some common materials

Material	Temperature (°C)	Penetration depth (D_p) (cm)
Water	25	1.4
Water	95	5.7
Ice	−12	1100
Paper, cardboard	25	20.60
Wood	25	8.350
Rubber	25	15.350
Glass	25	35
Porcelain	25	56
Polyvinyl chloride	20	210
Epoxy resin (Araldite)	25	4100
Teflon	25	9200
Quartz glass	25	16,000

water is 1.4 and 5.7 cm at 25°C and 95°C, respectively (Table 1.3). This means that during the experiments in a water solution on larger scales only some parts (outer layers) of the reaction mixture interact with microwaves to generate heat, which is then transported into the rest of the mixture conventionally (i.e., by the heat convection mechanism and/or the conduction mechanism). On the other hand, microwaves can penetrate and pass through the samples of the materials with low values of the loss factor (ε''). However, in the case of chemical processes, this should not be treated as a drawback, because due to the design of the microwave reactors, unabsorbed radiation passing through the sample is mostly reflected back and, finally, absorbed on later passes, which can provide enough energy to initiate a process [5].

Under microwave irradiation, the heating rate (i.e., temperature increase) of the material depends also on the shape and size of the sample. For instance, the study of the thermal behavior of alumina indicates that the temperature of the sample during microwave irradiation depends on the amount of the sample (Fig. 1.3). In a multimode cavity, the maximum temperature was obtained for ca. 200 g of alumina since it had a larger surface and in turn could absorbed more of the microwave energy per weight unit than, for example, 100 g of alumina. In contrast, the 500 g sample of alumina

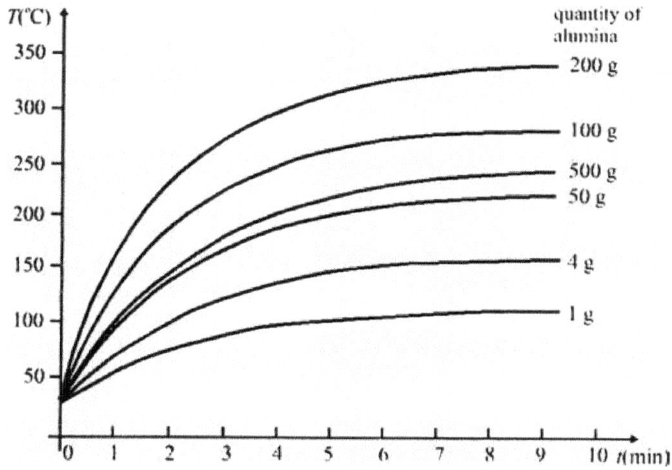

Figure 1.3 Thermal behavior of neutral alumina as a function of irradiation time and quantity.

was heated with a lower heating rate than the 200 g sample, but in this case the most probable explanation is that 500 g of alumina consumed the entire microwave energy provided by the reactor, which in fact could be less per weight unit than for 200 g of alumina. Most likely, the amount of sample that would be heated most efficiently by microwaves under such experimental conditions was somewhere between 200 and 500 g of alumina. In turn, 4 g of alumina appeared as the minimum amount for an appreciable thermal effect to be observed, while 1 g sample of alumina was unable to reach 100°C even after a relatively long irradiation time, ca. 20 min [6].

Recently, microwave technology has been applied to polymer and composite processing, and particularly its application to the polymerization process has become a subject of a great deal of academic and industrial research, with a lot of scientific and patent literature being generated [1, 7, 8]. Microwave technology is now widely accepted and has become a popular unconventional technology in polymer chemistry as an alternative to, and often improvement on, conventional heating. The relationship between the loss factor ε'' and the ability of some common materials to absorb microwave energy is shown in Fig. 1.4.

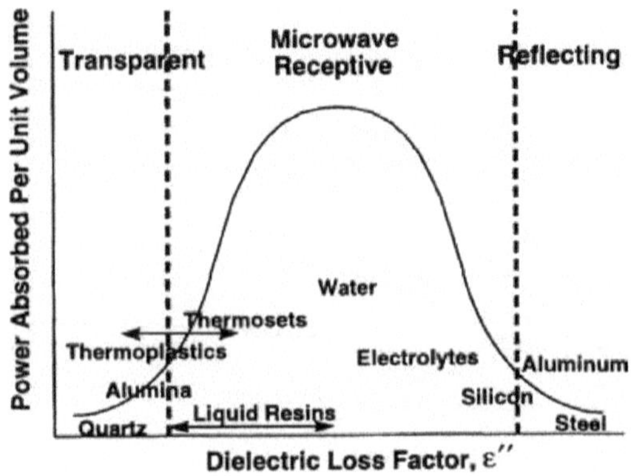

Figure 1.4 The relationship between the loss factor (ε″) and the ability of some common materials to absorb microwave energy.

Materials with high conductance and low capacitance (such as metals) have a high loss factor (ε″) and the penetration depth approaches zero. Thus, these materials are considered to be reflectors of microwave irradiation. On the contrary, materials with low values of the loss factor (ε″) have a large penetration depth. As a result, very little energy is absorbed by the material, and it is transparent to microwave irradiation, like quartz glass. Because of such behavior, microwaves transfer energy most effectively to materials that possess the lost factor (ε″) in the middle of the conductivity range (Fig. 1.4). In contrast, conventional heating transfers heat most efficiently to materials with high conductivity [9]. It is worth stressing that most of the polymers, that is, thermoplastics, thermosets, and liquid resins, are in the microwave-receptive region for the given frequency (Fig. 1.4), which is important during synthesis and processing of polymeric materials, polymer composites, and nanocomposites, as well as polymer hybrid materials [10].

1.2 Microwave Thermal Effect vs. Nonthermal Effect

The application of microwaves in chemistry is therefore so attractive that from the very beginning it was realized that a number of chemical processes can be carried out with a substantial reduction in the reaction time in comparison to conventional processes. Reactions that usually take many hours or days can be run in a considerably shorter time of several minutes or even seconds under the influence of microwave irradiation [11]. At the moment, there are two groups of hypotheses that are proposed to explain the reduction of the reaction time under microwave conditions in comparison to processes under conventional conditions.

According to the first group, in spite of the fact that the course of chemical processes under microwave conditions is considerably shorter than under conventional conditions, the kinetics and mechanism of the reactions are still the same. The reduction of the reaction time is the result of sudden and, sometimes, uncontrollable temperature growth of the reaction mixture under microwave irradiation, which in turn leads to the increase in reaction rates following common kinetic laws.

The second group supposes that during microwave irradiation of the reaction mixture there is a specific effect of microwave activation that causes an increase in the reaction rates, which the bulk temperature of the reaction mixture is inadequate to explain. Such an effect has been accepted to be called the nonthermal microwave effect or the specific microwave effect.

Critical reviews concerned with both groups of hypotheses have been published [12–16].

In this chapter, we are giving arguments that supported both groups of hypotheses, that is, nonthermal effects as well as thermal effects, and eventually, we are explaining that thermal gradients that are developed during microwave irradiation of different materials are the main factors for the reduction in reaction times. In particular, the development of high temperature gradients was observed for heterogeneous solventless systems, in which microwaves interact directly with reagents and, therefore, more efficiently drive chemical reactions. The possible accelerations of such reactions are expected to be optimal since they are not moderate or impeded by solvents, and the most successful examples of microwave applications were found to be related to heterogeneous systems.

1.2.1 *Nonthermal Effects*

Nonthermal microwave effects are often associated incorrectly with absorption of electromagnetic waves during rotational transitions of molecules (i.e., microwave spectroscopy), but the frequencies at which molecules can absorb electromagnetic waves to change their rotational states are higher than 2.45 GHz. For example, the first absorption lines of molecules such as carbonyl sulfide (OCS), CO, HF, and MeF occur at 12.2, 115, 1230, and 51 GHz, respectively. Moreover, molecular rotational spectra can be recorded only in the gas phase because in the condensed phase, rotation of molecules is suppressed by neighboring molecules. Internal bond rotations, referred to as torsional vibrations and oscillations in gas, liquid, and solid states, also require higher frequencies for excitation that falls into the infrared region. For instance, the excitation of molecules by infrared radiation at 100–400 cm^{-1} requires frequencies at 3000–12,000 GHz [17]. Thus, when the material absorbs microwave radiation at 2.45 GHz the increase in temperature is owing to macroscopic phenomena (e.g., dielectric heating mechanism), and this radiation does not bring molecules to higher rotational or vibrational energy levels. However, microwaves are converted into heat and quasi-accumulated to increase the internal energy of the material. It can be stated that microwaves pump energy into the material, and due to the influence of accumulated thermal energy within the material, it is possible to carry out a number of processes. Thus, it is considered that nonthermal microwave effects are due to such factors as [18]:

- Formation of hot spots during the interaction of microwaves with the material
- Change of reaction selectivity due to increase in heating rate of the reaction mixture
- Inversion of temperature pattern under microwave conditions in comparison to conventional conditions
- Development and/or increase in dipole moments of the reacting molecules in the transition state (TS) compared to the ground state (GS) and decrease in activation energy

Since microwaves are the source of volumetric heating of materials, which means that the increase in temperature of the reaction mixture can be faster than for conventional heating methods (i.e., electric mantle, oil bath, etc.), it is possible to find a different product distribution (selectivity) under microwave and conventional conditions. For instance, it was shown that for sulfonation of naphthalene in the presence of sulfuric acid, the rate of temperature growth of the reaction mixture influenced the ratio of the isomers 1- and 2-naphthalenesulfonic acid (1- and 2-NSA). By changing the rate of temperature increase it was possible to have either kinetic or thermodynamic control of the reaction to obtain 1-NSA or 2-NSA, respectively. Furthermore, the ratio of the isomers of NSA did not depend on the heating mode (i.e., microwave or conventional) but only on the heating rate [19]. For reaction mixtures that are large in size (volume), it is possible to obtain inversion of the temperature pattern under microwave conditions in comparison to conventional conditions. This means that the material under microwave irradiation can have a higher internal than external temperature, which is different than the bulk temperature of the reaction mixture. Therefore, it is possible to change selectivity or increase the yield of reactions under microwave conditions, while the mechanism and kinetics of the reaction stay unchanged.

For example, a calorimetric analysis (differential scanning calorimetry [DSC]) of the cure kinetics of glass/epoxy laminates with a thickness of 25 mm was conducted at a temperature range of 135°C–175°C for conventional and microwave experiments [20]. Both numerical and experimental results showed that microwaves promoted an inside-out cure of the thick laminates due to volumetric heating, which dramatically reduced the overall processing time. For the conventional process outside-in curing of composites was observed and resulted in visible matrix cracks, while cracks could not be seen in the microwave-processed composite. Thus, under conventional thermal conditions, to reduce thermal gradients, thick laminates were processed at a lower cure temperature and heated with slow heating rates, resulting in excessive cure times. It should be emphasized that the formation of cure gradients within the two composites cured under both microwave and conventional conditions was observed (Fig. 1.5).

Although cure gradients existed in both composites treated under microwave and thermal conditions, it can notified that

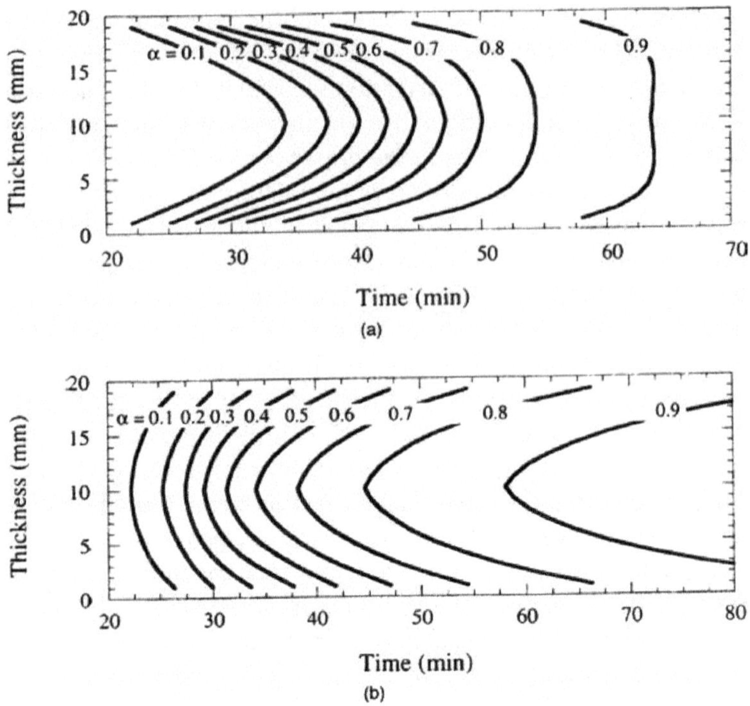

Figure 1.5 Formation of cure gradients with two laminates during (a) conventional cure and (b) microwave cure.

under microwave irradiation, the highest degree of conversions first appeared in central parts of the samples, while for conventional (autoclave) treatment the highest degree of conversion is first observed at the surface of the samples [21].

Finally, it has been postulated that the increase in the polarity of the reaction system (i.e., either development or increase of dipole moment) from the GS toward the TS can result in an acceleration of the reaction due to a stronger interaction of microwaves with the reagents during the course of the reaction. Thus, a nonthermal microwave effect can be expected for the reaction with polar mechanisms when the stabilization of the TS is more effective than that of the GS, which results in enhancement of the reactivity as a result of the decrease in activation energy (Fig. 1.6).

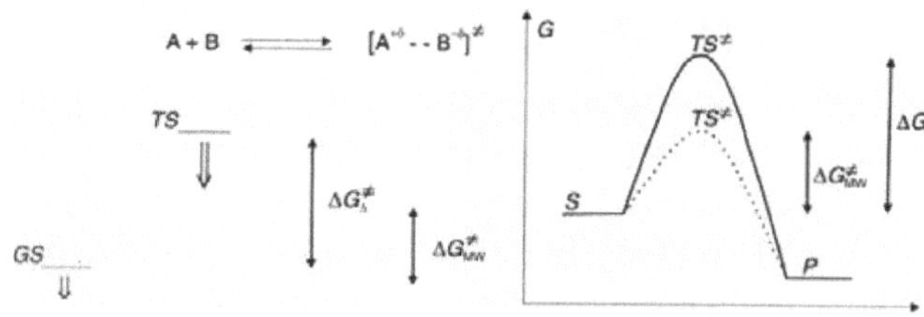

Figure 1.6 Relative stabilization of a more polar transition state (TS) vs. ground state (GS).

The most representative examples of this are concerned with unimolecular or bimolecular reactions between neutral molecules (i.e., dipole moments are developed in the TS) and anionic reactions of tight ion pairs (i.e., charge-localized anions), which leads to ionic dissociation in the TS [22]. All of these effects can be masked or limited by either protic (e.g., alcohols) or aprotic solvents (e.g., dimethylformamide [DMF], CH_3CN, dimethyl sulfoxide [DMSO]) that strongly interact with and absorb microwaves. In such cases the main interaction occurs between microwaves and polar molecules of the solvent that is used in excess. Thus particularly for samples large in size, microwaves do not penetrate the whole amount of the reaction mixture and a great amount of microwave energy is absorbed by the outer layers of the mixture (e.g., the penetration depth of water is ca. 1–2.5 cm; see Table 1.3). In turn, heat transfer to the internal parts of the reaction mixture is by the classical heat conduction mechanism from the outer layers, which is commonly observed in conventional heating. Therefore, reaction rates under microwave conditions in a polar solvent solution should be nearly the same as those observed under conventional conditions. However, this can lead to the conclusion that the most favorable systems for microwave activation are either nonpolar or solvent-free systems, so the application of polar solvents should be always taken into account; in particular, for the reagents that are weak microwave absorbers, a polar solvent can initiate temperature rise of the reaction mixture—a small amount of a polar substance added to a nonpolar medium can dramatically improve microwave absorption as well as temperature homogeneity of the sample.

1.2.2 *Thermal Effects*

Thermal effects that reduce reaction times and can be observed for reactions under microwave irradiation are related to different temperature regimes under microwave conditions in comparison to conventional conditions. Thermal effects can result in a seemingly faster course of chemical reactions; therefore, only proper temperature measurements and temperature analysis for the entire sample of the material (i.e., the bulk reaction mixture) lead to reaction rates that are comparable to reaction rates observed under conventional conditions. Three factors that can cause thermal effects are [33]:

- Overheating (superheating) of the reaction mixture that in particular can be observed in the presence of a large quantity of electric carriers (i.e., ions)
- Considerably faster heating rate of the reaction mixture, which means that the required temperature for a chemical reaction is achieved in a shorter time
- Increase of the rate of diffusive processes on the phase boundaries
- Overheating of the reaction mixture in sealed vessels—the pressure cooker effect

For instance, in one of the first published investigation, it was observed that a number of reactions can be carried out faster, provided that the reaction mixtures were placed in sealed vessels and irradiated in a microwave cavity. Moreover, the reactions needed even shorter reaction times if they were carried out in solvents with lower boiling points. Eventually, it was shown that the rates of the reactions were proportional to pressure developed in the reaction vessels during irradiation [23, 24]. In another example, a reduction in reaction time was observed during the synthesis of organometallic compounds of rhodium(I) and iridium(I) that are usually prepared by boiling the reagents in a mixture of alcohol and water for several hours. The reaction time during the microwave protocol was shorter than 1 min to afford products with satisfactory yield. Further more detailed investigation showed that the substantial reduction of the reaction time was attributed to the growth of pressure and temperature in sealed reaction vessels [25].

All the aforementioned examples prove that for the investigated reactions, the influence of microwave irradiation was limited to pure thermal effects (i.e., higher pressure and/or bulk temperature of the reaction mixture, development of a temperature gradient, overheating of solvents) that can be responsible for seemingly higher reaction rates under microwave conditions.

1.2.3 *Thermal Gradients*

As stated above, a careful consideration of all the factors that influence the reactions under microwave conditions can eliminate most of these effects. There are several reactions in which the influence of only thermal effects does not sufficiently explain the enhancement of yield and rate of the reactions. Moreover, temperature measurement under microwave conditions, particularly for reactions in dense and solventless media, is difficult [26, 27]. Therefore, it is a source of inappropriate interpretation of experimental results and divergence in descriptions of the same experiments. To illustrate, the four examples of the reactions under solventless conditions are considered. In a solventless reaction, three systems are usually considered:

- Reaction on a heterogeneous support
- Solvent-free phase transfer catalysis (PTC)
- Reaction of neat reagents

For all of these systems the experiments were performed under microwave irradiation, while the reaction temperature was additionally monitored by a thermovision camera to show temperature distribution on the surface of the reaction mixture. It is worth stressing that solventless systems are often dense, which causes difficulties with stirring and proper temperature measurements. On the other hand, as stated before, the best-observed results of microwaves applications are related to reaction in such viscous/solid systems; therefore the observation of temperature distribution at least on the surface of the reaction mixture might give a deeper view and explain the acceleration of these processes. For the reaction on the heterogeneous support the oxidation of alcohol as well as the oxidation of an activated methylene group with chromium dioxide (Magtrive^{TM}) were chosen [28, 29]. For the reaction of neat reagents and solvent-free PTC, synthesis of solid epoxy resins [30] and synthesis of benzofurans were, respectively, carried out [31]. Magtrieve^{TM} is DuPont's trademark for the oxidant based on tetravalent chromium dioxide (CrO_2), and it has been proven to be a useful oxidant in some reactions, including the oxidation of alcohols [32]. Moreover, it can be easily separated by employing its magnetic properties and then recycled and, in turn, reused for further processes. Magtrieve^{TM} as an oxidant is a very well-suited reagent for microwave synthesis, and it carries a benefit of efficient conversion of electromagnetic energy into heat according to the dielectric heating mechanism. For example, the irradiation of Magtrieve^{TM} (2.5 g) with the continuous power of a microwave reactor (100 W) in an open vessel (5 cm diameter) led to quick heating of the material up to 360°C within 2 min. As was expected, the temperature recorded by means of a thermovision camera had the highest value in the center of the reaction vessel (Fig. 1.7). Moreover, all the attempts to measure the temperature of Magtrieve^{TM} by means of a thermocouple after switching off the microwave power showed much lower temperature of ca.

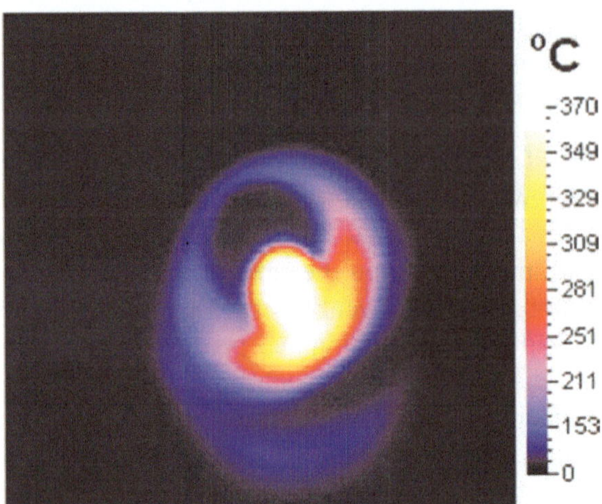

Figure 1.7 The temperature profile for Magtrieve^{TM} irradiated by microwaves (2 min of irradiation).

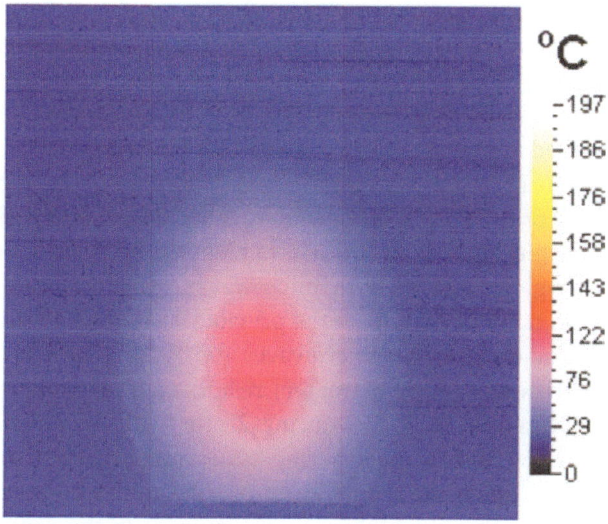

Figure 1.8 The temperature profile for Magtrieve™ in a toluene solution irradiated by microwaves (2 min of irradiation).

80°C–100°C, which leads to an important conclusion that the only valid temperature measurement must be done during microwave experiments while the material is irradiated with a continuous power.

When toluene, which is a weak microwave absorber, was introduced into the reaction vessel, the temperature of Magtrieve™ reached ca. 140°C within 2 min but became more uniformly distributed (Fig. 1.8).

In this case it is worth stressing that even though the temperature of the solid material was higher than the boiling point of toluene, we did not observe boiling in the reaction vessel. Moreover, a temperature gradient of the solid support can be observed, with the highest temperature at the center of the sample, so it is hard to say what the bulk temperature of reaction mixture is. It leads to another important conclusion that during reactions in heterogeneous systems (solid support/organic solvent), in which microwaves have been proved to be the most effective [33], the temperature of the solid material can be higher than the bulk temperature of the solvent and measurement of temperature with a pyrometer, thermocouple, or fiber optics thermometer does not give correct values. In such a case, the higher temperature of the solid support leads to higher conversion of reactants or/and higher reaction rates, which in turn might be a reasonable explanation for the so-called nonthermal microwave effect [15].

Another investigation in which temperature was monitored by means of a thermovision camera, that is, oxidation of active methylene groups, also confirmed that Magtrieve™ is characterized by a strong interaction with microwaves, which causes a rapid increase in temperature. As in the previous example, the irradiation of pure oxidant (2 g) by microwaves (with the power of 100 W) in an open vessel (5 cm diameter) results in a very quick heating of the material up to 350°C during 2 min of the experiment, but in this the results from the thermovision camera were turned into a 3D picture. Again the temperature-profile-recorded thermovision camera shows that the highest temperature is in the center of the reaction vessel but two other macro hot spots with temperatures of about 240°C can be seen within the sample (Fig. 1.9a).

These results showed that strong interaction of microwaves with Magtrieve™ was the main reason of the heterogeneity of heating, making it difficult to control. Therefore, to maintain a more uniform temperature of the reaction mixture, toluene, which is a very poor microwave absorber, was used. In Fig. 1.9b, it can be seen that the temperature of the Magtrieve™ surface may be higher than the boiling point of toluene, but boiling was not observed at all. Although the temperature was highest at the center of the sample, macro hot spots were not observed.

The next example of monitoring a chemical reaction with a thermovision camera is the reaction of neat reagents, that is, the polyaddition reaction of 4,4-isopropylidenediphenol (BPA) to a low-molecular-weight epoxy resin in the presence of a catalyst, which results in the synthesis of solid epoxy resins (SERs) (high-molecular-weight resins) [34, 35]. Usually, the reaction is carried out for a couple of hours in a temperature range of 140°C–180°C with a nitrogen blanket to minimize oxidative degradation reactions. The main advantage of the microwave process was a twofold reduction in the reaction time in comparison to conventional heating. It was found that the molecular weight distribution and the degree

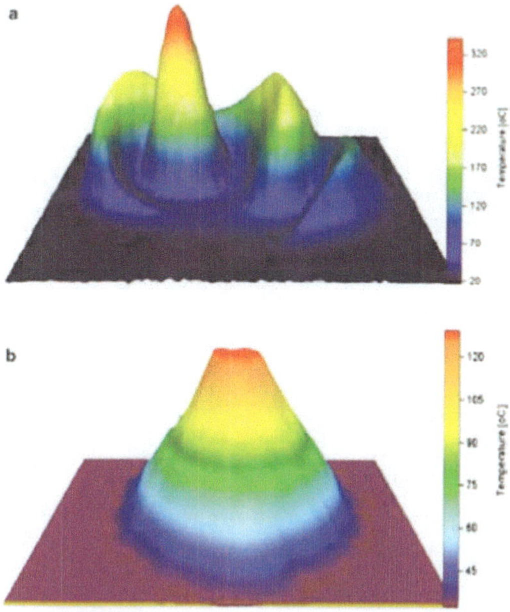

Figure 1.9 The temperature profile for Magtrieve^{TM} in a toluene solution irradiated by microwaves (2 min of irradiation).

of branching in resins synthesized under microwave irradiation were comparable with those obtained under conventional heating and were not influenced by the reduction in reaction time. The differences in the surface temperature homogeneity that were recorded by means of a thermovision camera under microwave and conventional conditions for both stirred and nonstirred reaction mixtures are presented in Fig. 1.10.

The experiments were carried out at 160°C, and as expected for an exothermic reaction, the highest temperatures were observed in the center of the reaction vessel for all the processes (Fig. 1.10a–d). Furthermore, for both microwave and conventional processes, the stirring of the reaction mixture resulted in more uniform temperature profiles (Fig. 1.10a,c), while for the nonstirred reaction mixture, the maximum surface temperature under microwave irradiation was much higher (200°C) than the bulk temperature. Moreover, for the nonstirred samples prepared under microwave

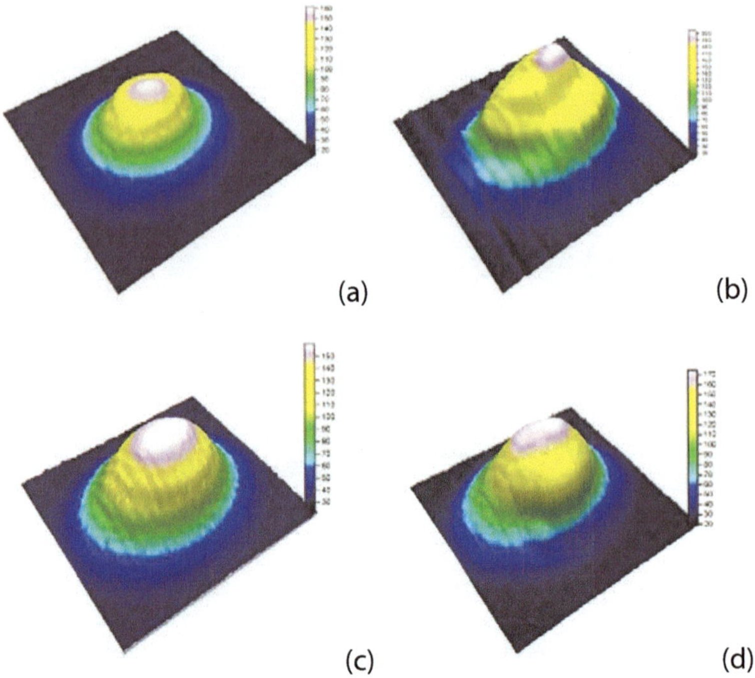

Figure 1.10 The surface temperature profiles of reaction mixtures recorded by means of a thermovision camera: (a) microwave stirred mixture, (b) microwave nonstirred mixture, (c) conventional stirred mixture, and (d) conventional nonstirred mixture.

irradiation, high temperature heterogeneity was observed (Fig. 1.10b), which, in turn, often led to crosslinking of resins. It was opposite to the reaction under conventional conditions, in which the temperature for the nonstirred reaction mixture was, however, more inhomogeneous than for the stirred mixture, but the maximum surface temperature reached only 170°C (Fig. 1.10d). It again leads to a conclusion that in the case of viscous reaction systems such as a mixture of a low-molecular-weight epoxy resin with BPA, in some parts of the sample, the local temperature under microwave irradiation can be higher than the overall bulk temperature.

Perhaps the most interesting example of the reaction under solventless conditions that was monitored with a thermovision camera is the reaction under solid–liquid PTC conditions of

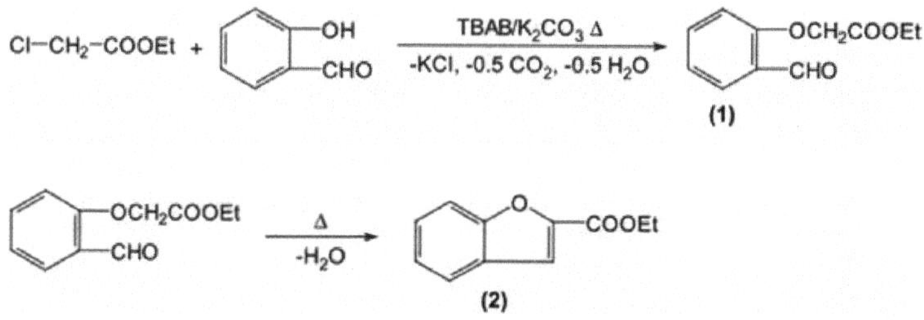

Figure 1.11 Preparation of benzofuran-2-carboxylic acid ethyl ester.

salicylaldehyde with ethyl ester of chloroacetic acid in the presence of K_2CO_3 and tetrabutylammonium bromide (TBAB) as a catalyst. The reaction resulted in the formation of (2-formylphenoxy)acetic acid ethyl ester (**1**), which after further intermolecular condensation gave benzofuran-2-carboxylic acid ethyl ester (**2**)—the final product (Fig. 1.11) [31].

When the reaction was run at lower temperatures, for example, 85°C (Table 1.4, entries 4 and 5) versus 110°C (Table 1.4, entries 2 and 3), the product distribution was much strongly affected by the microwaves, and the yield of **2** was higher in comparison with conventional experiments [36]. Thus, to observe the influence of microwave irradiation on chemical reactions if the yields were high under both microwave and conventional conditions, it was required not to increase but to decrease the reaction temperatures, which have been shown in reports [37–40].

It is worth stressing that crude compounds (**1**) and (**2**) adsorbed on the surface of K_2CO_3 gave different colors, that is, yellow (**1**) and brown (**2**), which were of utmost importance and permitted the researchers to visually follow the progress of the reaction (Fig. 1.12).

During the reaction, the temperature of the reaction mixtures and the temperature of the surface were determined by a fiber-optic thermometer and a thermovision camera, respectively. The reactions were carried out without a mechanical stirring of the reaction mixtures, and it was decided to carry out the reaction on a K_2CO_3 support at 110°C without a catalyst. The results of the microwave experiments are presented in Table 1.5.

Table 1.4 The distribution of the intermediate (**1**) and the final product (**2**) in the synthesis of benzo[b]furans under both conventional (Δ) and microwave (MW) conditions

No.	Temperature (°C)	Time (min)	Solvent (mL)	Yield (%) 1	2
1	110	10/MW	–	12	66
2	110	20/MW	–	–	100
3	110	20/ Δ	–	4	96
4	85	20/MW	–	15	85
5	85	20/ Δ	–	90	10
6	85	30/MW	Cyclohexane (1.5)	62	38
7	85	30/MW	EtOH (100%) (1.5)	0	100
8	85	30/MW	EtOH (96%) (1.5)	0	100
9	85	30/ Δ	EtOH (100%) (1.5)	47	53

The development of thermal gradients during microwave experiments on the surface of reaction mixtures was confirmed visually, that is, it was possible to observe the formation of brown spots in all the places where compound **1** was converted to compound **2** (Fig. 1.12). These observations were later proved by the analysis of the samples taken from different places from the surface of the reaction mixtures. In fact, the rotation of reaction vessels reduced overheating, the reaction mixture was exposed to a more homogenous microwave field, and, in turn, the yield of **2** was lower

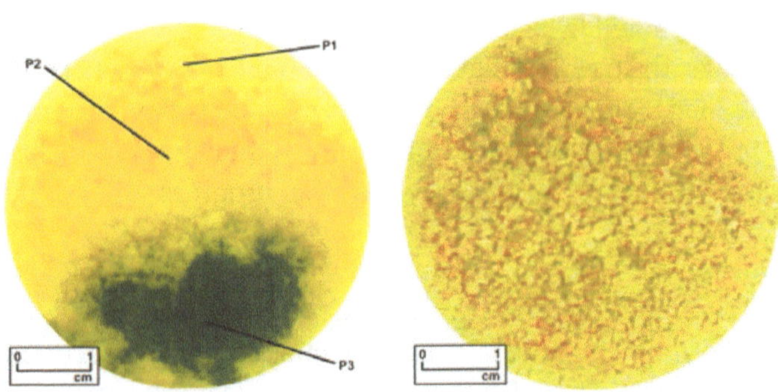

Figure 1.12 The surface of the reaction mixtures: (left) the reaction without rotation of the reaction vessel (Table 1.5, entry 1) and (right) the reaction with rotation of the reaction vessel (Table 1.5, entry 2).

Table 1.5 The distribution of the intermediate (**1**) and the final product (**2**) in the synthesis of benzo[b]furans under both conventional (Δ) and microwave (MW) conditions

No.	Selectivity[a] (%) 1	2	Description
1	33	67	No stirring
2	70	30	Vessel rotation
3	84	16	Vessel rotation and stirring
4	100	0	No stirring, addition of 6 mL of decane
5	67	33	No stirring, vessel rotation
6	77	23	No stirring, vessel rotation

[a]Average selectivity determined by gas chromatography/mass spectrometry (GC/MS) from the entire reaction mixture. All microwave experiments were carried out under the same conditions: maximal microwave power 240 W and maximal temperature 110°C.

(Table 1.5, entry 2) than for the experiments without rotation (Table 1.5, entry 1) but much closer (i.e., comparable) to the yield of the reaction under conventional conditions. Finally, the mechanical stirring with a quartz spatula placed within the reaction mixture improved the thermal homogeneity of the reaction mixture to a greater extent than the rotation of the reaction vessels (Table 1.5, entry 3); the results are comparable to those obtained under conventional conditions (the yields of **1** and **2** were 74% and 26%, respectively) [31].

Also, thermal gradients were revealed by means of a thermovision camera, which detected a strong temperature increase in the central part (on the side of the microwave waveguide in the reactor) of the reaction mixture in comparison to outer regions (Fig. 1.13). Eventually, the existence of thermal gradients was proved by the quantitative analysis (GC/MS) by taking samples from different places of reaction mixtures that exhibited different temperatures during the reactions (Fig. 1.12, left, and Fig. 1.13, entries P1, P2, and P3). The results are summarized in Table 1.6. It can be seen that the highest conversion of **1** to **2** was found for sample P3, which exhibited the highest temperature (ca. 200°C) during the microwave experiment (Table 1.6, entry 3).

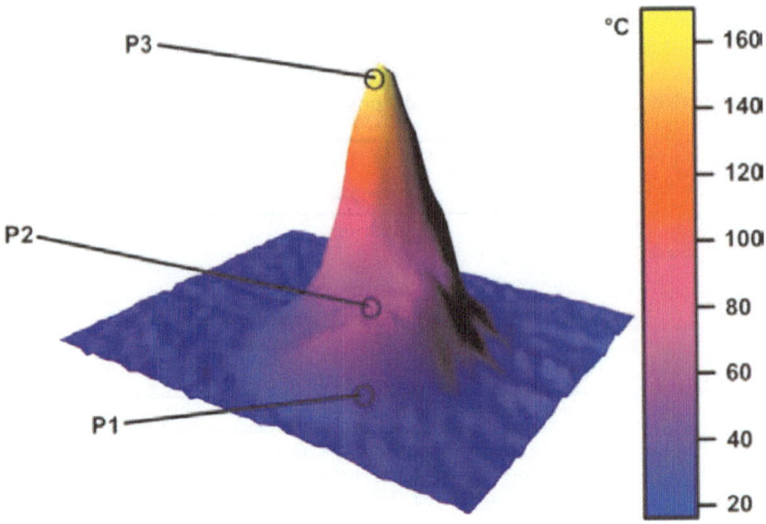

Figure 1.13 A thermovision photograph of the surface of the reaction mixture during the experiment.

Then for the conversion of **1** to **2**, 45% and 0% yields were observed for the samples P2 and P1, respectively, which gained lower temperatures than P3, that is, 125°C (Table 1.6, entry 2) and 70°C (Table 1.6, entry 1), respectively. These results are in good agreement with the experiments under conventional conditions, in which it was shown that only the samples that reached a temperature higher than 150°C can give high conversion of **1** to **2** without a catalyst (the yields of **1** and **2** were 11% and 89%, respectively) [31]. For the samples that reached temperatures lower than 100°C, no conversion of **1** to **2** was observed at all.

Table 1.6 Local composition of the reaction mixture (Figs. 1.12, left, and 1.13)

			Yield (%)	
Sample	**Color**	**Temperature (°C)**	**1**	**2**
P1	Light	70	100	0
P2	Brown	125	55	45
P3	Dark	200	0	100

The results presented above are important because one can expect that the overall reaction yields given in Table 1.5 should be similar for all the experiments and do not show significant differences during additional rotation of the vessels and/or stirring of the reaction mixtures. It can be assumed that the overall energy input given by microwaves is similar in all the cases; thus the reaction rate can be different in different regions of the reaction mixture because the temperature is different but the overall reaction rates should be comparable. In fact, we observed an opposite situation: the improvement of temperature homogeneity under microwave irradiation caused a decrease in the overall reaction yields and made them comparable to those obtained under conventional conditions.

1.3 Conclusion

In conclusion, in all the presented cases, proper temperature measurement in the case of a heterogeneous reaction mixture is very difficult. To maintain good temperature homogeneity and make some comparison with experiments under conventional conditions, effective stirring has to be provided, perhaps together with a small amount of an inert solvent. There is a general agreement that the application of fiber-optic thermometers is a reliable way to determine temperatures under microwave conditions.

Applying a thermovision camera, we found that for the reactions in heterogeneous systems under microwave irradiation, the temperature measurement with a fiber-optic thermometer can lead to serious errors like pyrometry, in particular for those experiments that are planned without any attention being paid to temperature homogeneity of the reaction mixture. In the latter case, a high temperature gradient within the reaction mixture generated by the microwaves leads to a higher conversion of reactants or/and reaction rates, which in turn might be a reasonable explanation of the so-called nonthermal microwave effects, that is, an increase in reaction rates that is inadequate for the temperature of the reaction medium. Therefore, before considering the increase in reaction rates by special microwave effects (thermal or nonthermal), first, we need to consider all the factors that might influence chemical reactions under microwave conditions, like reaction mechanisms, temperature profiles (gradients), and, in particular, proper design of our experiments.

References

1. Bogdal, D., and Prociak, A. (2007). *Microwave-Enhanced Polymer Chemistry and Technology* (Blackwell-Wiley, Oxford).
2. Kaatze, U. (1995). Fundamentals of microwaves, *Radiat. Phys. Chem.*, **45**, pp. 539–548.
3. Stuerga, D. (2012). Microwave-materials interactions and dielectric properties: from molecules and macromolecules to solids and colloidal suspensions. In *Microwaves in Organic Synthesis,* de la Hoz, A., and Loupy, A., eds. (Wiley-VCH, Weinheim), pp. 3–56.
4. Gabriel, C., Gabriel, S., Grant, E. H., Halstead, B. S., and Mingos, D. M. P. (1998). Dielectric parameters relevant to microwave dielectric heating, *Chem. Soc. Rev.*, **27**, pp. 213–224.
5. Stuerga, D., and Delmotte, M. (2003). Wave-material interaction, microwave technology and equipment. In *Microwaves in Organic Synthesis,* Loupy, A., ed. (Wiley-VCH, Weinheim), pp. 1–33.
6. Bram, G., Loupy, A., Majdoub, M., Gutierrez, E., and Ruiz-Hitzky, E. (1990). *Alkylation of potassium acetate in "dry media" thermal activation in commercial microwave ovens, Tetrahedron,* **46**, pp. 5167–5176.
7. Bogdal, D., Penczek, P., Pielichowski, J., and Prociak, A. (2003). Microwave assisted synthesis, crosslinking, and processing of polymeric materials, *Adv. Polym. Sci.*, **163**, pp. 193–263.
8. Hoogenboom, R., and Schubert, U. S. (2007). Microwave-assisted polymer synthesis: recent developments in a rapidlyexpanding field of research, *Macromol. Rapid Commun.,* **28,** pp. 368–386.
9. Thostenson, E. T., and Chou, T.-W. (1999). Microwave processing: fundamentals and applications, *Compos. Part A: Appl. Sci. Manuf.,* **30**, pp. 1055–1071.
10. Bogdal, D., Prociak, A., and Michalowski, S. (2011). Synthesis of polymer nanocomposites under microwave irradiation, *Curr. Org. Chem.,* **15**, pp. 178–188.
11. Loupy, A. (2002). *Microwaves in Organic Synthesis* (Wiley-VCH, Weinheim).

12. De la Hoz, A., Diaz-Ortiz, A., and Moreno, A. (2005). Microwaves in organicsynthesis. Thermal and non-thermal microwave effects, *Chem. Soc. Rev.*, **34**, pp. 164–178.
13. Kappe, C. O., Pieber, B., and Dallinger, D. (2013). Microwave effects in organic synthesis: myth or reality?, *Angew. Chem., Int. Ed.*, **52**, pp. 1088–1094.
14. Nüchter, M., Ondruschka, B., Bonrath W., and Gum, A. (2004). Microwave assisted synthesis: a critical technology overview, *Green Chem.*, **6**, pp. 128–141.
15. Perreux, L., and Loupy, A. (2001). A tentative rationalization of microwave effects in organic synthesis according to the reaction medium, and mechanistic considerations, *Tetrahedron*, **57**(45), pp. 9199–9223.
16. Gedye, R., Smith, F., Westaway, K., Laberge, L., and Rousell, J. (1986). The use of microwave ovens for rapid organic synthesis, *Tetrahedron Lett.*, **27**, pp. 279–282.
17. Strauss, C. R., and Trainor, R. W. (1995). Developments in microwave-assisted organic *chemistry, Aust. J. Chem.*, **48**, pp. 1665–1692.
18. Mingos, D. M. P. (2005). Theoretical aspects of microwave dielectric heating. In *Microwave Assisted Organic Synthesis* (Blackwell Publishing, Oxford), pp. 1–22.
19. Stuerga, D., Gonon, K., and Lallemant, M. (1993). Microwave heating as a new way to induce selectivity between competitive reactions application to isomeric ratio control in sulfonation of naphthalene, *Tetrahedron*, **49**, (28), pp. 6229–6234.
20. Thostenson, E. T., and Chou, T.-W. (1997). Microwave processing of thick section polymer composites, *Proc. 12th Annu. Meeting Am. Soc. Compos.*
21. Thostenson, E. T., and Chou, T.-W. (2001). Microwave and conventional curing of thick section of composites laminates: experiment and simulation, *Polym. Compos.*, **22**, pp. 197–212.
22. Perreux, L., Loupy, A., and Petit, A. (2012). Nonthermal effects of microwaves in organic synthesis. In *Microwaves in Organic Synthesis,* de la Hoz, A., and Loupy, A., eds. (Wiley-VCH, Weinheim), pp. 127–208.
23. Gedye, R., Smith, F., and Westaway, K. (1991). Microwaves in organic and organometallic synthesis, *J. Microw. Power Electromag. Energy*, **26**, pp. 3–17.
24. Loupy, A., Petit, A., Hamelin, J., Texier-Boullet, F., Jacquault, P., and Mathe, D. (1998). New solvent free organic synthesis using focused microwaves, *Synthesis*, **1998**(9), pp. 1213–1234.

25. Mingos, D. M. P., and Baghurst, D. R. (1991). Application of microwave dielectric heating effects to synthetic problems in chemistry, *Chem. Soc. Rev.*, **20**, pp. 1–47.
26. Kappe, C. O. (2013). How to measure reaction temperature in microwave-heated transformations, *Chem. Soc. Rev.*, **42**, pp. 4977–4990.
27. Mazubert, A., Taylor, C., Aubin, J., and Poux, M. (2014). Key role of temperature monitoring in interpretation of microwave effect on transestrification and esterification reactions for biodidiesel, *Bioresource Technol.*, **161**, pp. 270–279.
28. Bogdal, D., Lukasiewicz, M., Pielichowski, J., Miciak, A., and Bednarz, Sz. (2003). Microwave-assisted oxidation of alcohols using Magtrieve, *Tetrahedron Lett.*, **59**(5), pp. 649–653.
29. Lukasiewicz, M., Bogdal, D., and Pielichowski, J. (2003). Microwave-assisted oxidation of side chain arenes by Magtrieve^{TM}, *Adv. Synth. Catal.*, **345**(12), pp. 1269–1272.
30. Bogdal, D., and Gorczyk J. (2004). Synthesis and characterization of epoxy resins prepared under microwave irradiation, *J. Appl. Polym. Sci.*, **94**, pp. 1969–1975.
31. Bogdal, D., Bednarz, Sz., and Lukasiewicz, M. (2006). Microwave induced thermal gradients in solventless reaction systems, *Tetrahedron,* **62**(40), pp. 9440–9445.
32. Lee, R., and Donald, D. (1997). Magtrieve^{TM} an efficient, magnetically retrievable and recyclable oxidant, *Tetrahedron Lett.*, **38**, pp. 3857–3860
33. Strauss, C. R., and Rooney, D. W. (2010). Accounting for clean, fast and high yielding reactions under microwave conditions, *Green Chem.*, **12**, pp. 1340–1344.
34. Bogdal, D., and Gorczyk, J. (2003). Microwave-assisted synthesis and determination of chain branching in solid epoxy resins using ^{1}H-NMR spectrometry, *Polymer*, **27**, pp. 7795–7800.
35. Czub, P. (2008). Synthesis of high-molecular weight epoxy resins from modified natural oils and bisphenol-A or bisphenol A based epoxy resins, *Polym. Adv. Technol.,* **20**, pp. 194–208.
36. Bogdal, D., and Warzala, M. (2000). Microwave-assisted preparation of benzo[b]furans under solventless phase-transfer catalytic conditions, *Tetrahedron*, **56**(44), pp. 8769–8773.
37. Chatti, S., Bortolussi, M., and Loupy, A. (2000). Synthesis of new diols derived from dianhydrohexitols ethers under microwave-assisted phase transfer catalysis, *Tetrahedron*, **56**(32), pp. 5877–5883.

38. Cvengros, J., Toma, S., Marque, S., and Loupy, A. (2004). Synthesis of phosphonium salts under microwave activation: leaving group and phosphine substituents effects, *Can. J. Chem.*, **82**(9), pp. 1365–1371.
39. Hajek, M., and Radoiu, M. T. (2000). Microwave activation of catalytic transformation of t-butylphenols, *J. Mol. Catal. A*, **160**, pp. 383–392.
40. Perreux, L., and Loupy, A. (2003). Effect of cations, substrates and temperature on the microwave-assisted saponification of benzoic and mesitoic esters, *Org. Prep. Proced. Int.*, **35**, pp. 361–368.

Chapter 2

General Features of MW Interaction with Materials

S. Takayama,[a] M. Sato,[b] and J. Fukushima[c]

[a] *National Institute of Fusion Science, 322-6 Oroshi-cho, Toki-City, Gifu 509-5292, Japan*
[b] *Faculty of Engineering, Chubu University, 1200 Matsumoto-cho, Kasugai-City, Aichi 487-8501, Japan*
[c] *Department of Applied Chemistry, Tohoku University, 6-6-07 Aramaki Aoba, Aoba-ku Sendai-City, Miyagi 980-8579, Japan*
takayama.sadatsugu@nifs.ac.jp, satomoto@isc.chubu.ac.jp, fukushima@aim.che.tohoku.ac.jp

2.1 Introduction to General Features of MW Interaction with Materials: Microwave, Heat, and Materials

Since 1960, we have been successfully applying microwave energies for cooking food, drying wood, heating rubber, manufacturing medicines and textiles, sintering ceramics, etc.

As metals reflect microwaves on the surface, it was a surprise that in 1999, Roy et al. demonstrated the microwave sintering of powder [1]. In 2002, following the demonstrations, they wrote a scientific paper entitled "Definitive Experimental Evidence for Microwave

Microwave Engineering of Materials: From Mesoscale to Nanoscale
Edited by Erwann Guénin

ISBN 978-981-4669-42-9 (Hardcover), 978-981-4669-43-6 (eBook)
www.panstanford.com

Effects: Radically New Effects of Separated E and H Fields, such as Decrystallization of Oxides" [2].

Since then, microwave treatment has been paid more attention, for example, in refining of metal oxides and synthesis of organic or inorganic new materials, including for biological science and technologies [3–8].

It should be pointed out that a microwave is a wave, not just an oscillation, and the microwave energy losses are due to the interaction between the electromagnetic wave (vector) and the imaginary parts of permittivity. If the readers are familiar with the physics of electromagnetic waves and the permittivity tensor ε, etc., given by Maxwell's equations, they can skip to Eq. 2.29.

Microwaves are now opening the door to varieties of industrial applications.

In this chapter, the basic theories of microwaves will be outlined first. Next, the new phenomena discovered experimentally will introduce the readers to the cutting-edge technologies and the new science on material processing. A recent hypothesis will be introduced for describing the mechanism of interaction of microwaves and materials.

2.1.1 *Maxwell's Equations*

Microwaves are electromagnetic waves of 300 MHz to 300 GHz. Maxwell's equations express the relations between the electric and magnetic fields as follows:

$$\nabla \cdot \vec{B}(t, \vec{x}) = 0 \tag{2.1}$$

$$\nabla \times \vec{E}\,(t, \vec{x}) + \frac{\partial \vec{B}(t, \vec{x})}{\partial t} = 0 \tag{2.2}$$

$$\nabla \cdot \vec{D}(t, \vec{x}) = \rho\,(t, \vec{x}) \tag{2.3}$$

$$\nabla \times \vec{H}\,(t, \vec{x}) - \frac{\partial \vec{D}(t, \vec{x})}{\partial t} = j\,(t, \vec{x}) \tag{2.4}$$

where $\vec{B}$, $\vec{E}$, $\vec{D}$, $\vec{H}$, $\vec{j}\vec{j}$, and ρ indicate, respectively, the magnetic field, the electric field, the electric displacement, the magnetic field strength vectors, the current density, and the charge density. t and $\vec{x}$ are, respectively, time and space.

In vacuum, charge density ρ is equal to zero and the relations between $\vec{B}$, $\vec{E}$, $\vec{D}$, and $\vec{H}$ are written simply as

$$\vec{D} = \varepsilon_0 \vec{E} \tag{2.5}$$

$$\vec{H} = \frac{1}{\mu_0} \vec{B} \tag{2.6}$$

where ε_0 and μ_0 are, respectively, the permittivity and permeability in vacuum, and they are given by $\varepsilon_0 \mu_0 = c^2$.

Assuming that $\vec{E}$ and $\vec{B}$ vary exponentially as $\exp\left\{i\left(\vec{k}\cdot\vec{x} - \omega t\right)\right\}$, the solution of Maxwell's equations shows that the alternating electromagnetic field propagates as straight as an arrow in a direction perpendicular to the field vectors. The wave in vacuum has the following characteristics: (1) no mass, (2) propagation at the velocity of light, c, and (3) satisfaction of the superposition principle.

When electromagnetic waves encounter electric charges, the wave fields give forces to the charges that induce displacements and currents. According to Newton's third law of motion, the kinetic motions of charges give counterreactions to the wave. The electromagnetic field components $\vec{H}$ and $\vec{E}$ can be modified by $\vec{D}$ and $\vec{B}$ as follows:

$$\vec{D} = \varepsilon_0 \vec{E} + \vec{P}_{\mathrm{d}} \tag{2.7}$$

$$\vec{B} = \mu_0 (\vec{H} + \vec{M}) \tag{2.8}$$

where $\vec{P}_{\mathrm{d}}$ and $\vec{M}$ are called displacement and magnetization. Maxwell's equations (Eqs. 2.1–2.4) are given by

$$\nabla \cdot \vec{E} = \frac{1}{\varepsilon_0}(\rho - \nabla \cdot \vec{P}) \tag{2.9}$$

$$\nabla \times \vec{B} - \mu_0 \varepsilon_0 \frac{\partial \vec{E}}{\partial t} = \mu_0 \left(\vec{j} + \frac{\partial \vec{P}}{\partial t} + \nabla \times \vec{M} \right) \tag{2.10}$$

Equation 2.9 means that the polarization $\vec{P}$ reduces the electric field $\vec{E}$ in dielectric materials as

$$\vec{E} = \frac{1}{\varepsilon_0}\left(\vec{D} - \vec{P}\right) \tag{2.11}$$

Assuming that the polarization $\vec{P}$ varies proportionally with the applied electric field $\vec{E}$, they are written as

$$\vec{E} = \chi_e \varepsilon_0 \vec{P} \tag{2.12}$$

Then, the displacement $\vec{D}$ can be explained as

$$\vec{D} = \varepsilon_0 \left(1 + |\chi_e|\right) \vec{E} = |\varepsilon| \bullet \vec{E} \tag{2.13}$$

where χ_e and $|\varepsilon|$ are called polarizability and relative permittivity.

Similarly, introducing notations for currents

$$\vec{j}_P = \frac{\partial \vec{P}}{\partial t} \tag{2.14}$$

and

$$\vec{j}_M = \nabla \times \vec{M} \tag{2.15}$$

one can derive the relative permeability $|\mu|$ of magnetization. The current $\vec{j}_M$ and the time derivative of displacement $\partial \vec{D}/\partial t$ in Eq. 2.2 show reactions to the magnetic fields of microwaves in conductive and dielectric materials.

It must be noted that ε_0 and μ_0 are scalar numbers; however, $|\varepsilon|$ and $|\mu|$ have tensor forms. If they are complex conjugates, their imaginary parts correspond to the energy dissipations.

Once we obtain the permittivity tensor ε, Maxwell's equations can be written for plane waves showing the relationship between the alternating magnetic flux and its induced electric field:

$$\nabla \times \vec{B} = \mu_0 \varepsilon_0 \frac{\partial \vec{E}}{\partial t} + \mu_0 \vec{j} \tag{2.16}$$

$$\nabla \times \vec{E} = -\mu_0 \varepsilon_0 \frac{\partial \vec{B}}{\partial t} \tag{2.17}$$

2.1.2 *Waves in Materials and the Dispersion Relation*

Employing Fourier analysis $\exp\left\{i\left(\vec{k} \cdot \vec{x} - \omega t\right)\right\}$ Eqs. 2.16 and 2.17 combine to give the following wave equation in homogeneous materials containing charges and currents [9]:

$$\vec{k} \times \left(\vec{k} \times \vec{E}\right) + \varepsilon \cdot \vec{E} = 0 \tag{2.18}$$

where $\vec{k}$ is the wave vector, and the permittivity and electric field can be denoted as

$$\varepsilon \cdot \vec{E} = \begin{pmatrix} \varepsilon_{11} & -i\varepsilon_{12} & \varepsilon_{13} \\ i\varepsilon_{21} & \varepsilon_{22} & \varepsilon_{23} \\ \varepsilon_{31} & \varepsilon_{32} & \varepsilon_{33} \end{pmatrix} \begin{pmatrix} E_x \\ E_y \\ E_z \end{pmatrix} \tag{2.19}$$

For convenience, we introduce the dimensionless vector $\vec{n}$, as shown in Eq. 2.20:

$$\vec{n} = \frac{\vec{k}c}{\omega} \tag{2.20}$$

The magnitude $n = |\vec{n}|$ is the ratio of the velocity of light to the phase velocity. The reciprocal of n is the wave phase velocity divided by the velocity of light. Equation 2.18 can be written as

$$\vec{n} \times \left(\vec{n} \times \vec{E}\right) + \varepsilon \bullet \vec{E} = 0 \tag{2.21}$$

Introducing θ, the angle between $\vec{n}$ and the static magnetic field $\vec{B}_0 = \vec{z}B_0$, Eq. 2.21 is written as

$$\begin{pmatrix} \varepsilon_{11} - n^2\cos^2\theta & -i\varepsilon_{12} & \varepsilon_{13}\sin\theta\cos\theta \\ i\varepsilon_{21} & \varepsilon_{22} - n^2 & \varepsilon_{23} \\ \varepsilon_{31}\sin\theta\cos\theta & \varepsilon_{32} & \varepsilon_{33} - n^2\sin^2\theta \end{pmatrix} \begin{pmatrix} E_x \\ E_y \\ E_z \end{pmatrix} = 0 \tag{2.22}$$

The condition for a nontrivial solution is that the determinant of the square matrix is equal to zero. This condition gives the dispersion relation

$$An^2 + Bn + C = 0 \tag{2.23}$$

where

$$\begin{aligned} A &= \varepsilon_{11}\sin^2\theta + \varepsilon_{33}\cos^2\theta \\ B &= \frac{1}{4}\left(\varepsilon_{11}^2 - \varepsilon_{12}^2\right) + \varepsilon_{11}\varepsilon_{12}\left(1 + \cos\theta\right) \\ C &= \frac{1}{4}\left(\varepsilon_{11}^2 - \varepsilon_{12}^2\right)\varepsilon_{33} \end{aligned} \tag{2.24}$$

Equation 2.23 determines the propagating waves in materials in a simplified form of a quadratic equation. If the components of the permittivity tensor contain only real parts, the wave can propagate without energy dissipation. If they are complex conjugates, the energy transfers to the materials.

2.1.3 Energy Transfers from Microwaves to Materials

Energy transfers under microwave irradiation are analyzed. From Maxwell's equations (Eqs. 2.16 and 2.17), one can obtain Poynting's theorem, which is interpreted as a conservation theorem:

$$\nabla \cdot \left(\vec{E} \times \vec{B}\right) = \mu_0 \varepsilon_0 \left(\vec{B} \cdot \frac{\partial \vec{B}}{\partial t} + \vec{E} \cdot \frac{\partial \vec{D}}{\partial t}\right) \tag{2.25}$$

Poynting's vector $\vec{E} \times \vec{B}$ represents the flux of electromagnetic energy, and the right-hand quantity represents the rate of change of energy density.

Energies transferring from waves to charges balance with the counterreactions from the charges to the waves. If the relative permittivity and permeability are complex conjugates, the vector $\vec{E} \times \vec{B}$ might be also a complex conjugate. Suppose we have two vector quantities $\vec{A}$ and $\vec{B}$ and are interested in the product:

$$\begin{aligned}\vec{A}\vec{B} &= Re\left(A_0 e^{-i\phi}\right) Re\left(B_0 e^{-i\phi}\right) \\ &= \frac{1}{4}\left[A_0 B_0 e^{-2i\phi} + A_0^* B_0^* e^{-i\phi^*} + A_0 B_0^* e^{-i(\phi-\phi^*)} + A_0^* B_0 e^{-i(\phi^*-\phi)}\right]\end{aligned} \tag{2.26}$$

The asterisk indicates that the complex conjugate is to be taken. If we divide ϕ into its real and imaginary parts

$$\phi = \phi_r + i\phi_i \tag{2.27}$$

and if $\exp(\pm i\phi_r)$ actually periodic in space or time, we can average over a single period to obtain

$$\bar{AB} = \frac{1}{4}\left(A_0 B_0^* + A_0^* B_0\right) \exp 2\phi_i \tag{2.28}$$

provided ϕ_i is constant over this same interval. If $\exp(\pm i\phi_r)$ is almost periodic in space or time or if A_0, B_0, or ϕ_i is a slowly varying function of space or time, Eq. 2.28 is still correct in some mean value sense. Let us now assume that $\vec{E}$, $\vec{D}$, and $\vec{j}$ do not vary exactly as the real part of $E_0 \exp(\pm i\phi_r)$ but rather as the real part of $E_0 \exp\left\{-i \int_{-\infty}^{r} k(r')\,dr'\right\}$ or $E_0 \exp\left\{\int_{-\infty}^{t} \omega(t')\,dt'\right\}$.

Applying the averaged-value Eq. 2.28 to write Poynting's theorem on Eq. 2.25, the microwave energy transfer P_{MW} is written as follows:

$$P_{\mathrm{MW}} = \frac{1}{2}\sigma \left|\vec{E}\right|^2 + \pi f \varepsilon_0 \varepsilon'' \left|\vec{E}\right|^2 + \pi f \mu_0 \mu'' \left|\vec{H}\right|^2 \tag{2.29}$$

where σ is the conductivity; $|\vec{E}|$ and $|\vec{H}|$ are the absolute values of the amplitude of the electric field and the magnetic field, respectively; ε_0 and μ_0 are the permittivity of free space and the permeability of free space, respectively; ε'' and μ'' are the imaginary parts of relative permittivity and relative permeability; and f is the frequency of microwaves. The first term of the right-hand side corresponds to Joule heating in a conductive substance. The second term is due to dielectric losses. The alternating magnetic field $\vec{H}$ induces a magnetic loss by coupling with magnetization $\vec{M}$ in the third term of Eq. 2.29. The variety of knowledge on magnetism contributed toward an understanding of the mechanism of the loss at microwave frequencies; however, it has remained a frontier of physics. For example, although hysteresis loss is a well-known phenomenon of ferromagnetic materials, the microscopic mechanism has not been cleared yet for microwave irradiation. The new experimental results will be introduced in the next section. In this text, the notation μ'' describes these combined magnetic losses in Eq. 2.29, except for Joule loss by eddy currents that are induced by a temporal change of the magnetic flux, as explained by Maxwell's equations.

2.1.4 *Effect on Magnetization*

In microwave heating, magnetic materials as well as dielectric materials are heated. Could the materials be replaced by metals? But what about the metal state? In the case of bulk metal, it reflects microwaves. However, if it consists of powder metals, they can heat up to high temperatures by high-power microwave irradiation. The metal heating is derived only in powder form in which electromagnetic waves can penetrate deep into the body of the powder metal components. For example, in the experiments on microwave sintering of copper powders, the measured penetration depths were much larger than the penetration depth of the continuous metal (bulk metal). Here, the penetration depth δ is expressed as follows:

$$\delta = \frac{1}{\sqrt{\pi f \mu \sigma}} \tag{2.30}$$

where f indicates the microwave frequency and μ indicates the magnitude of permeability. For highly conductive materials (e.g., metal) the skin depth becomes very small at the microwave frequency. Copper is taken as an example of the heating of metal powders by microwaves [10]. The above experiment showed that permeability depends on the state of heterogeneity in micrometer order. The skin current shields the electromagnetic field on the bulk metal surface. If the nonconductive oxide layers could separate powder particles from each other, the skin current might circulate on each powder particle, and the wave fields could be transmitted into the matrix of powders. From the simulation [8], the density of the absorbed energy is maximum at some fixed particle radius, which equals 3.3 μm in the case of copper particles (the frequency of the microwave is 2.45 GHz). The study found out that single metallic particles are mostly heated by the magnetic component of the electromagnetic field.

2.1.5 *Internal Heating, Volumetric Heating, and Rapid Heating*

Microwaves penetrate into the bulk of condensed matter to be transformed into heat. This phenomenon is called internal heating.

In conventional heating, the surface temperature of condensed matter increases at first and then the thermal energy propagates into the bulk due to heat conduction. On the contrary, microwaves can supply the energy directly to the interior of materials. The phenomenon is called volumetric heating, which is advantageous for industrial applications, especially because conduction heating is very slow when a material is thick. As for a material with strong microwave absorbency, its temperature increases quickly, which is called rapid heating. The materials may fall into a thermal run-away when microwaves continue to be irradiated excessively, because ε'' becomes large with the increase of the materials' temperature. If thermal run-away occurs, it is quite difficult to control the temperature of materials.

2.1.6 *Selective Heating*

When a cup of milk is heated in a microwave oven, the temperature of the milk increases but the temperature of the cup does not increase. This feature of microwave heating is called selective heating. Selective heating occurs when absorption ratios of materials differ from each other. In selective heating, the local temperature difference is made in materials and a thermal nonequilibrium state occurs in the macrostate.

2.2 Nonthermal Microwave Effects

The macroscopic energy transfers depend on the electric and magnetic fields in the microwaves and on the permittivity, permeability, and conductivity of the materials. Instead of calculating the overall contributions of charged particles, which is practically impossible, we will embark on a statistical description in this chapter. Since 1990, it has been reported that a chemical reaction in microwave heating is different from that in conventional heating. The phenomena are well known as microwave effects. In addition to thermal microwave effects, as mentioned before, a lot of unknown phenomena were reported, for example, changes in the materials' structure in microwave **H**-field heating [2], sintering enhancement [11], etc. In the chapter, microwave effects are clearly divided into two, thermal microwave effects and nonthermal microwave effects. In the next section, examples of nonthermal microwave effects are discussed in detail.

2.2.1 *Decrystallization and Formation of Nanodomain Structures*

Roy et al. reported that Fe_3O_4 or TiO_{2-x} becomes a noncrystalline structure when the sample is heated at the maximum point of the microwave **H** field using a single-mode cavity (the phenomenon is called decrystallization) [2, 12]. Moreover, the authors reported that the decrystallized materials formed nanodomain structures [13, 14]. For example, the nanodomain structures of Fe_3O_4 are obtained

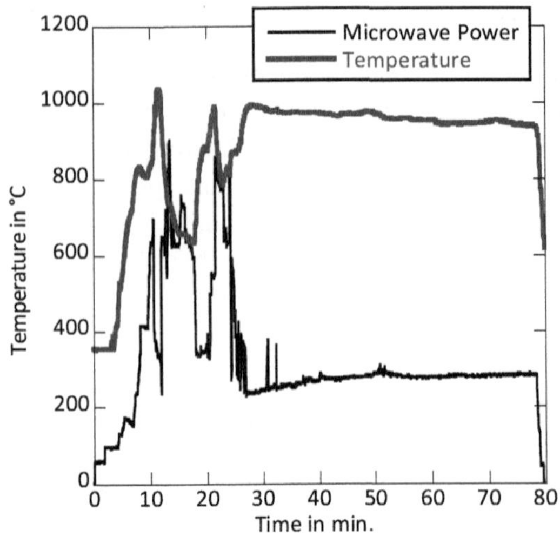

Figure 2.1 The typical progress of process temperature measured during microwave heating.

after microwave irradiation. The graphs in Fig. 2.1 show a typical progress of process temperatures measured during microwave heating of magnetite samples. The microwave power is controlled manually. The microwave power shows the remainder of input power and reflex power. For sintering experiments, the magnetite powder compacts were heated to the sintering temperature of 1000°C for 10 min, followed by 70 min of soak time. In the initial step of heating during the first 30 min, the sample temperature changed dramatically. Thereafter, the sample temperature settled a at microwave power as low at 270 W, which is one-third of the peak.

Figure 2.2a shows a magnetite sample sintered by using the **H** field of microwaves at 1000°C. The size of the sample varied from 6.2 to 8 mm in diameter after heating. The sintered sample was found to be oxidized only on the surface. The linear shrinkage of the sample was 17%, and no remarkable change in shape was observed. With a green density of 45%, a maximum sintered density of 82% theoretical density could be achieved. Figure 2.2b shows the X-ray diffraction profile of the cross section of magnetite after heating at 1000°C. It shows Bragg's reflections of both magnetite and hematite.

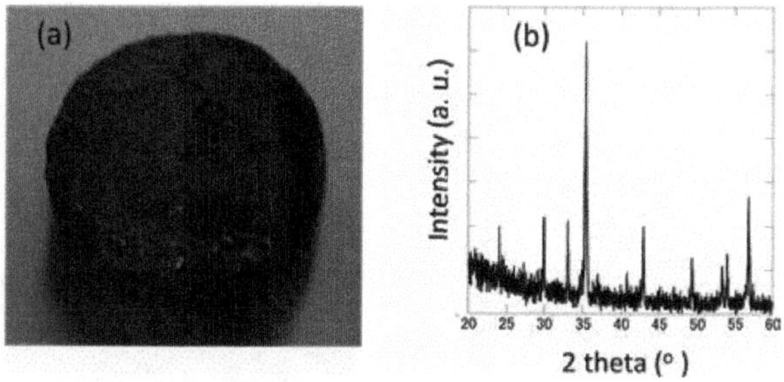

Figure 2.2 (a) Magnetite sintered by using the **H** field of microwaves at 1000°C. (b) X-ray diffraction of samples heated by the magnetic field at 1000°C.

The surface of the sample was oxidized because nitrogen gas was flowing in the cavity during heating.

The high-resolution transmission electron microscope (TEM) (JEOL JEM-3200) observed the images of the original Fe_3O_4 powder before heating and those excited in the selected magnetic field. The original crystal presents well-ordered lattice patterns over the whole crystal; therefore it has flat and homogeneous surfaces. Figure 2.3 shows the TEM image and the selected area electron diffraction pattern. The sample heated in the magnetic field exhibits the presence of a randomly oriented nanocrystal. The crystal size was an approximate average of 10 nm in the range. The randomness of the lattice orientation indicates that the particle–particle magnetic interaction is negligible in the heating–cooling process. It is very interesting that the temperature increase in the sample is accelerated at a temperature close to the Curie point and never rises above 1000°C but is well below the melting point of 1535°C.

The macroscopic magnetizations of samples were measured by Superconducting Quantum Interference Device (SQUID) (Quantum Design MPMS-XL7). Figure 2.4 shows the hysteresis loop of magnetite at 300 K. As can be seen in Fig. 2.4, the saturated magnetization (M_s) of sintered magnetite with a nanocrystal indicates alike M_s of original magnetite. The averaged remanence (M_r) and coercivity (H_c) of magnetite before heating were 11.74 emu/g and 108.9 Oe. On the other, the averaged remanence (M_r) and

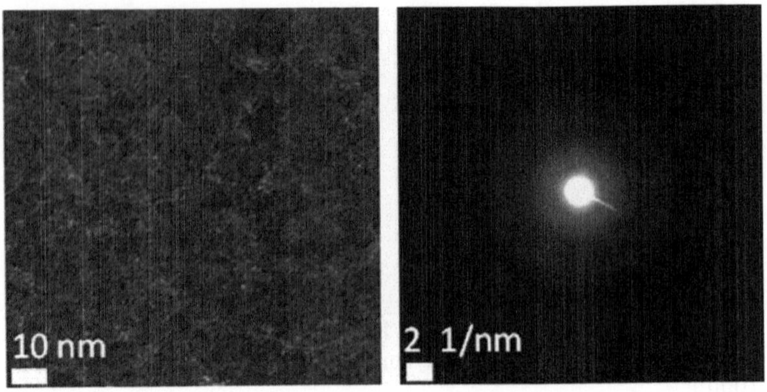

Figure 2.3 TEM image and electron diffraction pattern of magnetite sintered by using **H** field of microwaves at 1000°C.

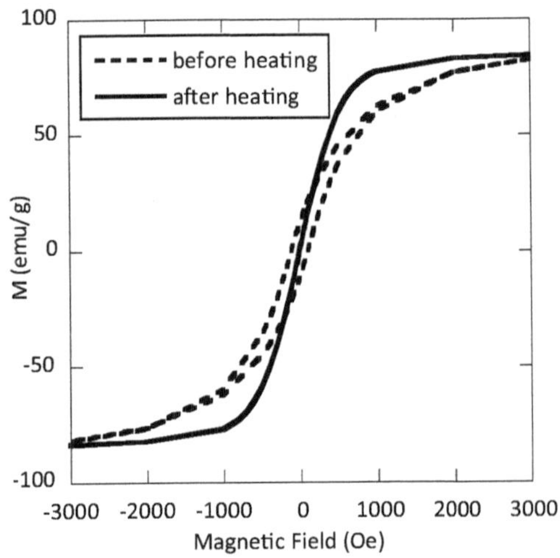

Figure 2.4 The hysteresis loop of magnetite sintered by using the **H** field of microwaves at 1000°C.

coercivity (H_c) of magnetite after heating were under the measuring limit of SQUID, that is, nearly zero.

In another example, the morphology of TiO_{2-x} after conventional heating and at the maximum point of **H**-field heating using the single-mode cavity is shown in Fig. 2.5. After conventional heating, TiO_{2-x} was crystallized because the Laue spots were round in

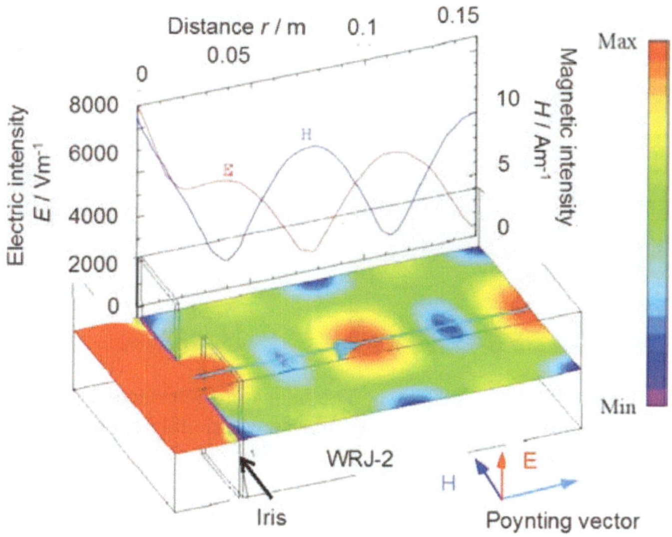

Figure 2.5 Distributions of **E**- and **H**-field intensities in a TE_{103} single-mode cavity.

shape. On the other hand, after **H**-field heating, the Laue spots showed concentric arcs, which illustrates that TiO_{2-x} crystals become amorphous. From the TEM image, TiO_{2-x} does not become amorphous but nanodomain crystalline structures are formed with 5–10 nm particle size, as shown in the areas encircled by red in Fig. 2.6. The volume of the nanodomain is less than 20% of the total volume of the sample [15].

2.2.2 *Enhancement of Nonthermal Reduction of CuO and TiO_{2-x} during Microwave Heating*

In this section, we indicate the study of reduction reaction of transition metal oxides as an example of nonthermal effect on enhancement of chemical reactions during microwave irradiation. A reduction behavior of CuO and TiO_{2-x} during microwave (**H**-field and **E**-field) heating and conventional (electric furnace) heating was investigated. For in situ qualitatively measuring reduction reaction, the authors developed the device that can analyze the outgas composition during microwave irradiation or conventional

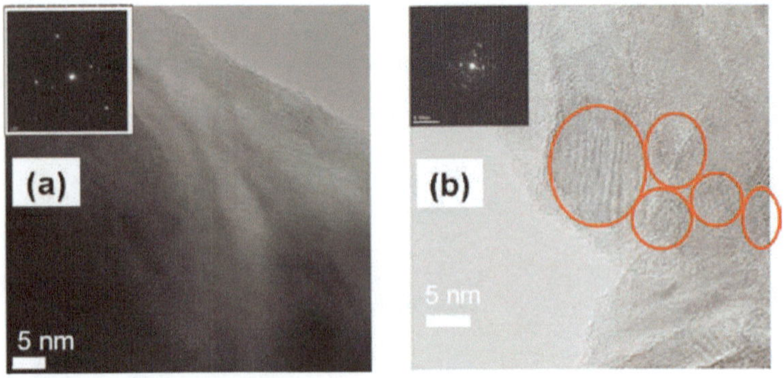

Figure 2.6 TEM image of TiO_{2-x}. The image of the selected area is shown at ttop left in both TEM images: (a) after conventional heating and (b) after **H**-field irradiation.

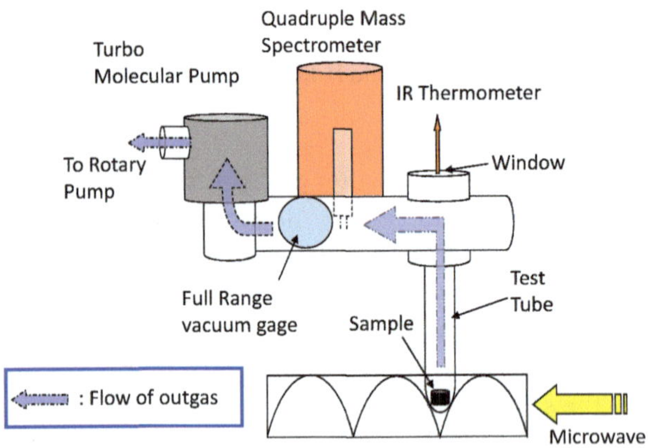

Figure 2.7 Schematic view of the experiment device. The blue arrow shows the flow of emitted oxygen. The schematic view shows when a sample is heated at the maximum point of microwave's **H** field.

heating. Figure 2.7 shows the in situ outgas analysis system and heating furnace. The vacuum level of the system was about 1.1×10^{-5} Pa. Single-mode cavity TE_{103} was used for the microwave-heating furnace. The previous report describes the device features in detail [15–17]. CuO and TiO_2 were selected as samples. The difference between CuO and TiO_{2-x} is the standard free energy of formation of oxides. It is easy to reduce CuO because the standard free energy of formation of oxides is small. On the other hand, it is relatively difficult to reduce TiO_2 because the standard free energy of formation of oxides is large. TiO_2 sample was heated using an electric furnace up to 900°C and held at the temperature for 1 h. After preheating, TiO_2 was reduced slightly to form stoichiometric TiO_2 (TiO_{2-x}). When TiO_2 becomes TiO_{2-x}, the unpaired electron in the $3d$ orbital appears. CuO powder has unpaired spin in the $3d$ orbital naturally. Thus CuO and TiO_{2-x} can be heated because the microwave's **H** field interacts with the unpaired electron [18]. The experimental results in **H**-field irradiation of CuO are shown in Fig. 2.8a. The **H** field means that the sample was heated at the **H**-field node. When the sample was heated in conventional heating, the rate of reduction increased gradually more than 500°C. When the sample was heated in **H**-field heating, the rate of reduction increased at 320°C, and it increased sharply at 360°C. Then it increased gradually to 620°C, and it increased sharply at 630°C and finally became stable from 640°C to 700°C.

These data clearly show that the enhanced reduction occurred when the sample was subjected to the **H** field. The temperature at which the reduction starts is low in **H**-field heating. In **H**-field heating, the reduction speed was at least 1 order faster than conventional heating. At 520°C, the rate of reduction in **H**-field heating was about 103 times larger than that in conventional heating.

Figure 2.9a shows the temperature dependence of the rate of reduction in **E**-field heating and that in conventional heating. When the sample was heated in an **E** field, the rate of reduction increased at 370°C, and it increased gradually to 590°C. However, the reduction speed was suddenly dropped at 590°C and decreased gradually to 620°C, and then it was slightly increased. In **E**-field heating, the oxygen emission was stopped after emitting a certain amount of oxygen.

Then we examined activation energy for reduction by thermal energy. The expression of the reduction rate is

$$k = A \exp\left(\frac{-\Delta E}{RT}\right) \tag{2.31}$$

$$\ln k = \frac{-\Delta E}{RT} + \ln A \tag{2.32}$$

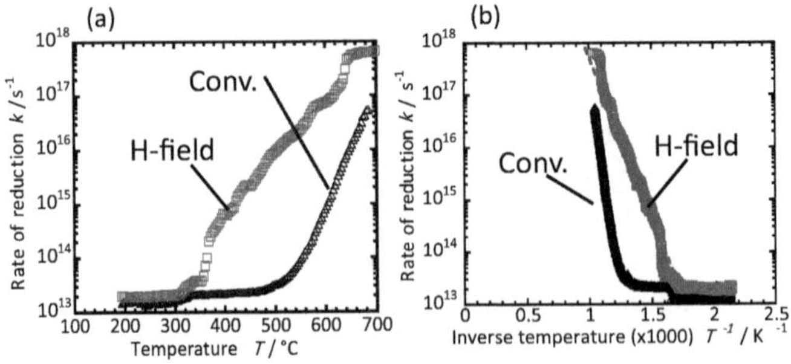

Figure 2.8 (a) Temperature dependence of the rate of reduction in an **H** field and in conventional heating of CuO. Open squares indicate the result in **H**-field heating, and open triangles indicate the result in conventional heating. (b) Arrhenius plot of the results.

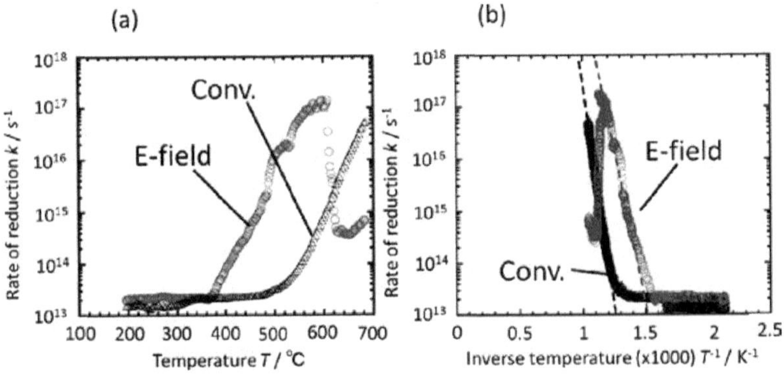

Figure 2.9 (a) Temperature dependence of the rate of reduction in an **E** field and in conventional heating of CuO. Open circles indicate the result in **E**-field heating, and open triangles indicate the result in conventional heating. (b) Arrhenius plot of the results.

where A is the pre-exponential factor and ΔE is the activation energy. When the logarithm of both sides of Eq. 2.31 is taken, the logarithm of the reduction rate is in proportion to the inverse temperature (Eq. 2.32). Figures 2.8b and 2.9b show the Arrhenius plot in each heating. The apparent activation energy (ΔE) is determined from the gradient of the Arrhenius plot. ΔE was 333 kJ/mol in conventional heating, 117 kJ/mol in **H**-field heating, and 228 kJ/mol in **E**-field heating.

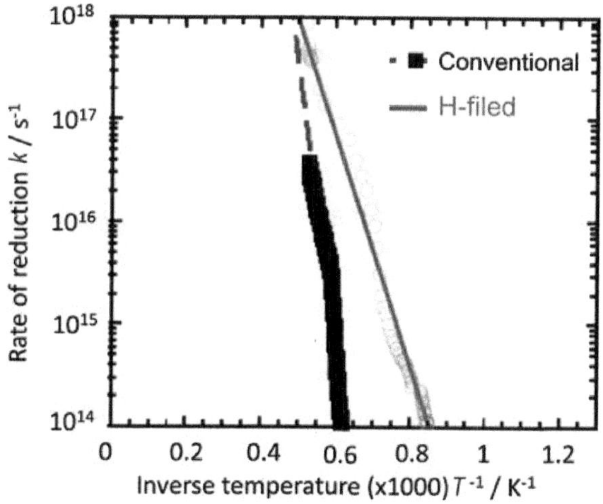

Figure 2.10 Arrhenius plot of the rate of reduction of TiO_{2-x} in conventional heating and in **H**-field heating. Solid squares indicate the result of conventional heating, and open circles indicate the result of **H**-field heating. Dashed lines (in conventional heating) and solid lines (in **H**-field heating) are determined by the least squares method.

Figure 2.10 shows the Arrhenius plot of the reduction rate of TiO_{2-x} in conventional heating and **H**-field heating. Oxygen was emitted in **H**-field heating at a lower temperature than in conventional heating. The results show that enhanced reduction of TiO_{2-x} occurred in **H**-field heating. ΔE of TiO_{2-x} in **H**-field heating decreased similarly as was observed for CuO. CuO; the values of apparent activation energies (E) were 517 KJ/mol in conventional and 214 KJ/mol in **H**-field heating. ΔE in conventional and **H**-field heating was 517 and 214 kJ/mol, respectively. Activation energy for reducing TiO_{2-x} in **H**-field heating decreased significantly compared to that in conventional heating. As the case of CuO, the activation energy of the TiO_{2-x} reduction reaction in **H**-field irradiation decreased significantly compared to that in conventional heating.

These phenomena introduced in Sections 2.3.1 and 2.3.2 cannot be explained, supposing all microwave electromagnetic field energies change to heat. That is, it is necessary to improve the first assumption that all microwave electromagnetic field energies change to heat. In the following section, the interaction of microwave electromagnetic waves and materials is improved once again and true "general features of microwave interaction with material" will be considered.

2.2.3 *Sintering Behavior of Metal Powders Involving Microwave-Enhanced Chemical Reaction [19]*

A sample was placed inside a container made of quartz, which is microwave transparent at room temperature. The quartz was used for microwave sintering when copper powders were sintered, because the quartz transmits infrared radiation. This transmission of infrared radiation is important for temperature measurements using a pyrometer.

The space between the container and the sample is filled with boron nitride powder to reduce the contact with air. The container was surrounded by thermal insulation made of silica–alumina fiberboard. The inner wall of the fiberboard was coated with silicon carbide (SiC) powder to a thickness of 2 mm. This SiC susceptor was used to achieve a high heating rate.

The samples were heated by microwaves. Microwaves from two magnetrons were introduced through two ports into the chamber. The magnetrons generate 1.5 kW each at 2.45 GHz. Each magnetron has small differences in frequency and the beat of waves occurring in the chamber. In addition, mode stirrers made the electromagnetic field uniform. Therefore, these beats were moved by two mode stirrers, and we were able to heat the sample uniformly because the beats channel a standing wave in the chamber. The temperature of the container was measured using a pyrometer, because metal samples cannot be measured directly using a pyrometer due to varying emissivity with temperature and the metal effect on the surface.

The samples were pressed uniaxially. The starting copper powder was made by dissociation. The copper powder type was PC100 with a purity of 99.7 wt% from Fukuda Metal Foil Powder, Japan. The powder was characterized by a dendritic particle shape, and 99.5 wt% of the particles were less than 45 mm in size. The starting metal powders were oxidized on the surface because they are usually kept in air. The color of the metal powder was reddish

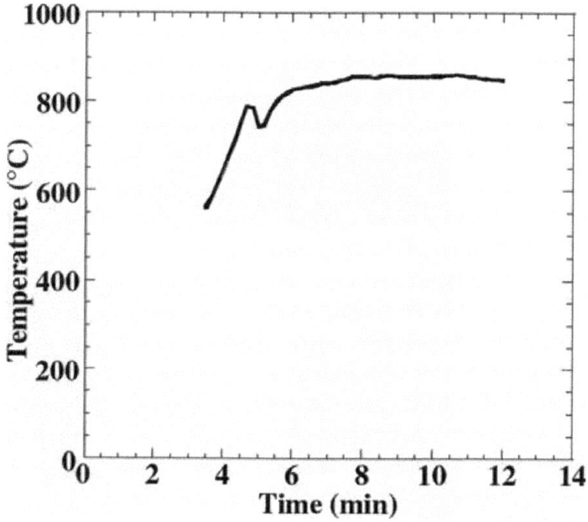

Figure 2.11 Heating profile for sintering copper powder sample at 850°C by microwaves.

brown. The compaction pressures were 5 and 1 ton/cm^2. The size of the samples compacted at 5 and 1 ton/cm^2 were 40 × 10× 6 and 100× 20× 6 mm^3, respectively.

Solid copper tubing, which was filled with copper powder up to 45 wt% density, was prepared for comparison. It has a rectangular shape and dimensions identical to those of the compacted powder samples. The 0.8 mm thickness of the tubing was sufficient to prevent microwave penetration. Both ends were loosely closed with solid copper plugs. Gases, including oxygen, could be transferred through the opening. The plugs blocked the microwaves.

Figure 2.11 shows the heating profile for sintering copper samples at 5 ton/cm^2 by microwaves. The temperature reached 850°C in 7 min and was held there for 5 min. The microwave power in the oven was 3 kW up to 850°C and was reduced from 2.5 to 1 kW while holding the temperature. The microwaves were continuous wave and were controlled manually. Figure 2.12a shows a picture of the copper sample sintered by microwaves. The samples were deoxidized on the surface. This sample shrank 5 wt%, and its bulk density reached 8.68 g/cm^3. Its relative density was 97 wt%.

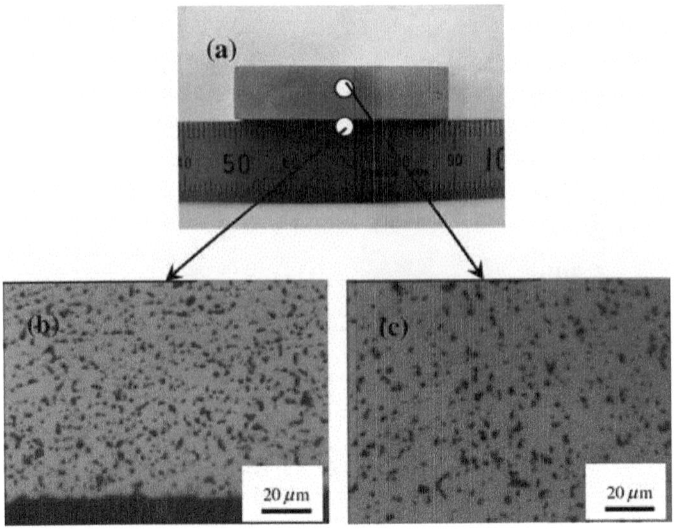

Figure 2.12 Copper powder sample sintered at 850°C by microwaves (a), (b) microscopic view of edge of cross section, and (c) microscopic view of inside of cross section.

Figures 2.12b and 2.12c show the microscopic view of the edge and the inside of the cross section of the sintered copper sample. The sample was well sintered throughout the cross section. In particular, the inside of the copper sample was denser than the surface, as is evidenced from scanning electron microscopy (SEM) images.

Figures 2.13a and 2.13b show energy-dispersive X-ray spectroscopy (EDX) data from the surface of a copper powder sample sintered at 900°C in air by microwaves and by conventional heating, respectively. The conventional method used electric heaters. The microscopic views in Figs. 2.13a and 2.13b show the areas analyzed by EDX on the surface of each sample. The only difference between Figs. 2.13a and 2.13b is the method of heating. In the case of microwave heating, we can observe the sintering behavior. However, in conventional heating, the oxide layer diffused into the sample surface without sintering. The ratio of oxygen in the copper powder sample sintered by conventional heating was above 50 at%. This means that sufficient air for oxidation remained in the container. However, in the copper powder sample sintered by

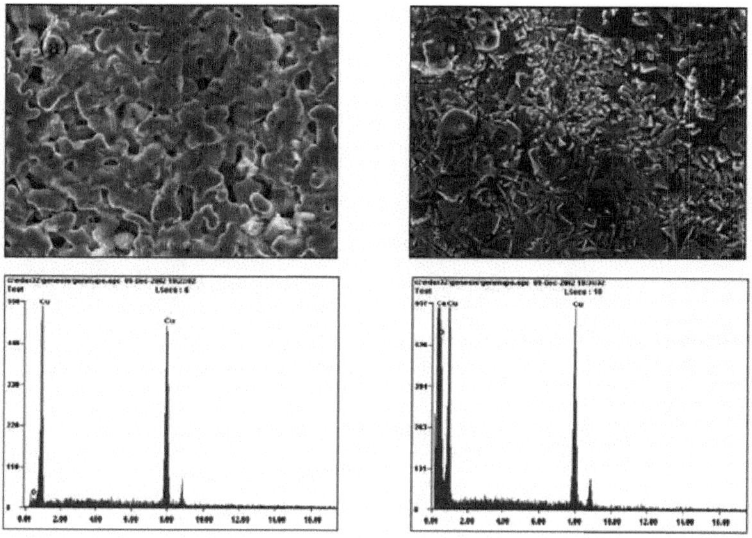

Element	wt%	at. %
Ok	02.21	08.24
Cuk	97.79	91.76

Element	wt%	at. %
Ok	22.93	54.16
Cuk	77.07	45.84

Figure 2.13 EDX data of the surface of copper powder sample sintered at 900°C in air.

microwaves, the ratio of oxygen was below 10 at%. This suggests that microwaves have accelerated the process of deoxidation. Figure 2.14a,b shows X-ray diffraction (XRD) patterns of the cross section of samples sintered at 900°C by microwave and conventional heating, respectively. The Cu_2O peak appears in the sample sintered by conventional heating. There are almost no Cu_2O peaks in the sample sintered by microwaves. The Cu_2O peak in Fig. 2.14b is smaller than the peak in the EDX data in Fig. 2.13b, because oxygen cannot enter the sample due to the oxide layer on the sample surface. The Cu_2O peak in Fig. 2.14b must be the oxide in the starting copper powder.

The solid copper tubing sample was heated to 850°C for 3 min and held for 4 min, as shown in Fig. 2.15. The microwave power was 3 kW during heating up to 850°C and was reduced to 2.5 kW to maintain the temperature at 850°C. Thus the power

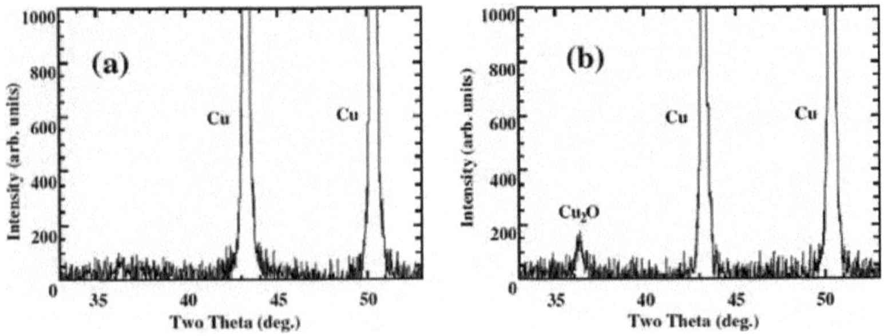

Figure 2.14 XRD patterns on cross section of copper powder sample sintered at 900°C (a) by microwaves and (b) by conventional heating.

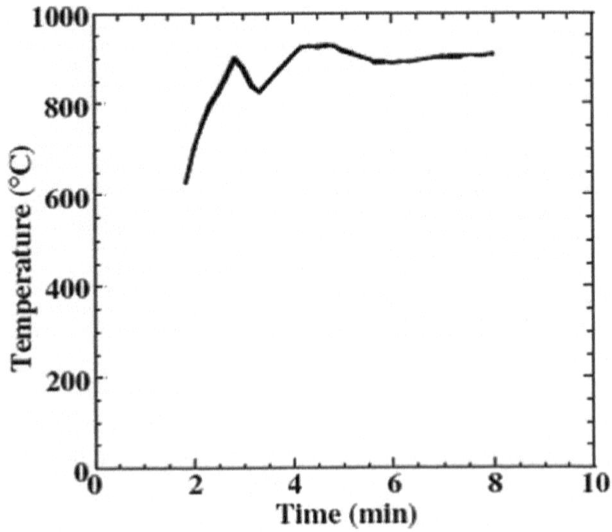

Figure 2.15 The heating curve of solid copper tubing for sintering at 850°C by microwaves.

required to maintain the temperature at 850°C decreased gradually. In the end only 1 kW was needed. Microwaves are reflected by copper melt. Consequently, the sample was heated by ohmic resistance on the sample surface and by radiant heat from the SiC susceptor. Therefore, the copper powders were heated by only thermal conduction without microwaves. Figure 2.16 shows the solid copper tubing before heating and after heating by microwaves.

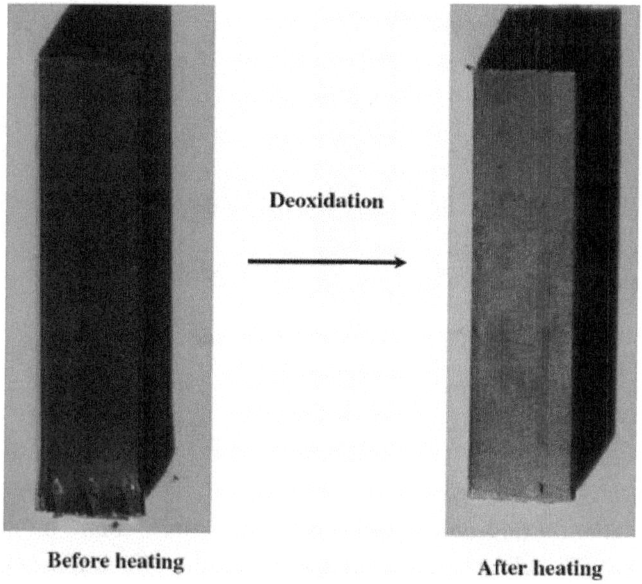

Figure 2.16 Solid copper tubing before heating and after heating at 850°C by microwaves.

The solid copper tubing looked dark reddish brown before heating, indicating copper oxide on the surface. However, after heating, the color changed to a metallic color. This suggests that the surface of the solid copper tubing was deoxidized during microwave heating. However, the colors of the copper powder in the solid copper tubing before and after heating were dark reddish brown and black, respectively. The copper powder in the tubing did not sinter. This suggests that the copper powders were not deoxidized. Exposure to the microwave field must have played a key role in direct microwave heating.

Figure 2.17a shows the microscopic views of a copper powder sample sintered by microwaves, and Fig. 2.17b shows the microscopic view of the copper powder in the solid copper tubing after microwave heating. The powder in the solid copper tubing retained a large amount of oxide components in the grains of particles and did not shrink from the green body to the sintered state. Samples without the copper tubing deoxidized and sintered to a high density

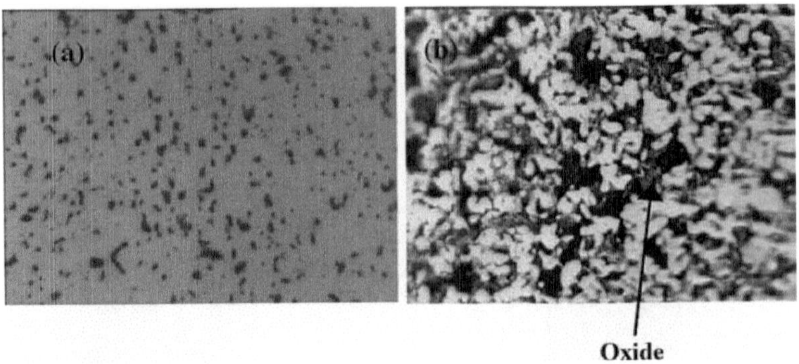

Figure 2.17 (a) Microscopic view of copper powder sample sintered at 850°C by microwaves; (b) microscopic view of copper powders in solid copper tubing after microwave heating at 850°C.

throughout. Good deoxidation, as good as those found on samples without the solid copper tubing, was found on the outer surface of the solid copper tubing that was exposed to the microwave field.

The results of the EDX data, as shown in Fig. 2.13a,b, indicate that the ratio of oxygen in the copper sample heated in air by microwaves is below 10 at.%. However, it is above 50 wt% in the sample heated in air by conventional heating. There is a very large difference between the sample heated by microwaves and that heated by conventional heating. We believe that microwaves have accelerated the deoxidation reaction. One possible explanation for this is that some arcing happens between the metal powders because the oxide layers on the metal powder surfaces are very thin and the input microwave power is very high at the kilowatt level. As the results of the XRD patterns shown in Fig. 2.14a,b indicate that Cu_2O peaks appear on the cross section of the sample heated by the conventional method. There are almost no Cu_2O peaks on the cross section of the sample heated by microwaves. These results show that microwaves affect deoxidation reaction during the sintering process. Of course, although microwave sintering of metal powder compacts receives growing interest, the mechanisms that allow heating metal powder samples homogeneously by microwaves to sintering temperatures and accelerating chemical reactions, such as deoxidation, are not well understood. However, the electromagnetic field of a microwave may be directly responsible for deoxidation; it may act as a catalyst for the deoxidation reaction. At the least, the results show that the microwave has activated the process of deoxidation.

2.3 Energy Transfer Mechanism on Microwave Processing

2.3.1 *Introduction*

A microwave is an electromagnetic wave. It is not heat. However, we know that materials become hot in colder circumstances by irradiation of microwaves. In this section, we will introduce mechanisms of the interactions of electromagnetic waves and dense materials according to the latest theoretical investigations.

Electromagnetic waves propagate in dielectric materials, depending on Maxwell's equations. The phase velocity of the electromagnetic wave reduces to $c/\sqrt{\varepsilon}$, where, c and ε are the speed of light in vacuum and the dielectric constant of the material. As ε is in the order of 1–100, the phase velocity of the microwave is almost 3–0.3 × 10^8 m/s. The wavelength at 3 GHz is 100 mm in vacuum and reduces to from 30 to 3 mm in the material. The wavelength is still much longer than the microscopic structures in the materials. It means that microwaves are considered to be plane waves in the microstructures. The electromagnetic wave impacts on the electrons and ions in the material in accordance with the alternation of each cycle of oscillation.

As the phases of the impacts are synchronized, the energies flow back and forth between the kinetic energies of charged particles and the electromagnetic field. The collisions between the charged particles scattered the ordered motions and transferred energy from the electromagnetic field to the material. The randomized phases reduced finally to the highest entropy called thermal equilibrium. It is a well-known mechanism of microwave heating.

However, in around 1980, a number of unexpected behaviors were reported under highly intense microwave irradiation. For example, when the activation energy of microwaves was sufficiently dropped in materials, chemical reactions become faster and higher in microwave irradiation than those under conventional heating. These phenomena are called microwave effects or nonthermal effects.

Stuerga and Gaillard wrote a paper entitled "Microwave Athermal Effects in Chemistry: A Myth's Autopsy, Part I and II" [20] in 1996. The paper suggested that the microwave effects could be the orienting effect of the electric field: microwave irradiation could induce molecular organization different from that induced by a heating mode. The numbers of dipoles will arrange between θ and $\theta + d\theta$ to the applied electric field, as described by a statistical average of angular position $\langle \cos\theta \rangle$, as written by

$$\langle \cos\theta \rangle = \frac{\int_0^{\infty} \cos\theta \sin\theta \exp\left(\frac{-W}{kT}\right) d\theta}{\int_0^{\infty} \sin\theta \exp\left(\frac{-W}{kT}\right) d\theta} \tag{2.33}$$

The anisotropic orientations of particles result in reductions of entropy by microwaves. The anisotropy will relax to random motions via a collisional process. This cannot be explained by a simple mechanism, depending on the collisional relaxations mentioned before.

In 2002, Roy et al. demonstrated that the **E** and **H** fields of microwaves showed different effects on materials [2]. The intense **H** field irradiations at microwave frequencies reconstructed the polycrystalline structures of metal oxides consisting of the micron meters of crystalline to the very fine nanocrystalline structure under a temperature as low as one-third to three-fourths of the melting point. It was clear evidence that the microwave energy couples to the crystal structures via specific channels, which might be translations, rotations, and vibrations, depending on the state of molecules and lattice.

In this section, we will investigate in detail the energy conversion paths in which the kinetic impacts, given by the alternating electromagnetic fields, are accumulated energies to the order of sufficient energy levels without a collisional process.

Generally speaking, materials are not homogenous. The discontinuity of surfaces, vacancies, magnetic domains in crystals and groups, and radicals on large molecules gives heterogeneity to the microscopic structures of materials. When alternating electric or magnetic fields are given by an external microwave to heterogeneous materials, it creates a force (stress) on the surface of the crystals. The wavelength of the microwave is much longer than the characteristic length of microscopic heterogeneity. The excited vibration is collective. Such chains of stress–strain of ordered motions of lattices, ions, and molecules can propagate with the phase velocities determined by the dispersion relations.

2.3.2 *Chemical Reactions and Phase Transitions under Microwave Irradiation*

In 1935, Eyring [21] published the famous transition state theory that derived the rate of chemical reactions by statistical methods. The theory introduced the hypothesis that the chemical reactions proceeded from the initial state A to the final state B by overcoming the potential peak, named activation complex, with an enthalpy $X^{\ddagger}$. The rate of chemical reactions was the result of the summation of each of the four reactions on the directivity to A$\rightarrow X^{\ddagger}$, A$\rightarrow X^{\ddagger}$ and B$\rightarrow X^{\ddagger}$, B$\rightarrow X^{\ddagger}$. Followed by the conventional transition state theory, Laidler [22, 23] and many other researchers extended it to the general principle for understanding of photochemical reactions on solids, liquids, surfaces, and catalysts.

In 2009, Towe hypothesized that reported physical changes in certain materials resulting from their low thermal exposure to intense gigahertz microwave radiations might be explained through generation of electromagnetically induced ultrahigh-frequency ultrasound (hypersonic) waves within their structures [24].

What is the common ground of a hypersound wave and microwave in a dense matter? Electromagnetic waves can propagate in vacuum by exchanging the energies between the alternating electric and magnetic fields. When electromagnetic waves are injected into a dense matter, the electric and magnetic fields react with the charged particles. The energy conservation law restricts that the summation of energies in the alternating electromagnetic fields and in kinetic waves shall be constant. It means that the electromagnetic fields in a dense matter are not the same in the vacuum.

The vacuum wavelength of microwave is in the range of 1–100 mm that is much longer than the lattice spacing of solids and the distances of atoms in molecules of liquids. It ensures that the microwave excites a forced oscillation of a plane wave on the boundary (surface) of vacuum and the matter. The wave fronts are the parallel planes of constant peak-to-peak amplitude normal to the phase velocity vector. The theorists tried to explain that the direct interactions of microwaves and electrons and ions in a molecule produced chemical reactions. The photon energy of the microwave frequency is of the order of 10^{-5} eV, which is too low to break the chemical bonding. What is the mechanism to accumulate the energy of the order of 1 eV? The question remained unanswered. No one had paid attention to the massive motions of particles excited by microwaves. Towe's suggestion introduced energy stimulations by ordered motions of particles.

In this chapter, we focus on the mechanisms in which the coherent motions excited, conserved, and stimulated energy to the level for breaking chemical bonds and transferring lattice structures.

2.3.3 *Excitation of Ultrasound Waves by External Microwaves*

Materials are heterogeneous in nature consisting of grain boundaries, magnetic domains in polycrystals, powders, and clusters. The electrons can move more widely on the surface than in the body of crystals. Figure 2.18 shows a schematic description of a sound wave

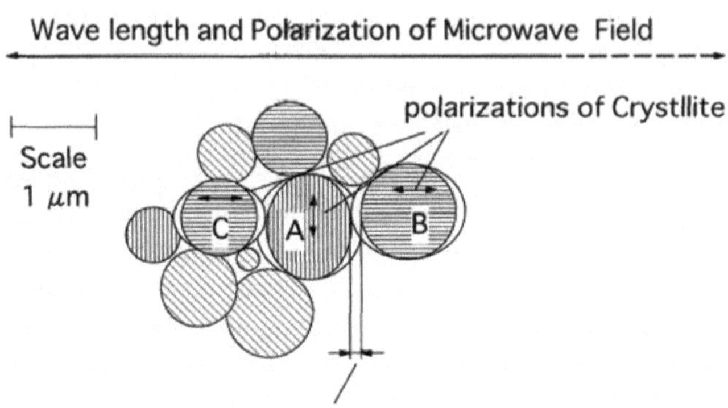

Figure 2.18 An image of excitation of sound in polycrystallin media by microwave the hatched lines show the direction of polarization in each crystallite.

coupling with a microwave on the surface of a permittivity material. The set of equations of motion of nonisotropic elasticity (depends on Hooke's law) and alternating electromagnetic field (Maxwell's equations) gives the chains of compression and decompression. Steele and Vural [25] showed the relations between stress and strain

$$\rho \frac{\partial^2 u_i}{\partial t^2} = \frac{\partial T_{ij}}{\partial x_j} \tag{2.34}$$

$$T_{kl} = C_{ijkl} S_{ij} \tag{2.35}$$

where T_{kl} are the components of the second-rank stress tensor T. In the absence of piezoelectricity, T is related to the second-rank strain tensor S through the fourth-rank elastic tensor C. If a unit cell of a crystal contains at least two different atoms, that crystal can exhibit piezoelectricity. In such material, stress can produce electricity, or vice versa, and applying a voltage can produce stress. The relationship between elastic and electrical variables requires the introduction of the piezoelectric tensor e, which is a third-rank tensor. In the acoustic-electric effect, the pertinent relations are given by

$$T_{kl} = C_{ijkl} S_{ij} - \bar{e}_{mkl} E_m \tag{2.36}$$

$$D_n = \varepsilon_{mn} E_m + \bar{e}_{nij} S_{ij} \tag{2.37}$$

where ε_{mn} is a lattice dielectric tensor. Equation 2.34 gives a well-known relation between the electric displacement $\vec{D}$ of anisotropic media and the electric field $\vec{E}$. If there are mobile electrons in the solid, a current subjected to an **E** field will be produced. The current is expressed as $J_{ij} = \sigma_{ij} E_j$, but it is now out of our concern in the processing of dielectric materials. For the wave, all variables are assumed to be of the form of $\exp\left\{i\left(\vec{k}\cdot\vec{x} - \omega t\right)\right\}$. The acoustic portion of the problem is completely described by the displacement u, which is a function of ρ, C, ω, k, e, and $\vec{E}$.

Let us consider the longitudinal sound propagation. Since the shear wave has only a u_x displacement,

$$S_{13} = S_{31} = \frac{1}{2}\frac{\partial u_x}{\partial z} \tag{2.38}$$

With only $E_z \neq 0$ for the longitudinal waves considered, the stress and electrical displacement components given by Eqs. 2.34 and 2.35 become

$$T_{13} = T_{31} = 2C_{44}S_{13} - \bar{e}E_z \tag{2.39}$$

and

$$D_z = \varepsilon E_z + 2\bar{e}S_{13} \tag{2.40}$$

Using Eqs. 2.34 and 2.40, the equation of motion of lattice becomes

$$\left(-\rho\omega^2 + C_{44}k^2\right)u_x = ik\bar{e}E_z \tag{2.41}$$

For $\bar{e} = 0$ (no piezoelectricity), Eq. 2.41 describes the usual sound wave equation of shear mode in a cubic crystal, with the sound velocity given by $v_s^2 = C_{44}/\rho$. With $\bar{e} \neq 0$, the sound is coupled to the electron motion through the field E_z. It is a hybrid mode with sound and electrostatic waves.

Now the coupling between the electron motion and the sound is calculated with the help of Maxwell's equations. This leads to a triplet of equations relating to the components of the self-consistent electric field

$$\sum_{j=1}^{3} \alpha_{ij}E_j = 0 \tag{2.42}$$

where α_{ij} contains both electrons and sound parameters. The determinants of three equations described by Eq. 2.40 yield the dispersion relation

$$D(\omega, k) = \mathrm{Det}\left(\alpha_{ij}\right) \tag{2.43}$$

The solutions of $D(\omega, k) = 0$ give the allowed modes of the composite system. These combined modes are called electrokinetic waves. As illustrated in Fig. 2.18, the microwave transmits in the z direction and incidents normal to a crystal surface. The electric field E couples to the electric dipole. The motion of electrons excites the shear sound wave at microwave frequency and the perturbations transmit in the z direction with a phase velocity given by Eq. 2.39. The E_x field gets a Debye shield on the surface; instead of it, the E_z field appeared in an electrostatic mode. If the electric field of an external microwave is perpendicular to the surface, dipole motions excite a longitudinal sound mode.

The sound waves dissipate energy along the propagation. The damping coefficient increases in proportion to the square of the frequency. The estimated length is less than a wavelength at 1 GHz in homogeneous materials. However, in heterogeneous materials with the characteristic size a of the particles, the sound absorption coefficient varies as ω^2 at a very low frequency ($\omega \ll \chi/a^2$, where χ is the thermometric conductivity); for$\chi/a^2 \ll \omega \ll v_{\rm ph}/a$, it varies as $\omega^{1/2}$; and for $\omega \gg v_{\rm ph}/a$, it again varies as ω^2. Typical ceramics such as alumina have $\chi = 1.24 \times 10^{-5}$ m^2/s and $a \approx 5$ μm; the dumping length is estimated as 0.6 m at 2.45 GHz. Sound waves can propagate in powder particles almost free from thermal diffusions [26].

It is clear that the external microwave energy must be converted to waves, including the kinetic motions of the atoms, molecules, and crystals. In thin materials such as air, the kinetic energy is negligibly small. However, in dense materials, such as solids and liquids, most of the energy couples to the ordered kinetic motions of the particles. The form of electromagnetic wave in vacuum is converted to electrokinetic or magnetokinetic waves according to the dispersion equation (Eq. 2.41). If the frequencies are of the order of the electron plasma frequency, only the electrons react with the field. If they are of the order of the ion plasma frequency, electron motions couple with ions through coulomb forces.

2.3.4 *Collisionless Energy Transfers from Electrokinetic Waves to Materials*

A wave transfers energies to materials in two different processes. The first is the collisional damping that the collective phonon mode scatters to thermal mode immediately. The velocity distribution function always keeps the Maxwell–Boltzmann form, where the Maxwell–Boltzmann velocity distribution is determined by

$$f_0(v) = \frac{\sqrt{m^*}}{\sqrt{2\pi kT}} \exp\left(\frac{-m^* v^2}{2kT}\right) \tag{2.44}$$

The other mechanism is the collisionless transfer. A qualitative explanation is as follows. As shown in Fig. 2.11, the external microwave field E_x does not penetrate into the material. If the collisions between particles in the unit cell of a crystal are negligibly small, the kinetic energy conserves in the chains of stress and strain.

Now, we are going to look into the analytical formula to deliver the rate of chemical reactions. The microwave accelerates particles in their translations in the z axis, that is, one freedom of motion in the six motions, namely three translations and three rotations belongs to the x, y, and z axes. In the following discussion, the velocity v denotes the translation velocity v_z. The Maxwell–Boltzmann velocity distribution function $f(v)$ is perturbed, as shown in the second term in the following equation:

$$f(v) = f_0(v) + f_0(v)\left(v - v_{\text{ph}}\right) g(v) \tag{2.45}$$

If the microwave is quite monochromatic, the perturbation takes the form of a delta function in the velocity space. However, the delta fluctuation will spread in the velocity space by many kinds of disturbances. The delta function is approximated by the sync function described as $f_x(x) = \sin(kx)/\pi x$, which is a well-known form in the quasi-linear approximation of Landau damping. However, the cross terms of $f_0(v)$ and $g(v)$ will need very complicated calculations to solve Eq. 2.43. The other approximation of a delta function is the normal function shown as follows:

$$g(v) = \frac{1}{\sqrt{2\pi\xi^2}} \exp\left(\frac{-\left(v - v_{\text{ph}}\right)}{2\xi^2}\right) \tag{2.46}$$

The form of Eq. 2.46 is symmetric at $v = v_{\text{ph}}$. The variance ξ^2 corresponds to the velocity spread by the accumulation of ordered motions near the median v_{ph} that synchronizes to the phase velocity of electrokinetic waves. Both functions of Eqs. 2.44 and 2.46 are of the form of a normal function; the cross term in Eq. 2.43 can be reduced to a summation term in an exponential function. The width of perturbations in velocity space is limited in the variance ξ^2. Figure 2.19 illustrates the perturbations to Maxwell–Boltzmann distribution. The disturbed part decreases the entropy of the system. It will cause the microwave nonthermal effects, as described in the following sections.

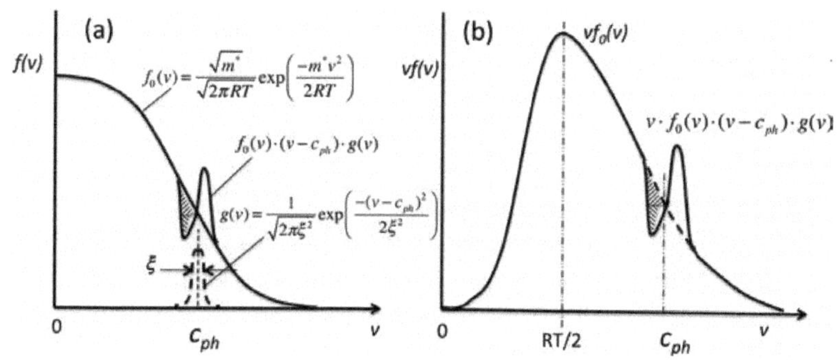

Figure 2.19 Perturbation to Maxwell–Boltzmann distributions by intense microwaves.

2.3.5 *Rate Equations with Perturbations by Microwave in Velocity Space*

The rate of chemical reactions are described by Arrhenius plots as a function of temperature:

$$k = \alpha \exp\left(\frac{-E_0}{RT}\right) \tag{2.47}$$

The rate of chemical reaction$k$ is related to velocities passing by an activation complex

$$\mathrm{A} + \mathrm{B} \rightleftharpoons X^{\ddagger} \rightleftharpoons \mathrm{C} + \mathrm{D}$$

The transition-state theories formulate the average velocity by the numbers of particles transmitting the activation complex $X^{\ddagger}$ as

$$\overline{V}_{\mathrm{f}} = \frac{\int_0^{\infty} vf(v)\,dv}{\int_0^{\infty} f(v)\,dv} \tag{2.48}$$

Followed by Eyring, the rate constant k^* between two molecules is

$$k^* = \frac{1}{2}\frac{\overline{V}_{\mathrm{f}}}{\delta}k^{\dagger} \tag{2.49}$$

Assuming the complexes are lying within the arbitrary length δ and the translational partition function, the rate functions $k^{\dagger}$ is derived as

$$k^{\dagger} = \frac{q^{\dagger}}{q_{\mathrm{a}}q_{\mathrm{b}}}\frac{\sqrt{2\pi m^* RT}}{h}\delta \exp\left(\frac{-E^*}{RT}\right) \tag{2.50}$$

where q_a, q_b, and $q^\dagger$ are the partition functions. Substituting Eqs. 2.46 and 2.48 into Eq. 2.47, the famous rate constant k^* is given by

$$k^\dagger = \frac{q^\dagger}{q_a q_b}\frac{RT}{h}\exp\left(\frac{-E^*}{RT}\right) \tag{2.51}$$

where R, m^*, and T are the molar gas constant, the reduced mass of the system, and the thermodynamic temperature, respectively.

Under intense microwave irradiations, the average velocity of particles across the peak of activated complex is calculated by substituting Eq. 2.45 into Eq. 2.48:

$$V_f = \frac{\int_0^\infty vf(v)\,dv}{\int_0^\infty f(v)\,dv} = \frac{\int_0^\infty vf(v)\,dv + \int_0^\infty vf_0(v - v_{ph})\,g(v)\,dv}{\int_0^\infty f(v)\,dv + \int_0^\infty f_0(v - v_{ph})\,g(v)\,dv} \tag{2.52}$$

The first integral in the denominator is the integration of Gaussian function as

$$\int_0^\infty f_0(v)\,dv = \int_0^\infty \frac{\sqrt{m^*}}{\sqrt{2\pi kT}}\exp\left(\frac{-m^*v^2}{2kT}\right)dv = \frac{1}{2} \tag{2.53}$$

The second integral is deformed to

$$\begin{aligned}
&\int_0^\infty f_0(v)(v - v_{ph})\,g(v)\,dv \\
&= \int_0^\infty \frac{\sqrt{m^*}}{\sqrt{2\pi RT}}\exp\left(\frac{-m^*v^2}{2RT}\right)\frac{(v - v_{ph})}{\sqrt{2\pi\xi^2}}\exp\left(\frac{-(v - v_{ph})}{2\xi^2}\right) \\
&= \frac{\sqrt{m^*}}{2\pi\sqrt{RT}\xi}\int_0^\infty (v - v_{ph})\exp\left(\frac{-v^2(m^*\xi^2 + RT) + 2RTvv_{ph} - RTv_{ph}^2}{2RT\xi^2}\right)dv
\end{aligned} \tag{2.54}$$

The stress induces electric field E_z that propagates in the z direction with an acoustic velocity. If the thermal motions of the particles are a little slower than the acoustic velocity, the static electric field E_z accelerates the charged particles in the z direction. On the other hand, faster particles push the wave. As the thermal velocity is slower than the sound velocity in a crystal, the numbers of particles getting energy is larger than that loosing energy. By assuming the kinetic energy $m^*\xi^2/2$ in the induced vibration by microwave is much smaller than the thermal energy $RT/2$ in the system, the mathematical equations describing the terms in parenthesis of exponential function is reduced to $-(v{-}v_{\mathrm{ph}})^2/2\xi^2$ and the integral becomes

$$= \frac{\sqrt{m^*}}{2\pi\sqrt{RT}\xi}\int_0^{\infty}\left(v - v_{\mathrm{ph}}\right)\exp\left(\frac{-\left(v - v_{\mathrm{ph}}\right)^2}{2\xi^2}\right)dv$$

Replacing $(v - v_{\mathrm{ph}}) \equiv y$, the integrals can be written as

$$= \frac{\sqrt{m^*}}{2\pi\sqrt{RT}\xi}\int_{-\infty}^{\infty} y\exp\left(\frac{-y^2}{2\xi^2}\right)dy$$

where the lower limit of integral $-v_{\mathrm{ph}}$ is replaced by $-\infty$, because if $\xi \ll v_{\mathrm{ph}}$, the integral between $-\infty$ and $-v_{\mathrm{ph}}$ is almost zero. The integral of an odd function is zero, and then the denominator can be determined by Eq. 2.50 as 1/2. By repeating similar procedures for the numerator in Eq. 2.50 by multiplying v in Gaussian integrations, we get a form of average velocity of particles across the peak of the activated complex under intense microwave as

$$\overline{V}_{\mathrm{f}} = \frac{\sqrt{m^*}}{\pi\sqrt{RT}}\left(\frac{RT}{m^*} + \xi^2\right) \tag{2.55}$$

Substituting Eq. 2.53 into Eqs. 2.46 and 2.47, the rate function under intense microwave is concluded as

$$k^* = \frac{q^{\dagger}}{q_{\mathrm{a}}q_{\mathrm{b}}}\left[\frac{RT}{h} + \frac{m^*\xi^2}{h}\right]\exp\left(\frac{-E^*}{RT}\right) \tag{2.56}$$

In Eq. 2.56, the first term RT/h within [ ] is the thermal reaction. The second term $m^*\xi^2/h$ is the kinetic energies of the ordered motions of molecules and lattices. It is not thermal energy. Equation 2.54 clearly shows that the rate of reaction is managed both by thermal energy and by the ordered motion induced by intense microwaves.

The physical images are illustrated in Fig. 2.19. From a thermodynamic point of view, the second term suggests that reduced entropy gives an acceleration of the particles passing by a peak of activated complex.

2.3.6 *Energy Balance Estimations*

Equation 2.56 suggests an unexpected new energy path in which the longitudinal wave couples to the chemical reactions and phase changes. When the temperature increases, the surface energy of the particles increases. For example, formation of necking and it is released via grain growth. The excess kinetic energy $m^*\xi^2$ enhances the formation of necking at lower temperatures than that in conventional heating. Why sintering occurs at lower temperatures is explained through Eq. 2.56.

We begin with a finite system in which longitudinal waves are propagating in the presence of an external alternating electric field corresponding to an **E**-field experiment in a single-mode cavity. The energy balance equation to our model is

$$\frac{\partial RT}{\partial t} = p_\mu - k^*\Delta H^0 - A\sigma T^4 - \frac{\partial (NRT)}{\partial t} \tag{2.57}$$

where p_μ, H_0, A, σ, and $\partial N/\partial t$ are the microwave power normalized to molar numbers, the difference of enthalpy between initial and final states, and the surface area and energy loss associated with particle diffusions caused by evaporations and reductions. Concerning steady-state reactions, we can neglect the time derivative term on the left-hand side. Substituting Eq. 2.56 into Eq. 2.57,

$$p_\mu = \frac{q^\dagger}{q_a q_b}\left[\frac{RT}{h} + \frac{m^*\xi^2}{h}\right]\exp\left(\frac{-E^*}{RT}\right)\Delta H^0 - A\sigma T^4 - \frac{\partial (NRT)}{\partial t} \tag{2.58}$$

Here we will notice that the partition functions q_a, q_b, and $q^\dagger$ do not vary by microwave perturbations. According to the quantum treatments, the Hamiltonian does not vary its energies with and without a microwave electric field of 10^{-5} eV. As a wave function is the other explanation of a partition function, it suggests that the partition functions remain unchanged by the microwave perturbations.

In the classic explanations, a microwave perturbation is the first order to the thermal energy of the Maxwell–Boltzmann velocity distributions. Figure 2.20 illustrates the energy flow diagram of microwave processing. The term $m^*\xi^2$ is the kinetic energy carried by longitudinal mode transmitted, depending on Eq. 2.39. We described that the longitudinal kinetic motions increase by the accumulation of collective motion described in the perturbed part in the velocity space. An oscillating longitudinal ordered motion is expressed by multiplying $m * v$ and the distribution function

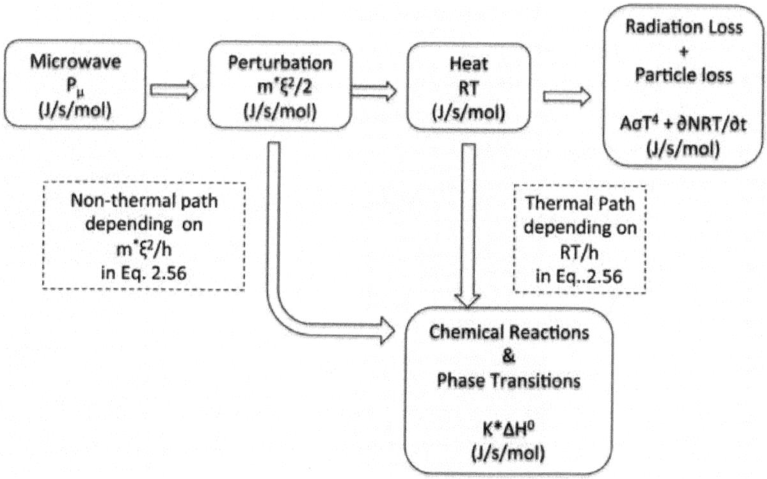

Figure 2.20 Energy flow diagram of microwave processing.

$vf_0(v)(v - v_{\text{ph}})g(v)$ and integrating as follows;

$$\tilde{w}_{\text{longitudinal}} = \int_0^\infty m^* v\left(vf_0(v)(v - v_{\text{ph}})\right) g(v)\, dv$$
$$= m^* \frac{\sqrt{m^*}}{2\pi\sqrt{RT}\xi} \int_0^\infty v^2 (v - v_{\text{ph}}) \exp\left(\frac{-(v - v_{\text{ph}})^2}{2\xi^2}\right) dv \tag{2.59}$$

Again, assuming that $m * \xi^2 \ll RT$, we get

$$= m^* \frac{\sqrt{m^*}}{2\pi\sqrt{RT}\xi} \int_{-\infty}^\infty (y + v_{\text{ph}})^2 \exp\left(\frac{-y^2}{2\xi^2}\right) dy$$
$$= \frac{m^*\xi^2 v_{\text{ph}}}{\sqrt{\pi} v_{\text{th}}} \tag{2.60}$$

The microwave field gives an impulse to lattices in a half cycle as

$$m^* v^2 = \frac{P_\mu}{\omega} \tag{2.61}$$

where ω and P_μ are, respectively, the angular frequency and the power of microwaves.

Stacking of collective motions terminates when collisions result in scattering of the monochromatic oscillation. The phase mixing is not the real collisions; however, it makes a beat between two oscillations with the frequencies ω and $\omega + \Delta\omega$. When the phase difference reaches π, the oscillations cancel each other. It is a disturbance to the momentum accumulations similar to real collisions. The period of time τ in which the phases of two oscillations differ by π is

$$\Delta\omega\tau = \pi \tag{2.62}$$

If the microwave power has a Gaussian-like narrow band frequency spread, the Q value is given by

$$Q = \frac{\omega}{\Delta\omega} \tag{2.63}$$

In the case of Eq. 2.61, periods of disturbance t can be delivered by an average of spreads written by Fourier integrals. Assuming the impulse shall accumulate energy within a time period τ, the longitudinal wave gets energy as

$$\tilde{w}_{\text{longitudinal}} = \int_0^{\tau} m^* \tilde{v} dt = \int_0^{\tau} \frac{P_\mu}{\omega} dt = \frac{P_\mu}{\omega}\tau \tag{2.64}$$

The energy in the longitudinal wave can be written as

$$\tilde{w}_{\text{longitudinal}} = 2\pi Q \frac{P_\mu}{\omega} \tag{2.65}$$

Putting Eqs. 2.60 and 2.65 as equal, the kinetic energies in the longitudinal wave is formulated by

$$m^* \xi^2 = 2\sqrt{\pi} Q \frac{P_\mu}{\omega} \frac{v_{\text{th}}}{v_{\text{ph}}} \tag{2.66}$$

where $v_{\text{th}} = \sqrt{2RT/m^*}$ is the velocity of thermal phonon vibrations. Substituting Eq. 2.62 into Eq. 2.54, the reaction rate can be expressed by experimentally detectable parameters of Q, ω, P_μ, T, and acoustic velocity v_{ph} as

$$k^* = \frac{q^\dagger}{q_a q_b} \left[\frac{RT}{h} + 2\sqrt{\pi} \frac{v_{\text{th}}}{v_{\text{ph}}} \frac{P_\mu / \Delta\omega}{h} \right] \exp\left(\frac{-E^*}{RT} \right) \tag{2.67}$$

or

$$k^* = \frac{q^\dagger}{q_a q_b} \left[\frac{RT}{h} + 2\sqrt{\pi} \frac{v_{\text{th}}}{v_{\text{ph}}} Q \frac{P_\mu / \omega}{h} \right] \exp\left(\frac{-E^*}{RT} \right) \tag{2.68}$$

2.3.7 *Magnetic Effects*

Many experiments have reported that microwave effects were observed more clearly in magnetic matters [2, 16]. Roy et al. [2] summarized as follows:

1. A large number of oxides containing 3*d* ions, most of them ferromagnetic, undergo a solid-state phase transformation from a crystalline phase to a non-crystalline phase in a modest ($\approx$0.5 oersted) magnetic field at 2.45 GHz in a few seconds at temperatures far below the melting point. Ferromagnetism has been ruled out as a requirement for decrystallization.
2. These decrystallized phases have no parallel in past work. They are not true glasses, since they have not been through the normal (bulk) liquid state required for being classified as glasses. The nearest analogue may be solid state decrystallized phases made by bombardment with high energy neutrons or α particles. What is so striking is the difference in energies in the radiation causing the change.
3. The presence of the remarkable wave-like patterns in the SEMs certainly indicates a liquid-like behavior on a very small scale, suggesting the action of a shock wave of some kind. Perhaps the term nano-glasses is not inaccurate.
4. Radical differences in interaction of the 2.45 GHz E field and the H field with ferroic oxide solids have been observed as gross differences in the phases produced, as proved by XRD and SEM of the respective products.
5. It is not possible to ignore the E and H fields as intensive variables in the thermodynamics of solids as materials scientists choose to do in analyzing phase relations in condensed matter. In virtually all such treatments one considers only pressure and temperature as affecting the free energy function, G.

We shall solve the above remaining questions by Late Rustum Roy. Introducing the ordered motions by microwave magnetic fields, the transition-state theories would couple to thermodynamic behaviors in magnetic matters.

Tanaka et al. reported that the nonresonant response of electron spins plays an important role in the unfilled 3*d* shell [18]. He used the first energy principle and performed the Monte Carlo simulations. Small spin perturbation in response to an alternating external magnetic field results in a large internal energy change through exchange interactions. The studies in the time domain show a performing dissipative spin dynamics simulation and detecting the linear response of spins. The nonresonant response gives the disturbances in their magnetic moments resulting in perturbations of velocity distribution functions. If the nonresonant responses were large enough, it would also be necessary to modify the partition function $q^{\dagger}$. These would enhance chemical reactions similar to what is mentioned in this chapter. The analytical equations are now under study for qualitative explanations.

2.4 Conclusion

In this chapter, we reviewed the unforeseen experimental results given by microwaves. In spite of overwhelming evidence shown in laboratories that a variety of solid-state reactions were radically different when carried out under intense microwave irradiations compared to conventional heating, the next question is whether it can be applicable to the scales of industry.

We introduced our latest working hypothesis that the radiation field links to the charged particles on the microstructures and the excited electrokinetic waves, such as super sounds and spine waves transmitting on the surfaces and in the body of multicrystalline solids and/or clusters in the liquid. The transferred energy might accumulate in the phase space via collisionless dumping to sufficient levels for enhancing the chemical reactions and phase transfers.

References

1. Roy, R., Agrawal, D., Cheng, J., and Gedevanishvili, S. (1999). Full sintering of powdered-metal bodies in a microwave field, *Nature,* **399**, pp. 668–670.

2. Roy, R., Peelamedu, P. D., Hurtt, L., Cheng, J. P., and Agrawal, D. (2002). Definitive experimental evidence for *microwave effects*: radically new effects of separated E and H fields, such as decrystallization of oxides in seconds, *Mater. Res. Innov.,* **6**, pp. 128–140.
3. de la Hoz, A., Diaz-Ortiz, A., and Moreno, A., (2005). Microwaves in organic synthesis. Thermal and non-thermal microwave effects, *Chem. Soc. Rev.,* **34**, pp. 164–178.
4. Herrero, M. A., Kremsner, J. M., and Kappe, C. O. (2008). Nonthermal microwave effects revisited: on the importance of internal temperature monitoring and agitation in microwave chemistry, *J. Org. Chem.,* **73**, pp. 36–47.
5. Ishizaki, K., Nagata, K., and Hayashi, T. (2007). Localized heating and reduction of magnetite ore with coal in composite pellets using microwave irradiation, *ISIJ Int.,* **47**, pp. 817–822.
6. Rao, K. J., Vaidhyanathan, B., Ganguli, M., and Ramakrishnan, P. A. (1999). Synthesis of inorganic solids using microwaves, *Chem. Mater.,* **11**, pp. 882–895.
7. Perreux, L., and Loupy, A. (2001). A tentative rationalization of microwave effects in organic synthesis according to the reaction medium, and mechanistic considerations, *Tetrahedron,* **57**, pp. 9199–9223.
8. Saitou, K. (2006). Microwave sintering of iron, cobalt, nickel, copper and stainless steel powders, *Scr. Mater.,* **54**, pp. 875–879.
9. Stix, T. H. (1992). *Waves in Plasmas* (Springer, New York).
10. Suzuki, M., Ignatenko, M., Yamashiro, M., Tanaka, M., and Sato, M. (2008). Numerical study of microwave heating of micrometer size metal particles, *ISIJ Int.,* **48**, pp. 681–684.
11. Booske, J. H., Cooper, R. F., and Freeman, S. A. (1997). Microwave enhanced reaction kinetics in ceramics, *Mater. Res. Innov.,* **1**, pp. 77–84.
12. Roy, R., Fang, Y., Cheng, J., and Agrawal, D. K. (2005). Decrystallizing solid crystalline titania, without melting, using microwave magnetic fields, *J. Am. Ceram. Soc.,* **88**, pp. 1640–1642.
13. Takayama, S., Kakurai, K., Takeda, M., Matsubara, A., Nishihara, Y., Nishijo, J., Sano, S., Nishi, N., and Sato, M. (2009). Investigation of crystal structure formation under microwave heating, *Nucl. Instrum. Methods Phys. Res. Sect. A,* **600**, pp. 246–249.
14. Takayama, S., Fukushima, J., Nishijo, J., Saito, M., Sano, S., and Sato, M. (2012). Sintering of soft magnetic material under microwave magnetic field, *Phys. Res. Int.,* **2012**, Article ID 165849, 4 pages.

15. Fukushima, J., Kashimura, K., Takayama, S., and Sato, M. (2012). Microwave-energy distribution for reduction and decrystallization of titanium oxides, *Chem. Lett.,* **41**, pp. 39–41.
16. Fukushima, J., Kashimura, K., Takayama, S., Sato, M., Sano, S., Hayashi, Y., and Takizawa, H. (2013). In-situ kinetic study on non-thermal reduction reaction of CuO during microwave heating, *Mater. Lett.,* **91**, pp. 252–254.
17. Fukushima. J., Sato, M., Matsubara, A., Takayama, S., and Kashimura, K. (2009). *Development of the High Vacuum Device to in-situ Measurement of Partial Pressure of Oxygen on Microwave/Conventional Heating,* Annual report of National Institute for Fusion Science, p. 420.
18. Tanaka, M., Kono, H., and Maruyama, K. (2009). Selective heating mechanism of magnetic metal oxides by a microwave magnetic field, *Phys. Rev. B.,* **79**, p. 104420.
19. Takayama, S., Saito, Y., Sato, M., Nagasaka, T., Muroga, T., and Ninomiya, Y. (2006). Sintering behavior of metal powders involving microwave-enhanced chemical reaction, *Jpn. J. Appl. Phys.,* **45**, pp. 1816–1822.
20. Stuerga, D. A. C., and Gaillard, P. (1996). Microwave a thermal effects in chemistry: a myth's autopsy, part I and II, *J. Microw. Power Electromagn. Energy,* **31**, p. 113.
21. Eyring, H. (1935). The activated complex in chemical reactions, *J. Chem. Phys.,* **3**, p. 107.
22. Laidler, K. J., and Klng, M. C. (1983). The development of transition-state theory, **1725**, pp. 2657–2664.
23. Laidler, K. J. (1987). *Chemical Kinetics,* 3rd ed. (Harper Collins Publications, New York).
24. Towe, B. C. (2009). Induced ultra-high frequency ultrasonic vibration as the driving force for reported sub-thermal microwave effects on materials, *Mater. Sci. Technol.,* pp. 597–607.
25. Steele, M. C., and Vural, B. (1969). *Wave Interactions in Solid State Plasmas* (McGraw Hill, New York).
26. Landau, L. D., and Lifshitz, F. M. (1953). *Theory of Elasticity, Course of Theoretical Physics,* Vol. 7 (Pergamon Press, Moscow).

Chapter 3

Microwave-Assisted Synthesis and Modification of Polymers

Martin Fimberger[a,b] and Frank Wiesbrock[a]

[a] *Polymer Competence Center Leoben, Roseggerstrasse 12, 8700 Leoben, Austria*
[b] *Institute for Chemistry and Technology of Materials, Graz University of Technology, NAWI Graz, Stremayrgasse 9, 8010 Graz, Austria*
frank.wiesbrock@pccl.at

3.1 Introduction

The application of microwave-assisted heating in organic chemistry, and especially in polymer science, received widespread attention during the initial decade of the millennium. With the availability of dedicated laboratory equipment, which offers accurate control over reaction conditions and safety measures, microwave-assisted heating nowadays has become a well-established method for fast and efficient syntheses. In polymer chemistry, dielectric heating offers an excellent pathway for accelerated polymerization reactions, especially for polymer classes known for slow propagation rates under conventional conditions. Following a thorough review about the application of microwave reactors in polymer chemistry, published in 2011 [1], this chapter focuses on the developments in the field since 2010.

Microwave Engineering of Materials: From Mesoscale to Nanoscale
Edited by Erwann Guénin

ISBN 978-981-4669-42-9 (Hardcover), 978-981-4669-43-6 (eBook)
www.panstanford.com

Reviewing the literature on this topic, clearly poly(2-oxazoline)s have received outstanding attention, due to the versatility of this polymer class for various applications. The synthesis of poly(2-oxazoline)s significantly benefits from microwave heating and without this application would probably not be in the focus of scientists, as conventionally heated syntheses proved to be very time consuming. A second polymer class under intense investigation comprises polyesters and polyamides obtained from by ring-opening polymerizations. Notably, also in controlled radical polymerizations (CRPs), dielectric heating has recently received significant attention. In a thorough review of the early research results on microwave-supported reversible addition–fragmentation chain transfer (RAFT) polymerization, Brooks and Sumerlin found that significant acceleration of reaction rates could be achieved, while reasonable control over the polymerizations was maintained [2]. Nevertheless, the authors indicated the need of a more thorough understanding of the mechanism of microwave-supported RAFT.

The aspect of green chemistry is not to be overseen. Microwave-assisted heating, especially in conjunction with ionic liquids, continuously raises interest among scientists. Ionic liquids are seen as environmentally benign alternative to volatile organic solvents, which provide excellent interaction with microwave irradiation, opening the pathway for efficient heating. One concern of the application of microwave-assisted heating in organic synthesis, yet not fully resolved, remains the upscaling of reactions. Despite various approaches to overcome this issue, the application of microwave-assisted heating still focuses on lab-scale syntheses.

In this chapter, five major topics will be discussed in detailed manner: CRPs, step-growth polymerizations involving the formation of C–C bonds, the ring-opening polymerization of 2-oxazolines as well as cyclic esters and amides, and, finally, the application of ionic liquids and click chemistry in microwave-assisted reactions.

3.2 Controlled Radical Polymerization

Among CRPs, in particular three types are under constant and thorough investigation, namely RAFT, atom transfer radical polymerization (ATRP), and nitroxide-mediated (radical) polymerization (NMP) (Scheme 3.1). The performance of NMPs under microwave irradiation is still in its infancy and, concomitantly, has been neglected in this section. Emulsions polymerizations have been included in this section, too.

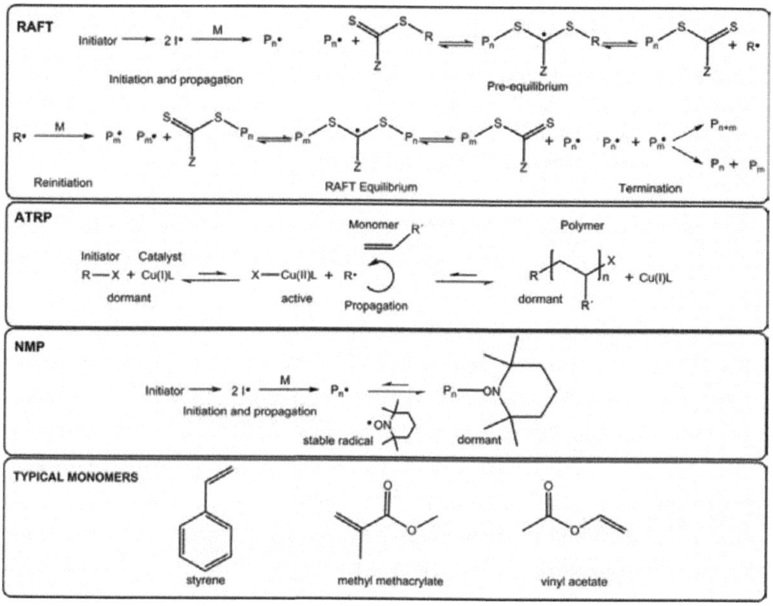

Scheme 3.1 General reaction schemes for reversible addition–fragmentation chain transfer (RAFT), atom transfer radical polymerization (ATRP), and nitroxide-mediated polymerization (NMP). Monomers commonly used in CRP are styrene, methyl methacrylate, and vinyl acetate.

3.2.1 *Reversible Addition–Fragmentation Chain Transfer*

Vivaldo-Lima and coworkers developed a kinetic model for the RAFT polymerization of styrene under microwave support using 2-cyanoprop-2-yl-1-dithionaphthalate as a chain transfer agent [3]. In its report, the group compared the kinetic data for the microwave-supported polymerization derived from a systematic model with experimental data. The correlation was found to exhibit deviations between the experimental and theoretical data (Fig. 3.1); however,

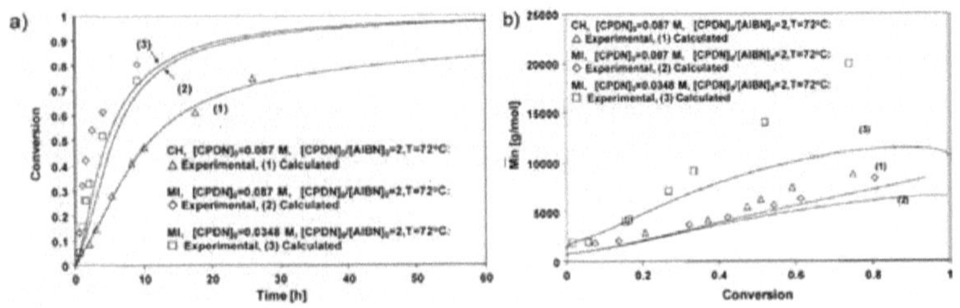

Figure 3.1 Experimental data compared to the calculated results from the kinetic models for monomer conversion and average molecular weight M_n, exhibiting mediocre correlation. Reprinted from Ref. [3]. Copyright © 2010, with permission from John Wiley and Sons.

it was shown that microwave support significantly increased the reaction rates of the RAFT polymerizations, while having no adverse effect on control.

Zetterlund and coworkers compared the thermally heated and microwave-supported polymerization of styrene [4]. They investigated both RAFT polymerization and conventional free-radical polymerization, while special care was taken for the control of the reaction temperatures; a CEM Discover microwave reactor with infrared (IR) temperature measurement was used for the microwave-assisted protocols. The polymerization rates were found to be independent of the heating source, indicating the absence of any nonthermal microwave effect (Fig. 3.2). Additionally, it was shown that radicals were not generated by microwave irradiation itself, as styrene could not be polymerized without an initiator or RAFT agent.

The successful syntheses of homopolymers and block copolymers of several vinyl esters by microwave-supported RAFT/macromolecular design via the interchange of xanthates (MADIX) polymerizations were reported by Sumerlin and Roy [5]. Poly(vinyl acetate) with molecular weights between 1 and 10 kDa could be synthesized within 15 min at a reaction temperature of 70°C. Ethylxanthogenic acid was used as a chain transfer agent and azobisisobutyronitrile (AIBN) as the initiator. The authors reported improved control of the polymerization compared to conventionally heated syntheses. Dispersity values between 1.16 and 1.25 and monomer conversions of up to 66% were reported. Conventionally

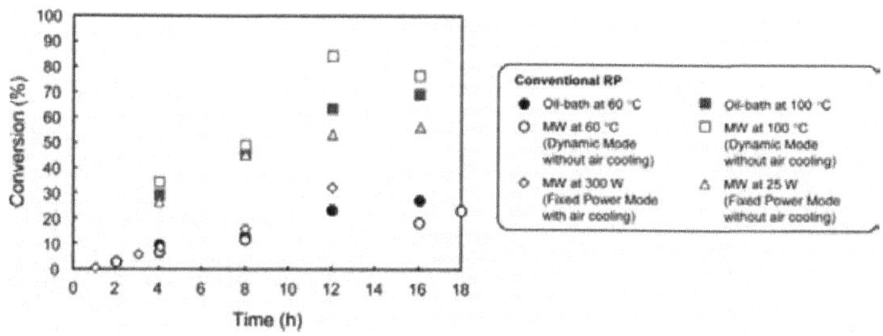

Figure 3.2 Conversion vs. time for an AIBN-initiated radical polymerization of styrene under conventional and microwave-assisted heating, exhibiting no significant difference in polymerization rates, which is indicative of the absence of any nonthermal microwave effects. Reproduced in part from Ref. [4] with permission of the Royal Society of Chemistry.

heated control experiments exhibited lower polymerization rates compared to microwave-assisted heating.

Zetterlund and Perrier reported the mechanistic investigation of RAFT polymerizations under microwave assistance [6]. The authors compared experimental kinetic data of the polymerization of styrene using cyano-^{iso}propyl dithiobenzoate and azo-bis-^{iso}butyronitril with simulated values obtained from PREDICI software. Three models were investigated; only one model (of increased propagation and addition rates) could be supported by the experimental data.

The microwave-assisted RAFT polymerization of *N*-vinylcarbazole was reported by Grigoras and Negru [7]. The polymerizations were performed at 70°C in a single-mode CEM Discover Labmate microwave reactor using 1,4-dioxane as the solvent and azo-bis-^{iso}butyronitril as the initiator. Trifunctional 1,3,5-benzyl tri(diethyldithiocarbamate) was used as a chain transfer agent, yielding a star-shaped polymer (Scheme 3.2). The initiator:monomer ratios of 1:1, 2:1, 4:1, and 5:1 were tested; the amount of the chain transfer agent was adjusted with respect to its trifunctional character. Number-average molecular masses between 2.3 and 32.3 kDa and $Đ_M$ values from 1.21 to 1.90 were reported. The linear relationship of the molecular masses with the monomer conversions was indicative of the controlled conditions during the polymerizations.

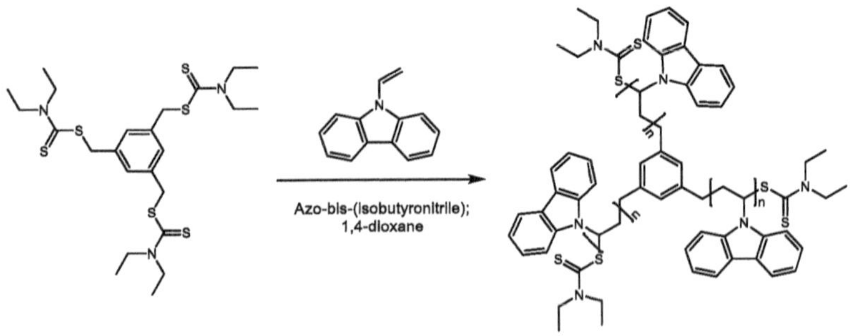

Scheme 3.2 A scheme of the RAFT polymerization of *N*-vinylcarbazole with the chain transfer agent 1,3,5-benzyl tri(diethyldithiocarbamate), yielding star-shaped polymers.

3.2.2 *Atom Transfer Radical Polymerization*

Similar to the previous work on the RAFT polymerization of styrene, Vilvaldo-Lima and Lopez-Dominguez also investigated the ATRP of methyl methacrylate and styrene by modeling tools [8]. Two models were under investigation: One included a proposed generation of radicals by microwaves, and the other focused on thermal effects on the polymerization. Both models were suitable to reproduce the experimental kinetic data, and it was concluded that further model studies are needed to distinguish between the models. Correspondingly, the existence of nonthermal microwave effects was not excluded by the authors at this stage of investigations.

The syntheses of copolyacrylates obtained by an ATRP from *N*-acryloyl-L-phenylalanine and methyl methacrylate were performed by Buruiana et al. [9]. Using 9-(chloromethyl)anthracene as initiator, polymers end-capped by anthracene were obtained. In a second approach, poly(*N*-acryloyl-L-phenylalanine-co-methyl methacrylate) was modified with 9-anthracenemethanol in order to obtain polymers bearing anthracene moieties in their side chains. While in the first case polymerizations were performed under conventional heating, the synthesis of poly(*N*-acryloyl-L-phenylalanine-co-methyl methacrylate) was performed in a CEM Discover microwave reactor. Dioxane and 1,1′-azo-bis-^{cyclo}hexanecarbonitrile were used as solvent and initiator, respectively.

Wang and Wu reported the copolymerization of styrene and acrylonitrile under microwave support [10]. $FeCl_3$/isophthalic acid was used as a catalyst and azo-bis-^{iso}butyronitril as the initiator. The microwave-supported copolymerization was significantly accelerated compared to conventionally heated polymerizations. A monomer consumption of 30% was reached after 38 min. A linear correlation of the molecular weights of the obtained polymers with increasing monomer consumption suggested living conditions of the polymerizations; the $Đ_M$ values <1.25 were found to be low.

The synthesis of HCl-doped tetraaniline end-capped conductive polystyrenes by microwave-supported ATRP was reported by Reynaud and coworkers (Scheme 3.3) [11]. The authors compared the microwave-supported synthesis with conventionally heated ones, reporting equivalent results in terms of material properties. However, a significant reduction of reaction times (under otherwise identical conditions) was stated. The polymerizations were performed in a CEM Discover single-mode microwave reactor in pulsed power mode. Additionally, heating while cooling was applied; this might have led to underestimation of the temperatures applied.

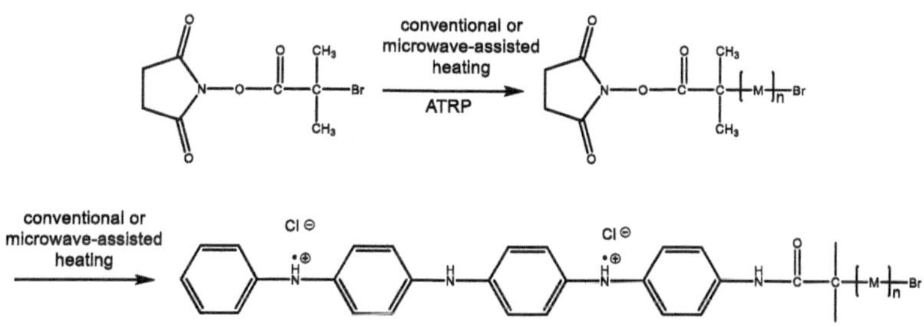

Scheme 3.3 A reaction scheme for the synthesis of the HCl-doped tetraaniline end-capped initiator used for the preparation of conductive polystyrenes.

3.2.3 *Emulsion Polymerization*

Sayer and coworkers reported the emulsion polymerization of methyl methacrylate initiated either by anionic or cationic initiators under microwave irradiation [12]. The syntheses were performed in a Synthos 3000 multimode reactor by Anton Paar.

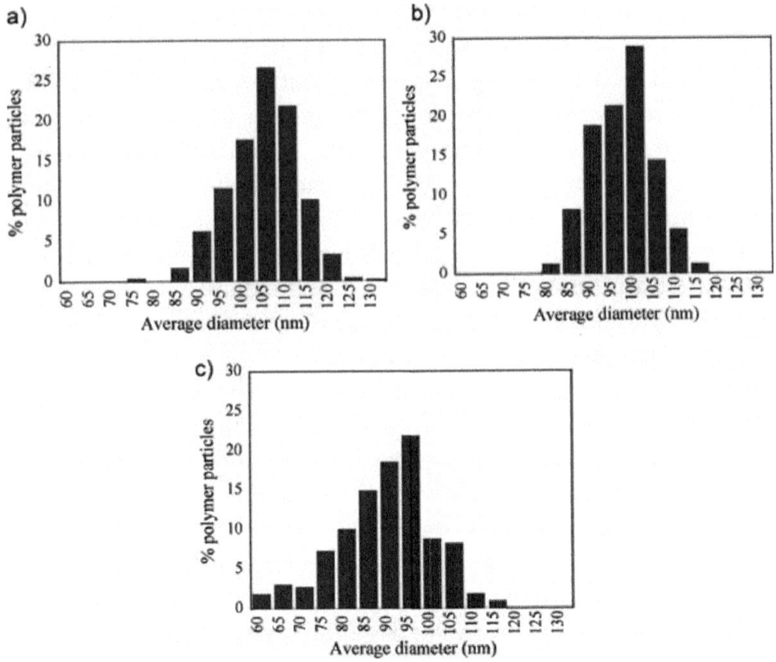

Figure 3.3 Size distribution of poly(methyl methacrylate) particles obtained from emulsion polymerizations using anionic potassium persulfate as initiator. (a) Conventional heating, (b) microwave-assisted heating at 80°C, and (c) microwave-assisted heating in pulsed mode. Reprinted from Ref. [12]. Copyright © 2011, with permission from John Wiley and Sons.

The initiators tested were potassium persulfate, 2,2′-azo-bis-2-methylpropionamidine dihydrochloride, and dodecyltrimethylammonium bromide. The authors reported the significant acceleration of the polymerization compared to conventional heating for both types of initiation systems (anionic/cationic) under reaction temperatures of 80°C. The acceleration was explained by an increased decomposition rate of the initiators, especially for potassium persulfate. Analogously, for pulsed heating cycles at full microwave power followed by active cooling, a significant acceleration was reported. In the case of the anionic initiators, the application of dielectric heated reportedly also influenced the size and size distribution of the obtained poly(methyl methacrylate) particles:

Smaller diameters were observed for the microwave-supported pathway (Fig. 3.3). Using the cationic initiator, the type of heating seemed not to have an influence.

Kappe and coworkers reported a comprehensive study about the existence or nonexistence of nonthermal effects in microwave-supported emulsion polymerizations [13]. Styrene and methyl methacrylate, respectively, were polymerized using potassium persulfate or 2,2′-azo-bis-2-methylpropionamidine dihydrochloride as the initiator. Microwave-assisted heating was provided by a single-mode Anton Paar microwave reactor (Monowave 300). The authors investigated the influence of microwave irradiation on the decomposition rates of the initiators as well as on the propagation rate of the polymerizations, concluding that there was no evidence for any nonthermal microwave effect. Observations of microwave effects reported by other groups were referred to nonprecise temperature measurements and local overheating of the reaction mixtures due to poor mixing.

3.3 Step-Growth Polymerizations

Suzuki coupling and Stille coupling, both of which are used for the formation of C–C bonds yielding (commonly) biphenyls as products, have evolved to standard reactions in organic synthesis (Scheme 3.4). Using (at least) bisfunctional compounds, polymers

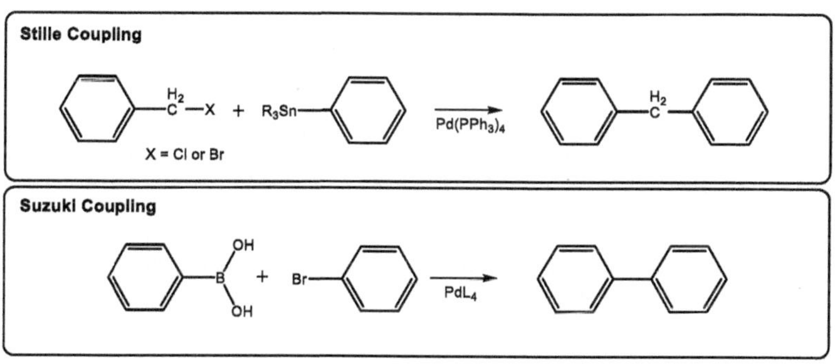

Scheme 3.4 General reaction schemes for the carbon–carbon bond-forming Stille coupling and Suzuki coupling. Both types of reactions are palladium catalyzed.

can be obtained from these coupling reactions, which will be described in a detailed manner in this section.

Schmücker and Kuckling reported the synthesis of poly (paraphenylene)s, capped with alkoxyamine end groups, by a microwave-supported Suzuki coupling [14]. The reactions were performed in a CEM Discover microwave reactor at reaction temperatures of up to 80°C. The polymers obtained were used as macroinitiators for the NMP of *N*-isopropylamide, yielding stimuli-sensitive block copolymers with a coil–rod–coil structure (Scheme 3.5). The times required for the synthesis of the macroinitiators were reported to be as short as 1 min. The molecular weights of the macroinitiator depended on the type of diboronic monomers used and ranged from 4.6 to 5.5 kDa for the polymer based on 1,4-benzenediboronic acid bis(pinacol) ester and from 7.3 to 9.6 kDa for 4,4′-biphenyl-diboronic acid bis(pinacol) ester.

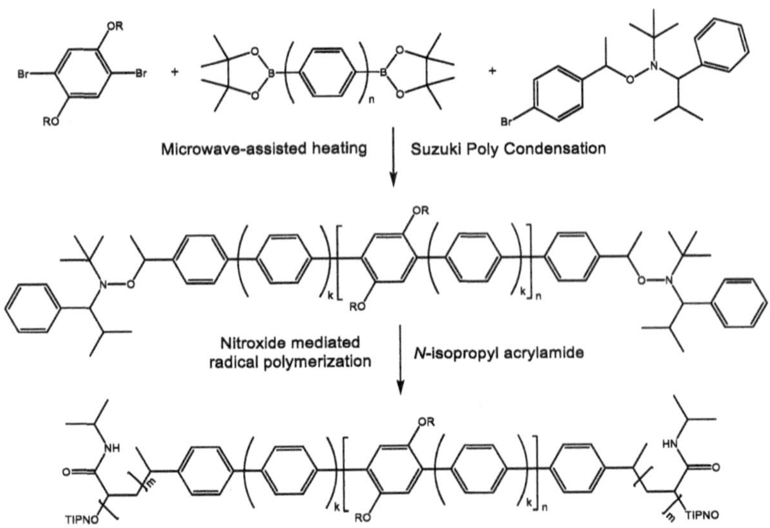

Scheme 3.5 Microwave-supported synthesis of the macroinitiator used for the NMP of *N*-isopropylamide.

Ma and coworkers contributed a comprehensive study on the microwave-heated synthesis of poly(9,9-dihexylfluorene)s as representatives of conjugated polymers [15]. The authors reported

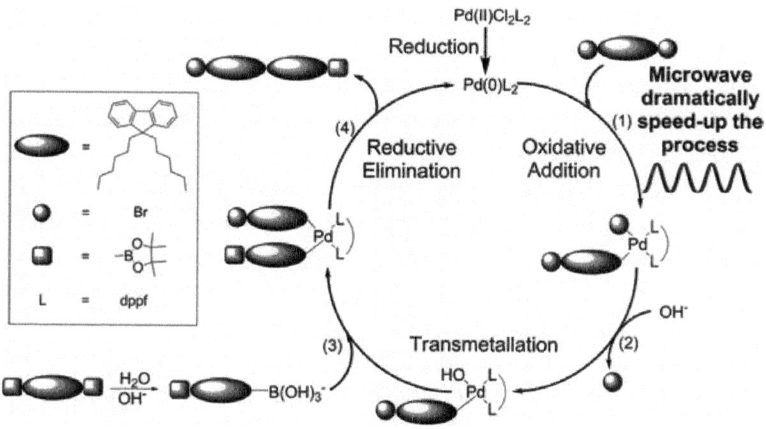

Figure 3.4 Possible mechanistic pathway for the microwave-heated catalytic cycle during palladium-catalyzed Suzuki coupling. Reprinted from Ref. [15]. Copyright © 2013, with permission from John Wiley and Sons.

a significant reduction of reaction times if dielectric heating of the palladium-catalyzed reaction was applied. Polymers with a molecular mass of 40 kDa were obtained within less than 15 min, which, under conventional heating, required 48 h. Monomer conversions of up to 68% and $Đ_M$ values of the polymers ranging from 1.21 to 2.15 were reported. Additionally, the authors provided mechanistic considerations of the catalytic reaction (Fig. 3.4).

Barbarella and coworkers delivered an early contribution to microwave support on Suzuki coupling reactions: they investigated the dielectrically heated preparation of thiophene oligomers under solvent-free conditions [16]. The reactions were performed on a Prolabo Synthewave 402 microwave system. Several palladium compounds were tested for catalytic activity.

Heeney and coworkers described the preparation of poly(3-alkyl-2,5-selenylenevinylene)s by a microwave-supported Stille coupling [17]. The work aimed at the preparation of selenophen-based compounds as low-band-gap electrically conductive materials. The polymerizations were performed in an Biotage Initiator microwave reactor operating at 100°C for 12 h. $Pd_2(dba)_2/Ph_3As$ was used as a catalyst and *N*-methyl pyrrolidone as the solvent. Two monomers, a hexyl-substituted and a decenyl-substituted selenophen, were tested. The hexyl-substituted selenophen yielded unsoluble polymers with a rather low M_n value of 7.6 kDa and an M_w value of 12 kDa. The decenyl-subtituted selenophen yielded polymers with M_n of 25 kDa and M_w of 62 kDa. A high regioregularity of the obtained products was reported.

A microwave-supported Wurtz coupling reaction for the polymerization of methylphenyl-dichlorosilane was reported by Sacarescu et al. [18]. The authors used a CEM Discover microwave reactor for the dielectrical heating of the solvent-free reaction mixture of methylphenyl-dichlorosilane and sodium as catalyst. The polymerization times could be reduced from several hours to only 3 min. A monomer conversion of over 80% and the monomodal distribution of the molecular weights of the obtained polymers were reported.

Hong et al. reported the preparation of several low-band-gap copolymers by a microwave-supported Stille coupling [19]. The polymerizations were performed in a Biotage Initiator microwave reactor. The reaction mixtures were prepared under inert conditions. Reaction times of 3 h under microwave irradiation were reportedly sufficient, which is a significant improvement compared to long-lasting syntheses under conventional heating. Yields of up to 90% were obtained, with molecular weights ranging from 13 to 21 kDa (M_n) and $Đ_M$ values in the range from 1.25 to 2.98.

3.4 Ring-Opening Polymerizations

The synthesis of polyesters, in particular biodegradable polyesters, and poly(2-oxazoline)s under microwave irradiation have been investigated in unparalleled degree, regarding the number of publications as well as the evolution of this research area manifested in, for example, material development and successful patent grants.

3.4.1 *Poly(ε-Caprolactone)s, Poly(Glycolic Acid), and Polylactides*

Among the polyesters, biodegradable polyesters such as poly(ε-caprolactone) (PCL), poly(glycolic acid) (PGA), and polylactide

(PLA) (Scheme 3.6) are the most extensively investigated ones, most likely due to their potential as materials in medical and medicinal applications.

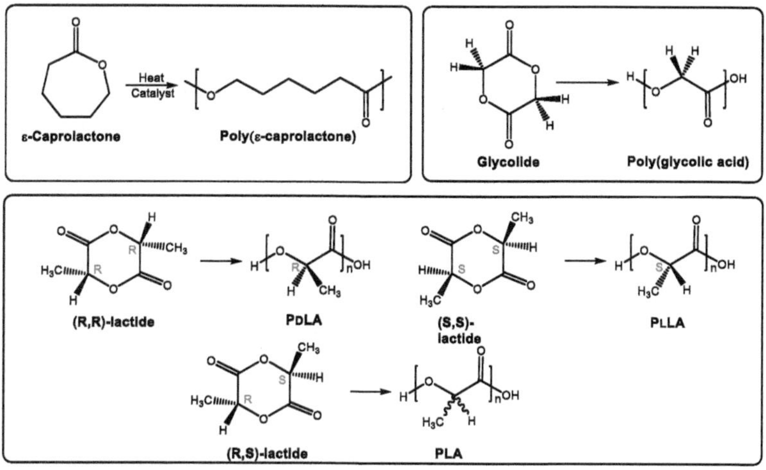

Scheme 3.6 General reaction schemes for the polymerization of ε-caprolactone, glycolide, and (R,R)-lactide, (S,S)-lactide, and *meso*-lactide.

3.4.1.1 Homopolymers of ε-caprolactone

Liu and coworkers reported a thorough investigation of the microwave-supported polymerization of ε-caprolactone [20]. The solvent-free polymerizations, catalyzed by zinc powder or stannous octanoate, were performed under conventional heating and microwave support. The polymers with the highest molecular weights were obtained using stannous octanoate and exhibited M_n values of up to 124 kDa. Zinc powder as catalyst yielded polymers with lower molecular weights. Experiments using conventional heating required significantly more time, yielding polymers with M_n values not higher than 60 kDa. In an additional study, the group reported the polymerization of ε-caprolactone, initiated by benzoic acid [21]. Also these experiments were performed under conventional heating and microwave support, with the latter yielding polymers with higher molecular weights.

The oligmerization of ε-caprolactone catalyzed by lanthanide halides was reported by Barbier-Baudry et al. [22]. Two types of microwave reactors, which provided power levels of up to 1200 W and internal temperature control by an IR sensor, were tested. PCLs with M_n values of up to 20.4 kDa were obtained; the maximum attainable molecular weight strongly depended on the water content of the monomers. Lower water contents resulted in higher molecular weights, which were ascribed to a possible co-initiation by water. Analogously, water coordinated to the catalyst seemed to have an influence on the obtained molecular weights.

The application of the enzyme Novozyme 435 as catalyst for the ring-opening polymerization of ε-caprolactone was reported by Ritter and Kerep [23]. The solvents investigated comprised tetrahydrofuran, dioxane, toluene, benzene, and diethyl ether. A CEM Discover lab-scale microwave reactor with IR temperature acquisition was used in open-vessel conditions. The results of these experiments provided controversial findings; the correlation of the performance of the polymerization and the polarity of the solvent was discussed.

In a further study, Ritter and Kerep used the same enzyme Novozyme 435 as catalyst for the microwave-supported synthesis of thiol-terminated PCLs [24]. The microwave-supported syntheses were compared to conventionally heated ones, revealing polymers with comparable $Đ_M$ values of 1.5 obtained from both pathways; however, higher yields were obtained using dielectric heating. Microwave-assisted heating also seemed to direct the synthesis toward thiol-terminated PCLs (Scheme 3.7). These polymers could

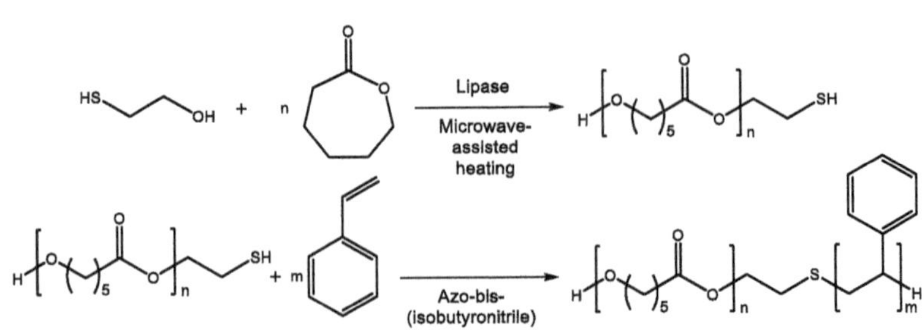

Scheme 3.7 Initiation of the ring-opening polymerization of ε-caprolactone by mercapto ethanol, yielding thiol-terminated poly(ε-caprolactone), a transfer agent for the radical polymerization of styrene.

be used as transfer agents for the copolymerization of styrene as second monomer.

Ritter and coworkers also reported the microwave-heated polymerization of ε-caprolactone using acrylic acid or methacrylic acid as initiator and stannous octanoate as catalyst [25]. A CEM Discover bench-top microwave reactor was operated at power levels limited to 50 W, resulting in temperatures of 180°C. Depending on the initiator:monomer ratios, polymers with M_n of up to 2.9 kDa and $Đ_M$ values in the range from 1.7 to 2.1 were obtained within reaction times of 90 min.

Liu and coworkers reported the application of carboxylic acids as well as maleic anhydride, succinic acid, and adipic acid for the metal catalyst-free synthesis of PCL [26]. Polymers with M_n values of up to 12 kDa were reported, exhibiting $Đ_M$ values ranging from 1.1 to 1.8. The power levels of microwave irradiation were varied between 270 and 720 W, revealing a linear correlation of the applied power and the obtained molecular masses. Extended reaction time, however, seemed to have an adverse effect on the molecular weights. Polymerizations in the presence of a model drug (ibuprofen) yielded a diffusion-controlled drug release system.

3.4.1.2 Copolymers of ε-caprolactone

Sosnik and coworkers reported the synthesis of PCL-*block*-poly(ethylene glycol)-*block*-PCL triblock copolymers [27]. The polymerizations were catalyzed by stannous octanoate and initiated by telechelic poly(ethylene glycol) with M_n of 400. Reaction times of 10 min and monomer-to-initiator ratios of 315:1 were found to yield polymers with the highest M_n values of 40 kDa. Polymer decomposition under prolonged irradiation time was observed.

The graft polymerization of ε-caprolactone onto poly(vinyl alcohol) was investigated by Liu and Yu [28]. Comb-shaped poly(vinyl alcohol)-*graft*-PCL polymers were obtained. A positive correlation of the substitution degree and applied microwave power as well as a positive correlation of the ε-caprolactone conversion with reaction times was reported.

The syntheses of poly(ε-caprolactam) and PCLs in an LT 502Xb microwave furnace operating at a variable frequency were performed by Scola and coworkers [29]. Temperature acquisition was provided by a calibrated thermocouple. The polymerizations of ε-caprolactam were performed in open Teflon vessels, using ω-aminocaprotic acid as initiator. The syntheses were conducted on several temperature levels. Yields ranging from 79% to 89% were obtained in reasonable reaction times of 3 h or less. PCL syntheses were initiated by butanediol or water, while the latter one could be introduced by the stannous octanoate catalyst. Several polymerizations were performed on temperature levels ranging from 120°C to 200°C.

3.4.1.3 Copolymers of poly(glycolic acid)

Xiong and coworkers reported the microwave-heated copolymerization (MAS-Π microwave furnace) of S,S-lactide and glycolide in the 10 g scale [30]. The polymerizations were catalyzed by stannous octanoate and initiated by octadecanol. The conventionally preheated reactants were irradiated by microwaves for up to 9 min at a reaction temperature of 120°C. The products with the highest viscosity of 0.88 dL/g were obtained after 5 min of irradiation. As observed also by other authors, prolonged irradiation times seemed to support polymer degradation and transesterifications. The glycolide content of the copolymers was higher than the initial feed ratio, which was ascribed to higher reaction rates of the glycolide.

The application of magnetic iron core nanoparticles coated by PLA-*stat*-PGA copolymers as drug delivery system was investigated by Carpenter and coworkers [31]. Pulsed microwave irradiation was used for localized heating of the microcomposites (Fig. 3.5), allowing for controlled release of a model dye. Three pulses lasting for 2 s were required for quantitative release of the fluorescent dye.

The microwave-heated synthesis of [PGA-*stat*-PLA]-*block*-poly (ethylene glycol)-*block*-[PGA-*stat*-PLA] copolymer was performed by Hadizadeh and coworkers [32]. The syntheses were initiated by hydroxy-terminated poly(ethylene glycol)s, which were added to mixtures of glycolide and R,S-lactide together with stannous 2-ethyl hexanoate as a catalyst. The polymerizations were performed in a Milestone MicroSYNTH reactor capable of 800 W of irradiation

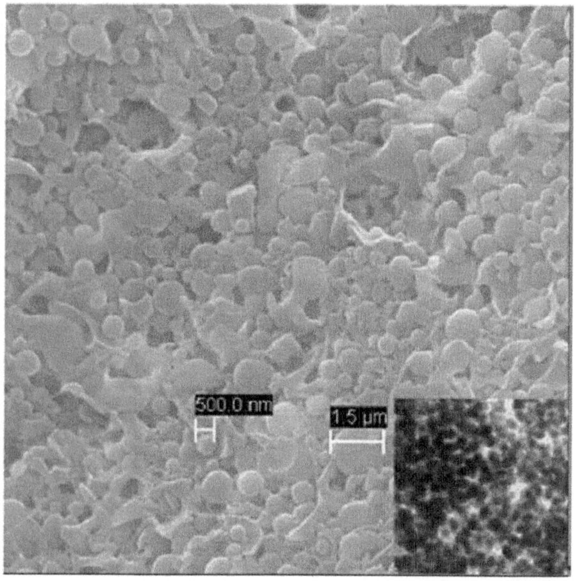

Figure 3.5 SEM image of the magnetic iron core nanoparticles coated by polylactide-*stat*-poly(glycolic acid). Reprinted with permission from Ref. [31]. Copyright 2008, AIP Publishing LLC.

power and were completed within 5 min. Prolonged irradiation times did not increase the overall molecular weights of the obtained polymers.

3.4.1.4 Homopolymers of lactic acid

Grande and coworkers reported the kinetic investigation of the polymerization of R,S-lactide [33]. The syntheses were performed either under conventional heating or under power- or temperature-controlled microwave-assisted heating. Neither the molar mass dispersity nor conversion rates were significantly influenced by the type of heating, indicative of the absence of any nonthermal microwave effect. First-order kinetics of the polymerizations were demonstrated for all of the applied heating methods.

The microwave-supported depolymerization of oligo(L-lactic acid) in order to obtain stereochemically pure S,S-lactide was investigated by Ohara and coworkers [34]. The oligomers were exposed to microwave irradiation at temperatures between 120°C and 180°C. Microwave support reportedly increased the yield of the lactide but to some extent also led to racemization of the product. Interestingly, the extent of racemization exhibited a correlation with the reduced pressure applied to the system.

Ohara and coworkers contributed a study of the synthesis of poly(L-lactide) following a two-step protocol [35]. Conventional heating was applied in the first step, yielding oligomers of L-lactic acid. The polymerizations were then continued using a CEM Discover bench-top microwave reactor under IR temperature surveillance. Compared to conventional heating, the polymers obtained under microwave support exhibited higher molecular weights. Additionally, racemization was observed to some extent.

Nikolic et al. reported the homopolymerization of R,S-lactide using a CEM Discover bench-top reactor [36]. The polymer syntheses were catalyzed by stannous 2-ethylhexanoate in various monomer:catalyst ratios ranging from 1000:1 to 10,000:1. The polymerizations were performed at 100°C within 30 min, while a residual monomer was found in all of the products. M_n values ranged from 26.7 to 112.5 kDa, and the molar weight dispersity ranged from 2.4 to 3.4. The obtained polymers were subsequently dissolved in tetrahydrofuran and sprayed into a water/poly(vinyl alcohol) mixture, resulting in microspheres with diameters of 50 μm.

The polymerization of R,S-lactide using stannous octanoate as a catalyst was performed by Peng and coworkers, who found that high monomer purity, especially the absence of water, was crucial to obtain polymers with high molar masses [37]. A domestic microwave oven was used for this study, heating a special SiC medium in close contact to the open reaction vessel. The highest intrinsic viscosities of polymers were obtained using lactides with 99.99% purity.

Usually, tin-based compounds are used to catalyze the polymerizations of lactides. To avoid the toxicity of the catalyst, Frediani et al. investigated the catalysis of the ring-opening polymerization of lactides by cone-shaped calix[4]arenes [38]. These macrocyclic titanium complexes exhibited thermal stability up to 230°C. The catalytic activities of the compounds were investigated under microwave-assisted heating revealing improved reaction rates (compared to conventional heating).

3.4.2 *Poly(2-oxazoline)s and Poly(2-oxazine)s*

2-Oxazolines feature the structural motif of a five-membered heterocycle (containing a nitrogen and an oxygen heteroatom) and undergo ring-opening polymerization after initiation by a cation (Scheme 3.8).

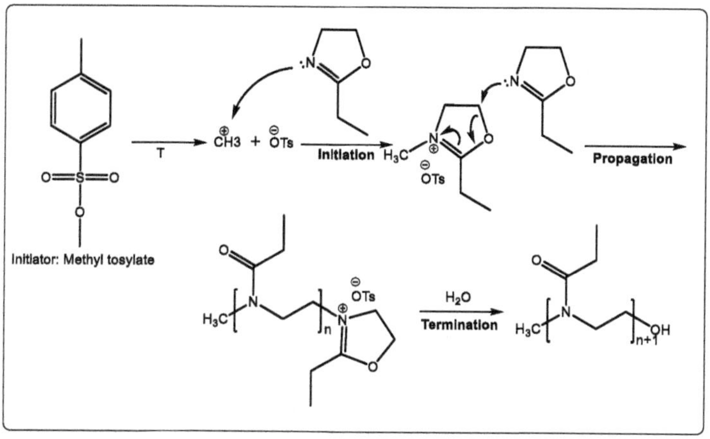

Scheme 3.8 Reaction scheme of the cationic ring-opening polymerization (CROP) of 2-ethyl-2-oxazoline, initiated by methyl cations.

Although poly(2-oxazoline)s have been investigated since the 1960s, their slow polymerization rates under conventional conditions have been a constant drawback. With the advent of dedicated lab-scale microwave reactors, this bottleneck could be overcome: Times required to synthesize poly(2-oxazoline)s (Scheme 3.8) were reduced from days to reasonable durations of several hours, which led to increased research activity on these materials.

Schubert and coworkers contributed a kinetic and mechanistic investigation of the microwave-supported cationic ring-opening polymerization (CROP) of various 2-oxazolines and oxazines [39]. The authors reported controlled polymerizations for 2-butyl-2-oxazoline,2-butyl-2-oxazine, 4-methyl-2-butyl-2-oxazoline,

4-ethyl-2-butyl-2-oxazoline, and 5-methyl-2-butyl-2-oxazoline, with decreasing rate constants of the polymerization in the above-mentioned order. Chain transfer reactions, especially during the polymerization of 4-methyl-2-butyl-2-oxazoline, were ascribed to trace impurities like water. The authors also reported copolymerizations, which revealed improved reaction rates of the sterically hindered species. Schubert and coworkers reported the microwave-supported preparation of poly(2-oxazine)s [40]. 2-Methyl-2-oxazine, 2-ethyl-2-oxazine, and 2-^{n}propyl-2-oxazine were polymerized with good reaction control. The reaction parameters were similar to prior studies concerning 2-oxazolines. At reaction temperatures of 140°C, polymerization rates of approximately four times slower than those of comparable 2-oxazolines were reported. Additionally, the authors reported a correlation of the cloud points of the poly(2-oxazine)s with the hydrophobicity of the materials.

Wiesbrock and coworkers reported the microwave supported synthesis of poly(2-oxazoline) copolymers and their subsequent usage as polymers in negative photoresists [41]. Two copolymers, namely poly(2-ethyl-2-oxazoline)-co-(2-but-3′-enyl-2-oxazoline) and poly(2-phenyl-2-oxazoline)-co-(2-dec-9′-enyl-2-oxazoline) were prepared with polymerization degrees of 100. Twenty percent of the monomers contained unsaturated side chains, enabling photochemically induced crosslinking by thiol-ene reactions. The statistical copolymers were prepared in a lab-scale Biotage Initiator 8 microwave reactor. Polymerization times of less than 100 min and a temperature of 140°C led to full monomer consumption. The obtained materials were soluble in environmentally friendly solvents and could be successfully applied as negative photoresists, revealing good adhesion to various materials and resolutions higher than 2 μm (Fig. 3.6).

To further expand the library of poly(2-oxazoline)s useful for photochemical reactions, Wiesbrock and coworkers investigated the application of 2-oxazolines with cyclic olefinic substituents [42]. 2-2′-Norbornenyl-2-oxazoline was polymerized successfully in the same Biotage benchtop microwave reactor. In this special case, an oligo(2-ethyl-2-oxazoline) macroinitiator was needed to start the polymerization, as the commonly used methyl tosylate was too reactive and yielded insoluble polymer (networks). The polymer

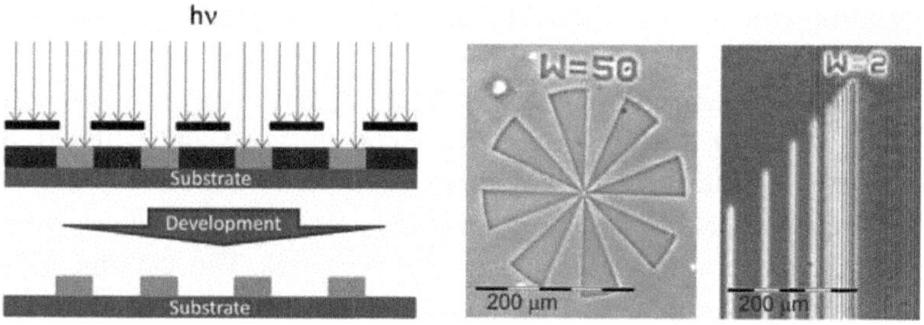

Figure 3.6 (Left) Principal procedure of photolithography with a negative photoresist, which becomes insoluble upon irradiation and after development reproduces the illumination pattern provided by a structured mask. (Right) Phase-contrast image of a CaF_2 sample coated with a poly(2-oxazoline)-based negative photoresist, which, after irradiation through a geometric pattern and development, exhibited resolutions of 2 μm. Reprinted from Ref. [41]. Copyright © 2012, with permission from John Wiley and Sons.

exhibited a trimodal distribution of molecular weights, which was ascribed to chain transfer reactions of the highly reactive monomer.

The synthesis of crosslinkable poly(2-oxazoline)s in microwave reactors was also described by Hoogenboom and coworkers [43]. Copolymers of 2-methyl-2-oxazoline and 2-dec-9′-enyl-2-oxazoline were synthesized under microwave support at 140°C. The reaction conditions followed a standard routine using acetonitrile as the solvent and methyl tosylate as the initiator. The obtained polymers were functionalized with the RGD peptide sequence and crosslinked, both by thiol-ene click reactions.

A different approach for the synthesis of hydrogels by Hoogenboom and coworkers aimed for the polymeranalogous crosslinking of poly(2-oxazoline)-based materials by an UV-induced thiol-ene reaction, applying bisfunctional thiols as crosslinking agents [44]. Copoly(2-oxazoline)s were prepared by a microwave-supported CROP of 2-dec-9′-enyl-2-oxazoline providing an olefinic moiety for the subsequent thiol-ene reaction and 2-methyl-2-oxazoline or 2-ethyl-2-oxazoline. Ratios of aliphatic to olefinic monomers ranged from 200:10 to 100:10. The subsequent photochemically induced crosslinking was monitored by photorheology. The swelling behavior of the obtained gels depended on the hydrophilicity/

hydrophobicity of the starting materials as well as the crosslinking degree.

Hoogenboom and coworkers also reported the bulk-(co)polymerization of 2-dec-9′-enyl-2-oxazoline [45]. An environmentally benign synthesis of the olefinic-substituted monomer from biologically derived starting materials and halogen-free solvents obtained from waste biomass was reported. The cationic ring-opening (co)polymerizations (eventually with 2-ethyl-2oxazoline) were initiated by methyl tosylate and performed under microwave support at 100°C. A polymerization degree of 60 was targeted by the initiator:monomer ratio. The unsaturated moiety was not affected by the reaction conditions. A kinetic investigation revealed first-order kinetics and a living character for the homopolymerization of the olefinic monomer. A $Đ_M$ <1.2 was determined for the homopolymer.

The preparation of a library of 32 different hydrogels based on crosslinked poly(2-oxazoline)s was reported by Wiesbrock and coworkers [46]. The study aimed for the detailed investigation of the stimuli-triggered release of a model compound (Eosin Y). The hydrogels were based on 2-ethyl-2-oxazoline and 2-phenyl-2-oxazoline, which were copolymerized in ratios of 150:0, 100:50, 50:100, and 0:150. For the in situ crosslinking, 1.5–30 equivalents of phenylene-1,3-bis-2-oxazoline were added. The copolymerizations were performed in a Biotage Initiator 8 single-mode microwave reactor at 140°C. Reaction times as short as 1 h were sufficient. Especially the materials with high contents of 2-ethyl-2-oxazoline revealed a pronounced hydrogel character. Cytotoxicity tests revealed no toxicity of the nondegraded materials.

The RGD functionalization of poly(2-oxazoline)-based hydrogels was investigated by Wiesbrock and coworkers [47]. The polymer networks were composed of 2-ethyl-, 2-nonyl-, 2-dec-9′-enyl-2-oxazoline, and bisfunctional 2,2′-tetramethylene-bis-2-oxazoline, which acted as in situ crosslinkers. Swelling degrees of the polymer networks could be influenced by the ratio of hydrophilic and hydrophobic monomers as well as the crosslinking degree. 2-Dec-9′-enyl-2-oxazoline provided the olefinic moiety for the thiol-ene reaction with a pentapeptide containing the RGD motif and a thiol

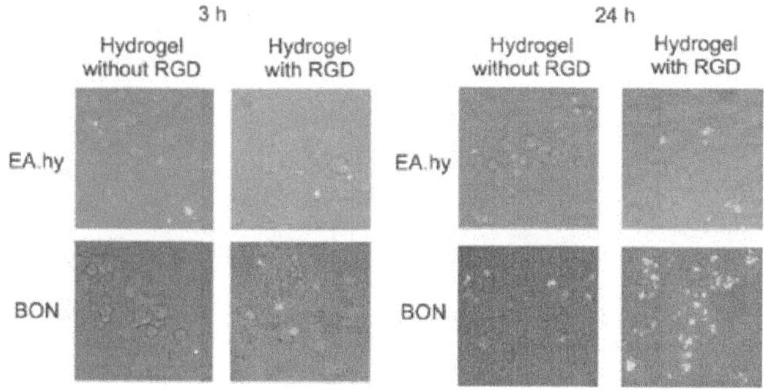

Figure 3.7 Transmission and fluorescence images of (top) noncancerous EA.hy cells and (bottom) cancerous BON cells incubated with an RGD-functionalized microgel exhibiting significant cell adhesion only to the malign cells. Reprinted from Ref. [47] with permission from MDPI AG.

moiety. Microparticles of this material selectively adhered to BON cancer cells (Fig. 3.7).

Schubert and coworkers reported the thermal and mechanical properties of different poly(2-oxazoline)s, which were obtained by a microwave-supported CROP [48]. 2-(3-Ethylpentyl)-2-oxazoline was polymerized as either homopolymer or copolymer with 2-ethyl-2-oxazoline. The reactions were performed at 140°C in an Initiator 8 bench-top microwave reactor with methyl tosylate as the initiator and acetonitrile as the solvent. A kinetic investigation revealed first-order kinetics of the monomer consumption.

The application of an alternative initiator system for the CROP of 2-alkyl-2-oxazolines was investigated by Schachner and coworkers [49]. The commonly used methyl tosylate was replaced by the cationic Brønsted acid $[H(OEt_2)_2]$-$[Al\{OC(CF_3)_3\}_4]$. Several polymerizations of 2-ethyl-2-oxazoline and 2-^{tert}butyl-2-oxazoline, as well as copolymerizations of these monomers, were performed. The kinetic survey of the polymerization of 2-ethyl-2-oxazoline revealed pseudo first-order kinetics and propagation rates of 128 mmol L^{-1} s^{-1} at a temperature of 140°C. Acetonitrile was used as the solvent. The syntheses were performed in an Initiator Sixty microwave reactor by Biotage.

Hoogenboom and coworkers reported the synthesis of S-2-butyl-4-ethyl-2-oxazoline, R-2-butyl-4-ethyl-2-oxazoline, and racemic

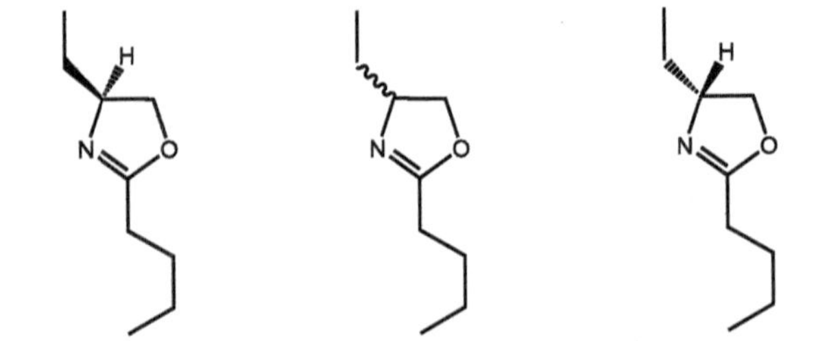

(4S)-2-butyl-4-ethyl-2-oxazolin **(4RS)-2-butyl-4-ethyl-2-oxazolin** **(4R)-2-butyl-4-ethyl-2-oxazolin**

Scheme 3.9 Structures of chiral 2-butyl-(2-oxazoline)s bearing an ethyl moiety on position 4.

2-butyl-4-ethyl-2-oxazoline (Scheme 3.9) as representatives of chiral 2-alkyl-2-oxazolines, which were subsequently polymerized by a microwave-supported CROP with a targeted degree of polymerization of 50 [50]. Circular dichroism measurements revealed helical structures for the polymers obtained from enatiomerically pure monomers and random coil structures for the polymer obtained from the racemic monomer.

In a subsequent study, Schubert and coworkers reported the synthesis of a library of chiral poly(2-oxazoline)s based on five chiral monomers [51], namely R-2-ethyl-4-ethyl-2-oxazoline, R-2-butyl-4-ethyl-2-oxazoline, R-2-octyl-4-ethyl-2-oxazoline, R-2-nonyl-4-ethyl-2-oxazoline, and R-2-undecenyl-4-ethyl-2-oxazoline. The polymers were obtained from a microwave-supported CROP using acetonitrile as the solvent. The polymerizations were performed at 180°C and reaction times of 60–90 min.

Schubert and coworkers also investigated the microwave-supported CROP of 2-^{cyclo}propyl-2-oxazoline [52]. Several polymers with targeted polymerization degrees ranging from 25 to 150 were synthesized at a polymerization temperature of 140°C. Reaction times ranging from 1 to 5 min were sufficient for monomer conversions of approximately 80%. A kinetic investigation revealed first-order kinetics as well as the living character of the polymerization. The observed polymerization rate was among the highest detected for 2-oxazolines bearing aliphatic side chains, which was ascribed to an electron-withdrawing effect of the cyclic substituent.

Wiesbrock and coworkers reported the application of (partially) hydrolyzed poly(2-oxazoline)s as contact biocides in a polypropylene matrix [53]. 2-Ethyl-2-oxazoline and 2-nonyl-2-oxazoline were polymerized in a Biotage Initiator 8 bench-top microwave reactor. The poly(2-oxazoline)s were subsequently hydrolyzed with hydrochloric acid also under microwave irradiation. A kinetic investigation revealed first-order kinetics of the hydrolysis step. Blends composed of 5 wt% of these polymers and 95 wt% of polypropylene were tested for antimicrobial activity against gram-positive and gram-negative microbes as well as fungi. Poly(2-nonyl-2-oxazoline)s with hydrolysis degrees of 50% and 75%, respectively, exhibited antimicrobial activity against all microbes tested.

Schubert and coworkers described the microwave-supported synthesis and selective partial hydrolysis of poly(2-ethyl-2-oxazoline)-*block*-poly(2-H-2-oxazoline) aiming at the preparation of poly(2-ethyl-2-oxazoline)-*block*-poly(ethylene imine) [54]. The authors reported an improved synthetic pathway for the 2-H-2-oxazoline. Block copolymers with ratios 2-ethyl-2-oxazoline:2-H-2-oxazoline ranging from 20:0 to 40:10 were prepared; the 2-H-2-oxazoline units could be selectively hydrolyzed under alkaline conditions.

Schubert and coworkers described the dielectrically heated synthesis of amphiphilic triblock copolymers based on 2-ethyl-2-oxazoline, 2-(1′-ethylheptyl)-2-oxazoline, and fluorinated 2-(2,6-difluorophenyl)-2-oxazoline [55]. The polymerizations were performed in an Initiator 60 single-mode microwave reactor with integrated IR temperature acquisition, operated in temperature-controlled mode. Homopolymerizations of the three monomers were performed at 140°C with a targeted degree of polymerization of 60. Kinetic investigation revealed first-order kinetics of the monomer conversion and the living character of the polymerizations. The highest polymerization rate among the chosen polymers was detected for the fluorinated species, which was ascribed to a higher nucleophilicity of the monomer. Several triblock copolymers were synthesized by an initial polymerization of 2-(2,6-difluorophenyl)-2-oxazoline and subsequent addition of 2-(1′-ethylheptyl)-2-oxazoline and 2-ethyl-2oxazoline, followed by microwave irradiation after each monomer addition. The monomer ratios ranged from 3/10/20 to 25/50/25. The procedure yielded polymers with a monomodal distribution of molecular weights and $Đ_M$ values between 1.11 and 1.17. The incorporation of all monomers in the product was confirmed by matrix-assisted laser desorption/ionizationtime-of-flight (MALDI-TOF) analyses.

Schubert and coworkers reported the microwave-heated synthesis of poly(2-(3′-ethylheptyl)-2-oxazoline as well as copolymers of that monomer and 2-ethyl-2-oxazoline [56]. The homopolymerization of the monomer with the branched side chain was performed at a temperature of 140°C in acetonitrile using methyl tosylate as initiator. First-order kinetics and dispersities of the polymer $Đ_M$ <1.23 were determined by gas chromatography and size-exclusion chromatography. Additionally, copolymers with monomer ratios ranging from 80:20 to 20:80 were synthesized. Also for the copolymerizations, a living character, first-order kinetics, and low dispersities $Đ_M$ <1.3 were maintained. The glass transition temperatures of the copolymers exhibited a linear correlation with the content of 2-(3′-ethylheptyl)-2-oxazoline repetition units, enabling for tailor-made tuning of the glass transition temperatures.

3.5 Highlights

Ionic liquids are prime microwave absorbers and have a huge potential for the replacement of halogenated solvents, paving the way for microwave-assisted green chemistry. Due to the uniform and internal heating provided by microwaves, also the performance of solvent-free polymerizations is a strategy to avoid halogenated solvents. The other hot topic described in this section is the application of the thiol-ene and azide–alkyne click chemistry in polymer sciences.

3.5.1 *Ionic Liquids and Solvent-Free Polymerizations*

Ionic liquids (Scheme 3.10) as solvents for polymerizations received considerable attention in environmentally more benign synthetic pathways. Additionally, ionic liquids, due to their polar nature, are excellent microwave absorbers.

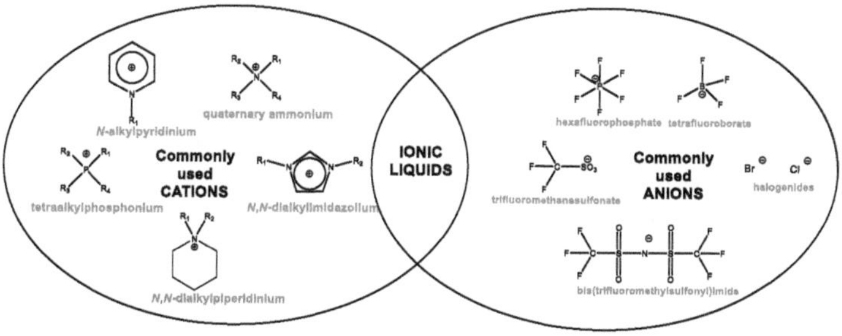

Scheme 3.10 Commonly used cations and anions in ionic liquids.

Sayer and coworkers investigated the application of the ionic liquid 1-^{n}dodecyl-3-methylimidazolium chloride as a surfactant during the microwave-heated emulsion polymerization of methyl methacrylate [57]. Irrespective of the heating source, comparable results of the polymerizations using dodecyltrimethylammonium bromide as surfactant were obtained in terms of particle size and monomer conversions. The authors also compared the results from microwave-supported polymerizations (Synthos 300 by Anton Paar) with those obtained under conventional heating, showing higher reaction rates in case of the dielectric heating. Polymerization rates were enhanced, especially in high-power pulsed-mode operation of the microwave reactor, which was ascribed to a specific effect of the microwave radiation on the ionic liquid.

Mallakpour and Dinari contributed to a study on the dielectrically heated preparation of chiral poly(amide-imide)s using tetrabutylammonium bromide as the ionic liquid [58]. The syntheses were performed in a domestic Samsung oven. 2-(3,5-Diaminophenyl)-benzimidazole was copolymerized with a monomer derived from the reaction of several different amino acids and pyromellitic dianhydride. The condensations were performed at a power level of 810 W for 3 min. Yields in the range of 50%–92% were obtained. The reactions were also performed applying conventional heating, which highlighted the advantages of microwave-assisted heating in terms of reaction rates

In another study, Mallakpour and Dinari used 1,3-diisopropyl imidazolium bromide as an ionic liquid reaction medium for the microwave-heated polycondensation of different aromatic diamines with (2-S)-4-[(4-methyl-2-phthalimidylpentanoylamino)benzoyl-amino]-isophthalic acid [59]. The polycondensations were performed in a domestic Samsung microwave oven. The authors concluded their study underscoring the efficient, clean, and easy procedure derived from the combination of ionic liquids and microwave-assisted heating.

Huang and coworkers described the preparation of a monolithic polymer-based material for application as a stationary phase in chromatography [60]. A two-step synthetic pathway was supported by ionic liquids and microwave-assisted heating. In the first step, 1-hexyl-3-methylimidazolium tetrafluoroborate was used as a reaction medium for the vinylization of a fused silica-coated column with 3-(trimethoxysilyl) propyl methacrylate, aiming to act as a reaction center for the subsequent monolith synthesis. The reaction mixture was irradiated with microwaves for 5 min in an unspecified domestic microwave oven. For comparison purposes, this step was also performed under conventional heating, which required 17 h for completion. The second step comprised the synthesis of the monolithic column material (Fig. 3.8) by the copolymerization of divinylbenzene and butyl methacrylate. The authors reported a significant acceleration of the synthesis using the combination of ionic liquids and microwave irradiation: Increased reaction rates by a factor of 300 were reported. An effect of the ionic liquid on the microporous structure of the obtained materials was also stated.

Mallakpour and Zadehnazari described the synthesis and characterization of a series of chiral hydroyphenyl-functionalized poly(amide-imide)s using dielectric heating and tetrabutylammonium bromide as solvent [61]. The preparations were performed in a Samsung domestic microwave oven. Yields ranging from 88% to 96% were obtained after 8 min of irradiation at approximately 130°C. The inherent viscosities of the polymers ranged from 0.32 to 0.49 dL/g.

In another study, Mallakpour and Zeraatpishe reported the synthesis of several poly(amide-imide)s based on 4,4′-methylene-bis(3-chloro-2,6-diethyl trimellit imidobenzene) by microwave-supported polycondensation of this monomer with various diamines [62]. The type of microwave reactor was not mentioned. Tetrabutylammonium bromide was used as solvent for the polymerizations,

Figure 3.8 SEM images of the column material derived from the microwave-supported polymerization of divinylbenzene and butyl methacrylate. Reproduced in part from Ref. [60] with permission of the Royal Society of Chemistry.

which were performed within 260 s at a microwave power of 900 W. Yields of up to 93% were reported.

The application of a two-phase system of ionic liquids/water for the microwave-supported copolymerization of aniline and *p*-phenylene diamine was described by Shim and coworkers [63]. The two monomers were dissolved in 1-butyl-3-methylimidazolium hexafluorophosphate. In the water phase, HCl or *p*-toluene sulfonic acid and potassium persulfate were mixed. Both phases were transferred into a microwave vial. The mixture was irradiated for 30 min in an unspecified microwave reactor, and the product recovered from the aqueous phase formed nanorods with a diameter of 40–100 nm.

3.5.2 *Click Chemistry*

Click chemistry (Scheme 3.11) received intense attention in organic synthesis as this type of reactions provides a convenient pathway for efficient, clean, and controlled syntheses.

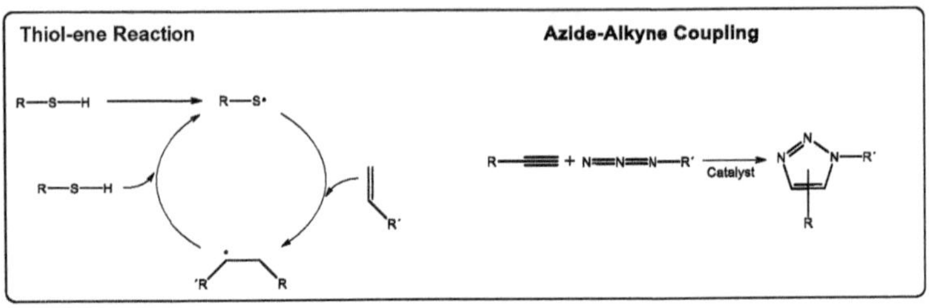

Scheme 3.11 The thiol-ene reaction and the azide–alkyne coupling reaction, representing two common types of click chemistry.

Jing and coworkers reported the preparation of biodegradable azide-functionalized polycarbonates [64]. The conventionally heated polymer syntheses started from 2,2-bis-azidomethyl-trimethylene carbonate, which was copolymerized with lactic acid. Subsequently, the polymers were modified by Huisgen 1,3-cycloaddition with three different azides. The polymeranalogous reactions were performed under microwave-assisted heating in a Yuhua Intruments MCR-3 reactor at 70°C for 5 min (Scheme 3.12). The micelle-forming polymers could be modified with anticancer drugs.

Kim et al. described the synthesis of well-defined self-assembling poly(3-hydroxy thiophen)-poly(2-vinyl pyrrolidone) copolymers [65]. The synthesis started from poly(3-hydroxy thiophen) that contained 5 mol.% of azide-functionalized repeating units, which were statistically spread over the polymer's backbone. Alkyne-terminated poly(2-vinyl pyrrolidone) were grafted onto this polymer by the copper-catalyzed azide–alkyne reaction performed in a Biotage Initiator microwave reactor at 65°C for 4 h. The grafted polymers

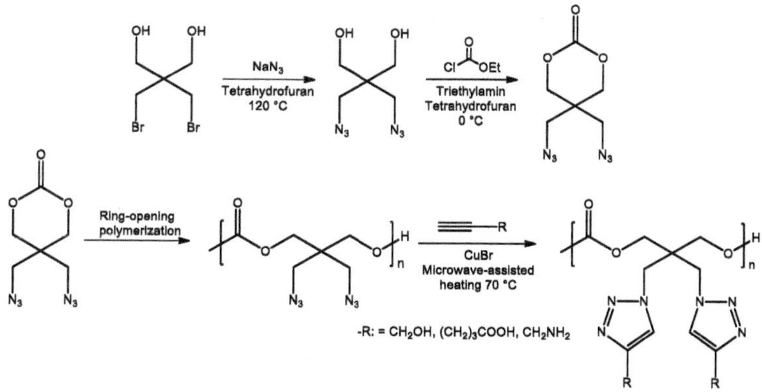

Scheme 3.12 A reaction scheme of the synthesis of azide-functionalized polycarbonates and the subsequent microwave-supported azide–alkine reaction.

revealed fibril-like, lamellar, cylindrical, and spherical structures, depending on the molecular weight of the grafted side chains.

In addition to polymer analogous modifications, also the synthesis of polymers by click chemistry was investigated. Huang and coworkers reported the microwave-supported click polymerization of aspartic acid–based or glutamic acid–based dialkynes with diazides [66]. The monomers were prepared from α-amino-functionalized aspartic or glutamic acid, which were modified at their acid functionalities either as an azide or as an alkyne. Homopolymers as well as copolymers were prepared by the CuBr-catalyzed azide–alkyne coupling reaction, which was performed in a Yuhua MCR-3 microwave reactor at 100°C for 30 min. The authors reported polymers with molecular weights ranging from 2.26 to 9.35 kDa and dispersity values in the range of 3.32–4.12. In a subsequent study, Huang and coworkers prepared polypeptides similar to the aforementioned ones, this time without the use of catalysts [67] using ^{tert}butyloxycarbonyl-protected aspartic acid or glutaric acid, which were deprotected after functionalization as azide or alkyne. The monomers were homopolymerized or copolymerized with dialkyne-functionalized poly(ethylene glycol) and terminated by monofunctional poly(ethylene glycol) alkynes. The azide–alkyne click polymerizations were performed at 120°C for 75 min. The obtained polymers were immobilized on crosslinked polystyrene microspheres functionalized with active ester groups; the aim of the work was to test the ability of the obtained material to remove endotoxins from proteins (bovine serum albumin [BSA] solutions).

3.6 Scale-Up

One of the major benefits of microwave reactors in organic synthesis is the efficient, quick, and homogeneous heat distribution inside the reaction vessel. A limitation to this fact is the low penetration depth of microwave irradiation. Scale-up of microwave-supported syntheses is therefore still challenging, as larger reaction vessels are not sufficiently penetrated by the microwave radiation, and efficient stirring is required for a uniform heat distribution. Two strategies to overcome these issues are currently focused on batch-mode reactors set up as single- or multimode microwave reactors (Fig. 3.9), as well as continuous-flow reactors.

Zhu and coworkers described the dielectrically heated batch-mode ring-opening polymerization of ε-caprolactone in the kilogram scale, performed in an industrial Nanjing WLD6S-01 multimode microwave reactor capable of a power output of 6 kW [68]. The polymerization was catalyzed by stannous octanoate. Five reaction vessels with a volume of 1.8 L each were used to hold a total mass of up to 2.5 kg of ε-caprolactone. Monomer conversion rates as high as 94% were reported, and polymers with

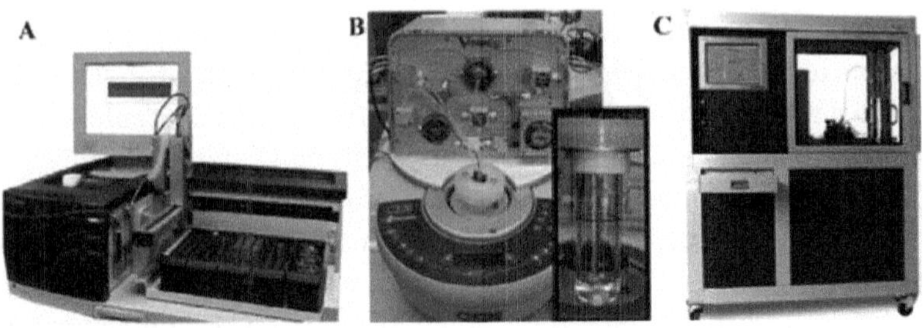

Figure 3.9 Three different microwave reactors providing reaction volumes in the range of (a) 5 mL to (c) 250 mL. Reprinted from Ref. [69]. Copyright © 2006, with permission from John Wiley and Sons.

M_n values of 100–1000 kDa were obtained. $Đ_M$ values were in the range of 1.72–2.75 according to size-exclusion chromatography. The reproducibility of the large-scale syntheses was demonstrated by a series of five experiments in the 1.5 kg scale performed at 850 W. Within this series, polymers with M_n values of 95 kDa were prepared at monomer conversion rates of more than 90%.

An earlier contribution by Schubert and coworkers described the up-scaled synthesis of poly(2-ethyl-2-oxazoline) in batch mode [69]. The polymerizations were performed in acetonitrile as the solvent and methyl tosylate as the initiator. Three different microwave reactors were used, depending on the scale of the polymerization, which ranged from 0.004 to 1 mol of the monomer (the latter representing almost 100 g of monomer), and a reaction volume of 250 mL. The authors reported excellent scalability of the polymerizations as the products were highly comparable, exhibiting $Đ_M$ values between 1.12 and 1.13 as well M_n values ranging from 9.5 to 9.7 kDa. Conversion rates larger than 96% were observed. Schubert and coworkers also investigated the scale-up of the CROP of 2-ethyl-2-oxazoline in continuous-flow mode [70]. A stock solution of the monomer and initiator in acetonitrile was pumped through a microwave-heated flow reactor with flow rates calculated for a residual time of approximately 15 min. Several reactor systems were tested, including a continuously stirred tank reactor, a tube reactor, and two flow coils. The recovered poly(2-ethyl-2-oxazoline)s exhibited a higher dispersity than those obtained from batch-mode synthesis. This was ascribed to a higher dispersity of residence time in the reactors, which is highly dependent on the reactor type.

3.7 Conclusion

This review of the proceedings of microwave-supported polymerizations and modifications of polymers clearly demonstrates the significance of microwave-assisted heating for facilitated, efficient, and also green synthesis strategies in polymer chemistry. Due to the rapidly growing interest of the community in this heating method during the first decade of this century, its application has been well established in the meantime. Especially the synthesis of poly(2-oxazoline)s clearly benefits from this technique. Also the ring-opening synthesis of polyesters and polyamides received intense attention, due to their potential applications in biomedic(in)al application. Controlled radical polymerizations (CRPs), which face slow reaction rates in favor of well-defined products, could also benefit from the fast and efficient heating technique.

3.7.1 *Safety Rules*

One general trend in the most recent research is the widespread application of microwave reactors dedicatedly designed for usage in laboratories; modified domestic ovens nowadays seem to vanish from laboratories. The application of dedicated laboratory equipment provides, aside from the safety infrastructure, excellent control over the reaction parameters—and beyond that highly precise temperature acquisition, which will contribute to the discussion about the existence/nonexistence of any nonthermal microwave effect.

Regardless of the serious benefits dielectrical heating provides for polymer chemistry, its major drawback is the scalability that still remains insufficiently addressed. Most contributions to that issue originate from the past decade. Recent investigations seem to focus on lab-scale amounts. In this sense, only fine-chemical polymers can be considered. Nevertheless, microwave-assisted heating remains a promising opportunity for the synthesis of smart and/or green materials, and especially the application of nonvolatile ionic liquids as replacement for volatile organic solvents (together with their prime microwave absorption) opens the field for further investigations.

Acknowledgments

This work was performed within the K-Project "PolyComp" at the Polymer Competence Center Leoben GmbH (PCCL, Austria) within the framework of the COMET program of the Federal Ministry for Transport, Innovation and Technology and the Federal

Ministry for Economy, Family and Youth, with contributions by the Institute for Chemistry and Technology of Materials at the University of Technology of Graz. Funding was provided by the Austrian Government and the State Government of Styria.

References

1. Ebner, C., Bodner, T., Stelzer, F., and Wiesbrock, F. (2011). One decade of microwave-assisted polymerizations: quo vadis?, *Macromol. Rapid Commun.*, **32**, pp. 254–288.
2. Brooks, W. L. A., and Summerlin, B. S. (2012). Microwave assisted RAFT polymerization, *Isr. J. Chem.*, **52**, pp. 256–263.
3. Hernandez-Ortiz, J. C., Jaramillo-Soto, G., Palacios-Alquisira, J., and Vivaldo-Lima, E. (2010). Modeling of polymerization kinetics and molecular weight development in the microwave-activated raft polymerization of styrene, *Macromol. React. Eng.*, **4**, pp. 210–221.
4. Sugihara, Y., Semsarilar, M., Perrier, S., and Zetterlund, P. B. (2012). Assessment of the influence of microwave irradiation on conventional and RAFT radical polymerization of styrene, *Polym. Chem.*, **3**, pp. 2801–2806.
5. Sumerlin, B. S., and Debashish, R. (2011). Block copolymerization of vinyl ester monomers via RAFT/MADIX under microwave irradiation, *Polymer*, **52**, pp. 3038–3045.
6. Zetterlund, P. B, and Perrier, S. (2011). RAFT polymerization under microwave irradiation: toward mechanistic understanding, *Macromolecules*, **44**, pp. 1340–1346.
7. Grigoras, M., and Negru, O.-I. (2012). Synthesis of star poly (N-vinylcarbazole) by microwave-assisted reversible addition-fragmentation chain transfer polymerization (RAFT), *Polymers*, **4**, pp. 1183–1194.
8. Lopez-Dominguez, P., and Vivaldo-Lima, E. (2013). Analysis of the microwave activated atom transfer radical polymerization of methyl methacrylate and styrene using modeling tools, *Macromol. React. Eng.*, **7**, pp. 463–476.
9. Buruiana, E. C., Murariu, M., and Buruiana, T. (2010). Copolyacrylates with phenylalanine and anthracene entities prepared by ATRP and microwave irradiation, *J. Lumin.*, **130**, pp. 1794–1801.

10. Wang, G., and Wu, H. (2011.) Microwave-assisted controlled copolymerization of styrene and acrylonitrile catalyzed by $FeCl_3$/isophtalic acid, *Polym. Bull.*, **67**, pp. 1809–1821.
11. Marcasuzaa, P., Reynaud, S., Grassl, B., Preud'homme, H., Desbrières, J., Trchovà, M., and Donard, O. F. X. (2010). Microwave synthesis: an alternative approach to synthesize conducting end-capped polymers, *Polymer*, **52**, pp. 33–39.
12. Costa, C., Santos, V. H. S., Sayer, C., Araujo, P. H. H., Santos, A. F., and Fortuny, M. (2011). Microwave effects due to anionic or cationic initiators in emulsion polymerization reactions, *Macromol. Symp.*, **302**, pp. 161–168.
13. Hayden, S., Studentschnig, A. F. H., Schober, S., and Kappe, C. O. (2014). A critical investigation on the occurence of microwave effects in emulsion polymerizations, *Macromol. Chem. Phys.*, **215**, 2318–2326.
14. Schmücker, S., and Kuckling, D. (2012). Enhanced preparations of alkoxyamine-functionalized poly(*p*-phenylene)s and their use as macroinitiators for the synthesis of stimuli-respinsice coil-rod-coil block copolymers, *Macromol. Chem. Phys.*, **213**, pp. 1725–1734.
15. Zhang W., Lu, P., Wang, Z., and Ma, Y. (2013). Microwave-assisted suzuki coupling reaction for rapid synthesis of conjugated polymers: poly(9,9-dihexylfluorene)s as an example, *J. Polym. Sci., Part A: Polym. Chem.*, **51**, pp. 1950–1955.
16. Melucci, M., Barbarella, G., and Sotgiu, G. (2002). Solvent-free, microwave-assisted synthesis of thiophene oligomers via Suzuki coupling, *J. Org. Chem.*, **67**, pp. 8877–8884.
17. Al-Hashimi. M., Baklar, M. A., Colleaux, F., Watkins, S. E., Anthopoulos, T. D., Stingelin, N., and Heeney, M. (2011). Synthesis, characterization, and field effect transistor properties of regioregular poly(3-alkyl-2,5-selenylenevinylene), *Macromolecules*, **44**, pp. 5194–5199.
18. Sacarescu, G., Simionescu, M., Harabagui, V., and Sacarescu, L. (2012). Microwave-assisted Wurtz coupling of methylphenyl-dichlorosilane in solvent-free conditions, *e-Polymers*, **12**, pp. 145–150.
19. Hong, Y.-R., Ng, J. Y., Wong, H. K., Moh, L. C. H., Yip, Y. J., Chen, Z.-K., and Norsten, T. B. (2012). Synthesis and characterization of a series of low-bandgap copolymers based on cyclopental[2,1-b:3,4-b′]dithiophene and thienopyrroledione for photovoltaic appliccations, *Sol. Energy Mater. Sol. Cells*, **102**, pp. 58–65.
20. Liao, Q., Liu, L. J., Zhang, C., He, F., Zhuo, R. X., and Wan, K. (2002). Microwave-assisted ring-opening polymerization of ε-caprolactone, *J. Polym. Sci., Part A: Polym. Chem.*, **40**, pp. 1749–1755.

21. Yu, Z. J., Liu, L. J., and Zhuo, R. X. (2002). Microwave-improved polymerization of ε-caprolactone initiated by carboxylic acids, *J. Polym. Sci., Part A: Polym. Chem.*, **41**, pp. 13–21.

22. Barbier-Baudry, D., Brachais, C.-H., Cretu, A., Loupy, A., and Stuerga, D. (2002). An easy way toward ε-caprolactone macromonomers by microwave irradiation using early lanthanide halides as catalysts, *Macromol. Rapid Commun.*, **23**, pp. 200–204.

23. Kerep, P., and Ritter, H. (2006). Influence of microwave irradiation on the lipase catalyzed ring-opening polymerization of ε-caprolactone, *Macromol. Rapid Commun.*, **27**, pp. 707–710.

24. Kerep, P., and Ritter, H. (2006). Chemoenzymatic synthesis of polycaprolactone-*block*-polystyrene via macromolecular chain transfer reagents, *Macromol. Rapid Commun.*, **28**, pp. 759–766.

25. Sinnwell, S., Schmidt, A. M., and Ritter, H. (2006). Direct synthesis of (meth-)acrylate poly(ε-caprolactone) macromonomers, *J. Macromol. Sci., Part A: Pure. Appl. Chem.*, **43**, pp. 469–476.

26. Song, Y., Weng, X., Zhuo, R., and Liu, L. (2002). Acid-initiated polymerization of ε-caprolactone under microwave irradiation and its application in the preparation of drug controlled release system, *J. Biomater. Sci. Polymer Edn.*, **14**, pp. 241–253.

27. Gotelli, G. A., Bonelli, P., Abraham, G. A., and Sosnik, A. (2010). Fast and efficient synthesis of high molecular weight poly(epsilon-caprolactone) diols by microwave-assisted polymer synthesis, *J. Appl. Polym. Sci.*, **121**, pp. 1321–1329.

28. Yu, Z., and Liu, L. (2007). Biodegradable poly(vinyl alcohol)-graft-poly(ε-caprolactone) comb-like polyester: microwave synthesis and its characterization, *J. Appl. Polym. Sci.*, **104**, pp. 3973–3979.

29. Fang, X., Simone. C. D., Vaccaro, E., Huang, S. J., and Scola, D. A. (2002). Ring-opening polymerization of ε-caprolactam and ε-caprolactone via microwave irradiation, *J. Polym. Sci., Part A: Polym. Chem.*, **40**, pp. 2264–2275.

30. Li, G., Zhao, N., Bai, W., Chen, D. L., and Xiong, C. D. (2010). Microwave-assisted ring-opening polymerization of poly(glycolic acid-*co*-lactic acid) copolymers, *e-Polymers*, **10**, pp. 557–562.

31. Naik, S., and Carpenter, E. E. (2008). Poly(D,L-lactide-*co*-glycolide) microcomposite containing magnetic iron core nanoparticles as drug carrier, *J. Appl. Phys.*, **103**, pp. 07A313-1–07A313-3.

32. Khodaverdi, E., Tekie, F. S. M., Mohajeri, S. A., Ganji, F., Zohuri, G., and Hadizadeh, F. (2012). Preparation and investigation of sustained drug

delivery systems using an injectable, thermosensitive, in situ forming hydrogel composed of PLGA–PEG–PLGA, *AAPS Pharm. Sci. Tech.*, **13**, pp. 590–600.

33. Ramier, J., Renard, E., and Grande, D. (2012). Microwave-assisted ring-opening polymerization of D,L-lactide: a probe for the nonexistence of nonthermal microwave effects, *Macromol. Chem. Phys.*, **213**, pp. 784–788.
34. Hirao, K., Masutani, K., and Ohara, H. (2009). Synthesis of L,L-lactide via depolymerization of oligo(L-lactic acid) by microwave irradiation, *J. Chem. Eng. Jpn.*, **42**, pp. 687–690.
35. Hirao, K., Masutani, K., and Ohara, H. (2009). Noncatalytic polycondensation of L-lactic acid under microwave irradiation, *J. Chem. Eng. Jpn.*, **42**, pp. 417–419.
36. Nikolic, L., Ristic, I., Adnadjevic, B., Nikolic, V., Jovanovic, J., and Stankovic, M. (2010). Novel microwave-assisted synthesis of poly(D,L-lactide): the influence of monomer/initiator molar ratio on the product properties, *Sensors*, **10**, pp. 5063–5073.
37. Jing, S., Peng, W., Tong, Z., and Baoxiu, Z. (2006). Microwave-irradiated ring-opening polymerization of D,L-lactide under atmosphere, *J. Appl. Polym. Sci.*, **100**, pp. 2244–2247.
38. Frediani, M., Sémeril, D., Matt, D., Rizzolo, F., Papini. A. M., Frediani, P., Rosi, L., Santella, M., and Giachi, G. (2010). L-lactide polymerization by calix[4]arene-titanium (IV) complex using conventional heating and microwave irradiation, *e-Polymers*, **10**, pp. 177–184.
39. Lambermont-Thijs, H. M. L., Fijten, M. W. M., (Ton) van der Linden, A. J., van Lankvelt, B. M., Bloksma, M. M., Schubert, U. S., and Hoogenboom, R. (2011). Efficient cationic ring-opening polymerization of diverse cyclic imino ethers: unexpected copolymerization behaviour, *Macromolecules*, **44**, pp. 4320–4325.
40. Bloksma, M. M., Paulus, R. M., van Kuringen, H. P. C., van der Woerdt, F., Lambermont-Thijs, H. M. L., Schubert, U. S., and Hoogenboom, R. (2012). Thermoresponsive poly(2-oxazine)s, *Macromol. Rapid Commun.*, **33**, pp. 92–96.
41. Schenk, V., Ellmair, L., Rossegger, E., Edler, M., Griesser, T., Weidlinger, G., and Wiesbrock, F. (2012). Water-developabel poly(2-oxazoline)-based negative photoresists, *Macromol. Rapid Commun.*, **33**, pp. 396–400.
42. Fimberger, M., Schenk, V., Rosegger, E., and Wiesbrock, F. (2013). UV-induces crosslinking of poly[2-(2′-norbornenyl)-2-oxazoline]s, *Period. Polytech. Chem. Eng.*, **58**, pp. 69–74.

43. Farrugia, B. L., Kempe, K., Schubert, U. S., Hoogenboom, R., and Dargaville, T. R. (2013). Poly(2-oxazoline) hydrogels for controlled fibroblast attachment, *Biomacromolecules*, **14**, pp. 2724–2732.

44. Dargaville, T. R., Forster, R., Farrugia, B. L., Kempe, K., Voorhaar, L., Schubert, U. S., and Hoogenboom, R. (2012). Poly(2-oxazoline) hydrogel monoliths via thiol-ene coupling, *Macromol. Rapid Commun.*, **33**, pp. 1695–1700.

45. Kempe, K., Hoogenboom, R., and Schubert, U. S. (2011). A green approach for the synthesis and thiol-ene modification of alkene funktionalized poly(2-oxazoline)s, *Macromol. Rapid Commun.*, **32**, pp. 1484–1489.

46. Kelly, A. M., Hecke, A. B., and Wiesbrock, F. (2011). Synthesis of poly(2-oxazoline)-based hydrogels with tailor-made swelling degrees capable of stimuli-triggered compound release, *Macromol. Rapid Commun.*, **32**, pp. 1815–1819.

47. Schenk, V., Rossegger, E., Ebner, C., Bangerl, F., Reichmann, K., Hoffmann, B., Höpfner, M., and Wiesbrock, F. (2014). RGD-functionalization of poly(2-oxazoline)-based networks for enhanced adhesion to cancer cells, *Polymers*, **6**, pp. 264–279.

48. Kempe, K., Rettler, E. F.-J., Paulus, R. M., Kuse, A., Hoogenboom, R., and Schubert, U. S. (2013). A systematic investigation of the effect of side chain branching on the glass transition temperature and mechanical properties of aliphatic (co-)poly(2-oxazoline)s, *Polymer*, **54**, pp. 2036–2042.

49. Rudolph, T., Kempe, K., Crotty, S., Paulus, R. M., Schubert, U. S., Krossing, I., and Schachner, F. H. (2013). A strong cationic Brønsted acid, $[H(OEt_2)_2]$-$[Al\{OC(CF_3)_3\}_4]$, as an efficient initiator for the cationic ring-opening polymerization of 2-alkyl-2-oxazolines, *Polym. Chem.*, **4**, pp. 495–505.

50. Bloksma, M. M., Rogers, S., Schubert, U. S., and Hoogenboom, R. (2009). Secondary formation of main-chain chiral poly(2-oxazoline)s in solution, *Soft Matter*, **6**, pp. 994–1003.

51. Bloksma M. M., Rogers, S. Schubert, U. S., and Hoogenboom, R. (2011). Main-chain chiral poly(2-oxazoline)s: influence of alkyl side-chain on secondary structure formation in solution, *J. Polym. Sci., Part A: Polym. Chem.*, **49**, pp. 2790–2801.

52. Bloksma, M. M., Weber, C., Perevyazko, I. Y., Kuse, A., Baumgärtel, A., Vollrath, A., Hoogenboom, R., and Schubert, U. S. (2011). Poly(2-cyclopropyl-2-oxazoline): from rate acceleration by cyclopropyl to thermoresponsive properties, *Macromolecules*, **44**, pp. 4057–4064.

53. Kelly, A. M., Kaltenhauser, V., Mühlbacher, I., Rametsteiner, K., Kren, H., Slugovc, C., Stelzer, F., and Wiesbrock, F. (2013). Poly(2-oxazoline)-derived contact biocides: contributions to the understanding of antimicrobial activity, *Macromol. Biosci.*, **13**, pp. 116–125.
54. Tauhardt, L., Kempe, K., and Schubert, U. S. (2012). Toward the design of lpei containing block copolymers: improved synthesis protocol, selective hydrolysis, and detailed characterization, *J. Polym. Sci. Part A: Polym. Chem.*, **50**, pp. 4516–4523.
55. Kempe, K., Baumgaertel, A., Hoogenboom, R., and Schubert, U. S. (2010). Design of new amphiphilic triblock copoly(2-oxazoline)s containing a fluorinated segment, *J. Polym. Sci., Part A: Polym. Chem.*, **48**, pp. 5100–5108.
56. Kempe, K., Jacobs, S., Lambermont-Thijs, H. M. L., Fijten, M. M. W. M., Hoogenboom, R., and Schubert, U. S. (2010). Rational design of an amorphous poly(2-oxazoline) with a low glass-transition temperature: monomer synthesis, copolymerization and properties, *Macromolecules*, **43**, pp. 4098–4104.
57. Costa, C., Santos, V. H. S., Dariva, C., Santos, A. F., Fortuny, M, Araujo, P. H. H., and Sayer, C. (2012). Ionic liquid as surfactant in microwave-assisted emulsion polymerization, *J. Appl. Polym. Sci.*, **127**, pp. 448–455.
58. Mallakpour, S., and Dinari, M. (2013). Straightforward and green method for the synthesis of nanostructure poly(amide-imide)s-containing benzimidazole and amino acid moieties by microwave irradiation, *Polym. Bull.*, **70**, pp. 1049–1064.
59. Mallakpour, S., and Dinari, M. (2010). A study of the ionic liquid mediated microwave heating for the synthesis of new thermally stable and optically active aromatic polyamides under green procedure, *Macromol. Res.*, **18**, pp. 129–136.
60. Shih, Y. H., Singco, B., Liu, W.-L., Hsu, C.-H., and Huang, H. Y. (2011). A rapid synthetic method for organic polymer-based monoliths in a room temperature ionic liquid medium via microwave-assisted vinylization and polymerization, *Green Chem.*, **13**, pp. 296–299.
61. Mallakpour, S., and Zadehnazari, A. (2013). Microwave-assisted synthesis and morphological characterization of chiral poly(amid-imide) nanostructures in molten ionic liquid salt, *Adv. Polym. Technol.*, **32**, p. 21333.
62. Mallakpour, S., and Zeraatpishe, F. (2013). Molten salt ionic liquid-assisted synthesis of nano-structured poly(amide imide)s based on 4,4′-methylenbis(3-chloro-2,6-diethyl trimellit imidobenzene) via mi-

crowave process as an environmentally friendly methodology, *Polym. Sci. Ser. B.*, **55**, pp. 271–279.

63. Haldoray, Y., Zong, T., and Shim, J.-J. (2011). Microwave-assisted facile and rapid synthesis of self-assembled conducting copolymer nanorods via aqueous/ionic liquid interfacial polymerization, *Mater. Chem. Phys.*, **127**, pp. 385–390.
64. Hu, X., Yan, L., Xiao, H., Li, X., and Jing, X. (2013). Application of microwave-assisted click chemistry in the preparation of functionalized copolymers for drug conjugation, *J. Appl. Polym. Sci.*, **127**, pp. 3365–3373.
65. Kim, H. J., Paek, K., Yang, H., Cho, C.-H., Kim, J.-S., Lee, W., and Kim, B. J. (2013). Molecular design of "graft" assembly for ordered microphase seperation of P3HT-based rod-coil copolymers, *Macromolecules*, **46**, pp. 8472–8478.
66. Guo, J., Wei, Y., Zhou, D., Cai, P., Jing, X., Chen, X.-S., and Huang, Y. (2011). Chemosynthesis of poly(ε-lysine)-analogous polymers by microwave-assisted click polymerization, *Biomacromolecules*, **12**, pp. 737–746.
67. Guo, J., Meng, F., Li, X., Wang, M., Wu, Y., Jing, X., and Huang, Y. (2012). PEGylated click polypeptides synthesized by copper-free microwave-assisted thermal click polymerization for selective endotoxin removal from protein solutions, *Macromol. Biosci.*, **12**, pp. 533–546.
68. Xu, Q., Zhang, C., Cai, S., Zhu, P., and Liu, L. (2010). Large-scale microwave assisted ring-opening polymerization of ε-caprolactone, *J. Ind. Eng. Chem.*, **16**, pp. 872–875.
69. Hoogenboom, R., Paulus, R. M., Pilotti, Å., and Schubert, U. S. (2006). Scale-up of microwave-assisted polymerizations in batch mode: the cationic ring-opening polymerization of 2-ethyl-2-oxazoline, *Macromol. Rapid Commun.*, **27**, pp. 1556–1560.
70. Paulus, R. M., Erdmenger, T., Becer, C. R., Hoogenboom, R., and Schubert, U. S. (2007). Scale-up of microwave-assisted polymerizations in continous-flow mode: cationic ring-opening polymerization of 2-ethyl-2-oxazoline, *Macromol. Rapid Commun.*, **28**, pp. 484–491.

Chapter 4

Microwave Processing of Ceramics and Glasses

D. C. Folz,[a] D. E. Clark,[a] and R. L. Schulz[b]

[a]*Department of Materials Science and Engineering, Virginia Polytechnic Institute and State University, 445 Old Turner Street, Blacksburg, VA 24061, USA*
[b]*Corning Incorporated, 1 Science Center Drive, Corning, NY 14830, USA*
dfolz@mse.vt.edu

The ceramics industry represents a major segment of the global economy and includes a diverse range of mature (i.e., whitewares, refractories, glass) and emerging (i.e., technical and advanced ceramics) products. The manufacture of nearly all ceramics products requires the use of both high (firing) and low (drying) temperatures that, up to the present, have been provided primarily by conventional electric resistance or gas heating. Microwave energy as a replacement and/or complement to traditional methods has the potential to reduce processing costs, improve product performance, and provide a greener solution for manufacturing ceramics.

Some companies already are investing in this technology but, for obvious reasons, are reluctant to discuss their successes. Implementation of microwave energy by industry has been slow due to the capital investments required, as well as their hesitation to replace familiar technologies. In this chapter, recent microwave

Microwave Engineering of Materials: From Mesoscale to Nanoscale
Edited by Erwann Guénin

ISBN 978-981-4669-42-9 (Hardcover), 978-981-4669-43-6 (eBook)
www.panstanford.com

research from universities and national laboratories is discussed, with an emphasis on ceramic and glass processing.

4.1 Introduction

4.1.1 *The Ceramics Industry*

The worldwide ceramics industry comprises many subindustries, such as whitewares, refractories, abrasives, cements, glass and glass-ceramics, structural clay products, and hi-tech ceramics. Together, these subindustries contribute hundreds of billions of dollars to the global economy. Most of the research today is on high-tech ceramics—those that are used in products that are fairly sophisticated in design and require a high degree of engineering skills and process control in order to manufacture them. Typical high-tech ceramic products include sensors, pollution control devices, biomedical implants, optical fibers, electronic circuitry, and composites, to name a few.

The manufacture of most ceramic products requires controlled temperature elevation to alter either the thermodynamics or the kinetics of the process. For example, when clays are formed into bricks, these bricks (structural clay products) must be heated to high temperatures in order to cause the clay products to decompose, react, and densify (sinter) into an environmentally stable and mechanically strong product. Elevated temperatures are used for a variety of ceramic processes, such as drying, binder burnout, decomposition reactions, alloying, clinkering, melting, sintering, nucleation and crystallization, and annealing. A diagram illustrating the approximate temperature ranges over which some of these important processes take place is shown in Fig. 4.1.

The required temperatures for most of these processes typically are achieved with electrical resistance or gas heating. However, significant research and development activities have demonstrated that either microwave energy alone or microwave energy combined with electric or gas heating can provide a more efficient and sometimes otherwise advantageous method for heating ceramics. The objective of this chapter is to review our present understanding

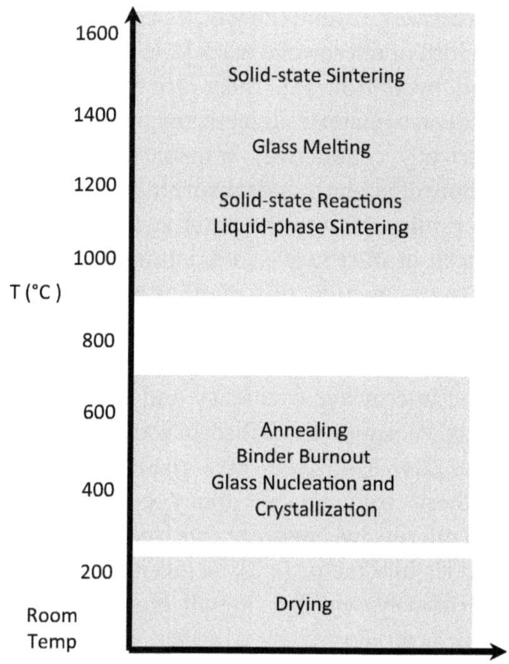

Figure 4.1 Approximate temperature ranges where important ceramic processes occur.

of how microwave energy offers economic and/or technical advantages over conventional methods. We also will discuss some of the challenges in implementing the use of microwave energy on a more wide-scale basis for processing ceramics.

4.1.2 *Brief Review of Microwave–Ceramic Interactions*

Microwave energy is recognized as a viable method for processing a variety of materials (polymers, metals, and ceramics) ranging from liquids to solids. Following publication of the 1987 article in *American Ceramic Society Bulletin*, "Microwave Processing of Ceramic Materials," by W. H. Sutton [1], research in the area of ceramics processing via microwave energy has grown substantially. Through the symposia and international conferences to follow, papers by researchers around the world provide the evidence of the depth and breadth of microwave work [1–11].

As outlined by Sutton [1], there are specific categories that describe microwave–material (electromagnetic field) interactions: opaque (electrically conductive), transparent (electrically nonabsorbing and nonreflecting), and absorbing (electrically lossy and insulating). Ceramics fall into the latter two categories. It is due to these differences in microwave interactions that ceramics can be used to assist in the manufacture of other materials, or they can be the products themselves.

The extent to which a ceramic is either transparent or absorbing depends on the microwave frequency and the temperature. While there are some ceramics that absorb well at room temperature at 915 MHz or 2.45 GHz (the most readily available equipment operates at these frequencies), many ceramics are essentially transparent to microwave energy at this frequency and temperature [1]. The dielectric loss factor ($\varepsilon''_{\text{eff}}$), as given in Eq. 4.1, of a material determines primarily whether it will absorb microwave energy [12]:

$$\varepsilon''_{\text{eff}} = \varepsilon''_{\text{c}} + \varepsilon''_{\text{p}} \tag{4.1}$$

This equation indicates that the effective dielectric loss factor is the sum of the losses at a specific frequency and temperature, where the subscripts c and p, respectively, represent losses due to long-range charge migration (conductivity) and short-range polarization processes (polar species relaxation). The dielectric loss factor $\varepsilon''_{\text{eff}}$ may also be expressed in terms of the loss tangent tan δ and the dielectric constant ε' (a measure of the electric polarizability of the materials)

$$\varepsilon''_{\text{eff}} = \varepsilon' \tan \delta \tag{4.2}$$

Metaxas and Meredith state that good candidate materials for microwave processing will have an $\varepsilon''_{\text{eff}}$ between 10^{-2} and 5, with 5 being a strong lossy material [12]. Although there is an abundance of papers that discuss dielectric properties and measurements [12–19], as well as online resources that list dielectric loss of materials at various frequencies and temperatures, this is an area where a much larger and more systematic database would be useful. A database that will be useful for all interested parties must not be too general in nature. For example, all alumina does not have the same value of $\varepsilon''_{\text{eff}}$ and thus will not heat the same in a microwave field. Likewise, different geometries of microwave applicators (and thus microwave fields) will not heat identical alumina samples in the same way.

There are several methods for obtaining dielectric data for materials at specific frequencies and temperatures. For example, one that is used quite often is cavity perturbation [13, 14]. Using this method, the electric field strengths in the microwave cavity are measured both with and without a sample present. On the basis of energy dissipation, the dielectric loss factor of the sample material is back-calculated for the specific frequency and temperature conditions in the cavity. The samples themselves are quite small, and the measurements and calculations depend heavily on the physical characteristics of the material, not just its pure dielectric behavior. For example, if 10 samples of identical size and shape are taken from a single bulk material, the dielectric loss factors obtained can vary on the basis of the structural characteristics of each sample (i.e., porosity, grain size and orientation, chemical or structural defects). The resulting dielectric data can provide a basis for initial decisions regarding the potential for microwave heating in the material, but it may not always give an accurate picture of the behavior of the actual large-scale, nonuniform production piece.

Another factor that can affect the extent of microwave–material interaction is the type of microwave applicator used. The energy dissipation within the sample can look entirely different in a single-mode versus a multimode cavity. Depending on the strength and uniformity of the field that impinges on the sample, it may behave quite differently, making the choice of equipment to be used for fundamental measurements and for manufacturing an issue. When the construction of a useful database is discussed, it is important to consider carefully what type of information will be needed by the end users, be they researchers or product manufacturers.

Table 4.1 shows some typical values of the dielectric properties of ceramics and water at a specific frequency. Dielectric properties are dependent on not only the composition, frequency, and temperature but also the purity and structure. For example, crystalline lithium disilicate has a lower $\varepsilon''_{\text{eff}}$ than does lithium disilicate glass with

Table 4.1 Dielectric properties for some typical ceramics and glasses. Dielectric properties can be very sensitive to the purity and phase of the material. Therefore, the data shown should be used only as a guide

Ceramic materials	Frequency (GHz)	Temperature (°C)	ε''	ε'	tan δ
Mullite[a]	0.001	25	0.012	5.8	0.002
Fused quartz[b]	2.45	25	0.0001	3.8	0.00003
Soda lime glass [20]	2.45	20	1.20	6.0	0.20
Pyrex[b]	2.45	25	0.005	4.0	0.0013
Boron nitride [12]	8.52	25	0.0131	4.37	0.00300
Aluminum nitride[c]	2.45	100	0.008	9	0.000889
Silicon nitride [12]	8.52	25	0.0199	5.54	0.00360
	2.45	25	0.015	0.68	0.022059
Lithium disilicate glass[d]	2.45	25	0.17	7.5	0.023
Lithium disilicate glass-ceramics[d]	2.45	25	0.02	5	0.004
99.7% Al_2O_3[d]	2.45	25	0.009	2.32	0.0039
99.7% Al_2O_3[d]	2.45	1016	0.022	2.5	0.008
Zircon refractory[d]	2.45	25	0.0045	9.89	0.001
Zircon refractory[d]	2.45	1160	0.5624	11.52	0.049
α–SiC[d]	2.45	25	1.29	10.51	0.123
α–SiC[d]	2.45	1015	0.914	13.36	0.068
Water[d]	2.45	25	10.2	78.00	0.131

[a]www.accuratus.com
[b]Data provided by J. Gerling, Gerling Applied Engineering, Inc., Modesto, CA.
[c]Committee on Microwave Processing of Materials: An Emerging Industrial Technology, Commission on Engineering and Technical Systems, National Research Council. (1994). *Microwave Processing of Materials* (National Materials Advisory Board, National Academy Press, Washington, DC).
[d]Dielectric data provided by Dr. R. Hutcheon, Microwave Properties North, Canada.
Source: Reprinted from Ref. [20] with permission of the American Ceramic Society, www.ceramics. org

the same composition, making the latter much easier to heat with microwaves.

Factors that affect the power absorbed per unit volume P_a and thus the extent of heating are shown in Eq. 4.3:

$$P_a = -\varepsilon_o \varepsilon''_{eff} E^2_{rms} \quad (4.3)$$

where is the angular frequency, ε_o is the permittivity of free space, and E_{rms} is the root-mean-square value of the electric field.

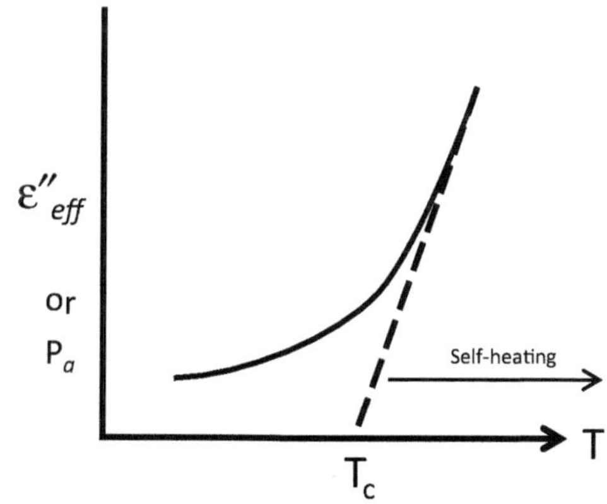

Figure 4.2 The effect of temperature or $\varepsilon''_{\text{eff}}$ on power absorbed. As the temperature of the sample is increased, a point is reached where ε'' increases significantly and the power absorbed results in self-heating. Reprinted from Ref. [19] with permission of the American Ceramic Society, www.ceramics.org All rights reserved.

Thus in addition to the dielectric loss factor, the ability to heat a ceramic depends on the electric field strength. Typically, single-mode applications have higher field strengths than do multimode applications, which are most commonly used. Therefore, samples that cannot be easily heated in a multimode applicator can be heated in a single-mode application.

4.1.2.1 Methods of heating

Many ceramics exhibit an abrupt increase in $\varepsilon''_{\text{eff}}$ with increasing temperature, as shown in Fig. 4.2. Since the power absorbed, P_a, is directly proportional to $\varepsilon''_{\text{eff}}$, there is a corresponding increase in the heating rate of the ceramic at T_c. The reason for this abrupt change in $\varepsilon''_{\text{eff}}$ most likely is due to easier dipole rotation at higher temperatures that moves the peak in microwave absorption closer to the microwave frequency being used. The temperature where the abrupt change in $\varepsilon''_{\text{eff}}$ occurs is referred to as the critical temperature, T_c. Below T_c, an external energy source is required to heat the ceramic. Above T_c, the sample becomes self-heating in the microwave field. Very few T_c values have been reported for ceramics thus far. The T_c for SiC is around room temperature, and thus, it can be directly heated with microwave energy, while materials such as Al_2O_3 and ZrO_2 must be heated to a fairly high temperature (>800°C) before they can absorb enough microwave energy to become self-heating.

There are two primary methods for using microwave energy to process ceramics and glasses (Fig. 4.3): (1) *direct microwave heating* and (2) *hybrid microwave heating.* In direct microwave heating, the sample will begin to absorb at ambient temperature without the aid of any secondary energy input. Good candidates for direct microwave heating are those with relatively high dielectric loss factors (i.e., silicon carbide, carbon). However, a significant number of technically important ceramics, such as alumina and zirconia, will not heat up from room temperature via direct microwave energy. It is here where an understanding of the relationship between $\varepsilon''_{\text{eff}}$ and T_c becomes important.

The purpose of hybrid microwave heating is to achieve a critical temperature where $\varepsilon''_{\text{eff}}$ is sufficiently high such that the material will begin to absorb microwave energy on its own. Hybrid microwave heating can be further broken down into two categories,*extrinsic* and *intrinsic*. In *extrinsic hybrid heating*, a secondary form of energy

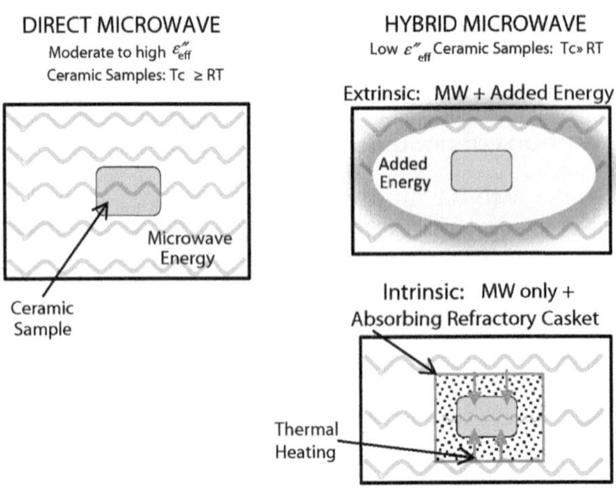

Figure 4.3 Simple schematic representation of direct and microwave hybrid heating of ceramic samples.

is used in conjunction with microwaves to achieve the T_c. Gas-fired or electric furnaces and ultrasonics are examples of additional energy sources documented in the literature [21–27]. Once T_c for the ceramic is achieved, the material absorbs microwave energy on its own and heats without the aid of the secondary energy source.

Intrinsic hybrid heating, on the other hand, uses a composite as a microwave-absorbing casket (sample holder) that is composed usually of a high-dielectric-loss-factor ceramic (known as a *susceptor*) in a refractory ceramic. The susceptor, in the form of granules, rods, or plates, absorbs microwave energy at low temperatures and heats the sample contained within [28–30]. As the sample temperature rises, it will reach T_c, where it then has $\varepsilon''_{\text{eff}}$ high enough to begin absorbing the microwave energy itself. The refractory phase of the sample holder then becomes important, as it will hold the heat generated by both the sample and the suscepting phase and lead to a uniform, volumetric heating environment inside the casket. One system that is used often in creating intrinsic hybrid heating caskets is silicon carbide (high dielectric loss, high thermal conductivity) embedded in alumina (very low dielectric loss at low temperatures, low thermal conductivity) [31, 32].

In using either intrinsic or extrinsic hybrid heating, the uniform thermal profile can lead to excellent mechanical and physical properties throughout the sample.

4.1.2.2 Benefits of microwave processing

Clark and Sutton produced a fairly comprehensive list of the benefits, challenges, and needs of the materials research and manufacturing community [33]. An abbreviated version of that table is provided below (Table 4.2)—abbreviated in that it addresses those issues most important in ceramics processing

While the applications for microwave processing of ceramics continue to grow slowly, it is a technology that is likely to be most useful in niche markets where either (1) it is the only method by which a product can be made or made efficiently or (2) the benefits of adding to or replacing all together the current processing methods

Table 4.2 Major benefits, challenges, and needs of microwave processing for ceramics and glass

Benefits
• Volumetric and uniform heating
• Selective heating
• Lower temperatures to achieve superior properties
• Rapid heating
Challenges
• Heating low dielectric loss materials
• Controlled rate of heating and maximum temperatures
• Integrating microwave processing into existing manufacturing operations
Needs
• Availability of microwave equipment and compatible component materials
• Process controls, electronic tuning/automation (smart processing)
• Improved fundamental understanding of microwave–material interactions
• Better communication among manufacturers, researchers, and end users

Source: Adapted from Ref. [33] with permission of the American Ceramic Society, www.ceramics.org

with microwave energy far outweigh the costs of implementing completely new production methods.

Throughout the remainder of this chapter, examples of microwave applications in each of the temperature regimes described in Fig. 4.1 are presented. How these benefits, challenges, and needs affect the manufacture of ceramics in each of the temperature regimes will be discussed.

4.2 Low-Temperature Microwave Applications: Drying

There are a number of low-temperature processes in ceramics manufacture, even including chemical synthesis, such as sol–gel formation. However, simple drying is the dominant process, especially in those materials where the primary liquid for removal is water, such as whitewares (clay-based materials). The reason that this is an obvious microwave application is that water has a moderately high dielectric loss factor [34] at 2.45 GHz and room temperature, while the ceramic does not, thus creating a composite material—a partially microwave transparent material—with the water as the primary absorbing phase.

In conventional heating, the surface layers are heated first and the thermal conductivity of the material dictates the rate at which each layer reaches the desired temperature. In the case of drying, it is the water present in each layer that must absorb the thermal energy in order to evolve out of the part, layer by layer. Given that the thermal conductivity for many ceramics is low, such as in the whitewares that are the topic of this discussion, the process for bringing the entire product up to the desired temperature in a uniform manner requires a significant amount of time and thermal energy.

The advantage of using microwave energy in drying is that it is associated with volumetric heating, in which the microwave energy is distributed throughout the cross section of the product uniformly rather than layer by layer. Because the microwave energy is absorbed by the water contained throughout the part and is not limited by the rate of thermal energy absorption by the ceramic to heat the water, drying takes place more rapidly at lower temperatures and more uniformly across the product cross section. As seen in the work discussed below, this uniformity in drying leads to a more efficient drying process (shorter times; lower temperatures) and more homogeneous physical and mechanical properties throughout the cross section of the products.

A recent study [35] compared microwave and conventional drying methods for heat-treating porcelain materials. This study used the flexural strength (σ_f) and Weibull modulus (m) to evaluate the success of both drying methods. From an industrial standpoint, the Weibull modulus is a crucial parameter in selecting processing methods. Essentially, it is a statistical determination of the maximum stress a material can experience before failure [36]. It is based on the premise of the "weakest link." Thus it allows a manufacturer to have confidence that his or her methodology is sufficient to produce a part that will survive a minimum required load for a specific application (probability of survival, P_s).

In the case of porcelain materials, the conventional drying time to produce parts with sufficient green (dried) strength requires 10–14 days [35]. This long drying time is the driving force behind the

Table 4.3 Weibull modulus for four compositions of porcelain ceramics dried using conventional and microwave heating

Composition	Conventional (m)	Microwave (m)
A	9.3	15.6
B	10.3	15.2
C	11.2	15.5
D	9.3	9.4

Source: Adapted from Ref. [35]. Copyright © 2008. Reproduced with permission of John Wiley & Sons, Inc.

search for a faster, more efficient method that can result in the same or better mechanical properties of the greenware (advantage: rapid heating).

In this study, the researchers used four different porcelain compositions. The goal of the conventional and microwave drying cycles was to reduce the total moisture content from 16% to 0.5% approximately. All of the samples were dried in air for 1 h, followed by a conventional 8 h drying cycle with a maximum 2 h soak at 120°C or a microwave drying cycle of 4 h at a maximum temperature of 70°C. The microwave power at 2.45 GHz was held constant. Three-point bend testing was used to calculate the flexural strength (σ_f) of the samples in MPa. The Weibull modulus (m) is calculated as the slope of the best-fit line plotted by $\ln[\ln(1/P_s)]$ versus $\ln(\sigma_f)$. The results of their work are summarized in Table 4.3.

Note that with the exception of composition D, the microwave-dried samples showed significantly higher statistical reliability than those dried conventionally. In other words, their probability of survival could be predicted with more certainty than those dried conventionally. The results obtained by Hemanthakumari and Satapathy show that microwave processing is indeed a viable method for low-temperature processing of ceramic materials [35].

Additional work in the area of low-temperature microwave heating that is promising is in the area of environmental applications. Reviews of some of this work are presented by Jones et al. [37] and Tyagi and Lo [38].

4.3 Moderate-Temperature Applications: Glass Nucleation and Crystallization

There are many different types of heat treatments that take place over the moderate temperature range, primarily between 400°C and 600°C (Fig. 4.1). Annealing to relieve stresses in glasses, binder burnout to remove a wide variety of primarily polymer binders that must be removed prior to high-temperature firing, and, in the case to be discussed here, nucleation and crystallization in glasses to form glass-ceramics are examples of low-temperature processes.

Glass-ceramics are polycrystalline ceramics that are formed from glass through the process of nucleation and subsequent crystal growth. They are valuable in forming ceramics with densities close to 100% of theoretical. By using traditional glass-forming techniques to cast the shapes required, the nucleation and crystallization steps can result in a highly dense polycrystalline glass-ceramic in near-net shape [39]. The microstructure of the final glass-ceramic is controlled by two factors, the number and distribution of the nuclei and the rate and uniformity of crystal growth [40]. One of the commercialized applications for this particular glass-ceramic composition is in dental materials [41].

In work performed by Mahmoud et al. [40, 42], the effect of microwave energy on crystal growth in lithium disilicate ($Li_2S_2O_5$, abbreviated as LS_2) glass-ceramics was studied. In this case, the frequency was not held constant at 2.45 GHz. Instead, a variable-frequency microwave (VFM) oven was used during the crystal growth phase (Fig. 4.4). The VFM uses a traveling-wave tube versus a fixed-frequency magnetron to generate electromagnetic energy. The advantage is the ability either to select one specific frequency or to raster over a range of frequencies around a selected central frequency in order to achieve a uniform field within the sample cavity and the samples themselves [43] (advantage: uniform heating).

The glass and glass-ceramic processing temperatures were determined using differential scanning calorimetry (DSC) that provided the glass transition temperature (T_g), the nucleation temperature (T_n), and the crystallization temperature (T_c). The initial glass samples were formed into cylinders using conventional

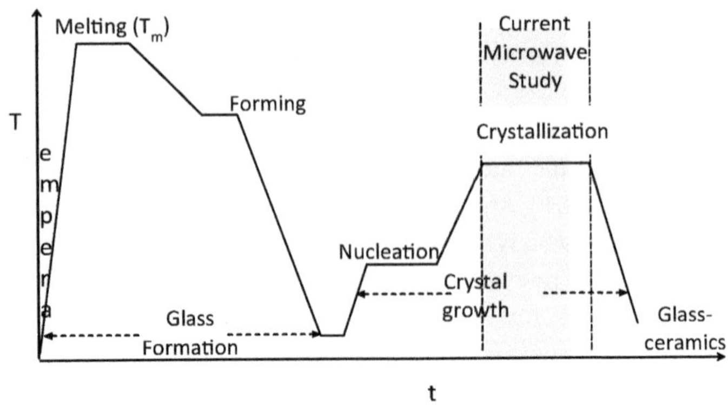

Figure 4.4 Temperature versus time plot illustrating the formation of glass and glass-ceramics. Adapted from Refs. [40, 42] with permission of the American Ceramic Society, www.ceramics.org All rights reserved.

glass melting and casting techniques [42]. They were annealed in electric resistant ovens for 10 h at 400°C. The samples were sectioned into their final sample sizes and nucleated conventionally for 2 h at 480°C.

The microwave research was performed on the crystallization phase of the heat treatment—the phase in which crystals grow from the nuclei to form the polycrystalline glass-ceramic. In this case, intrinsic hybrid heating was used in combination with VFMs. Lambda Technologies Microcure Model 2100 was used to raster a microwave field at a central frequency of 6.425 GHz, sweeping at a rate of 0.1 s over a 1.15 GHz bandwidth. Temperature was monitored and controlled using an infrared sensor, as well as a K-type thermocouple connected to a controller.

Using Fourier transform infrared spectroscopy (FTIR) to analyze the phases present in the samples, it was concluded that the VFM-processed glass-ceramics were fully crystallized in shorter times and lower temperatures than those processed conventionally to the same extent of crystallization. The process showed the same results for all 20 samples processed for each condition (advantages: lower processing temperatures, rapid heating). Microstructural analyses of the samples confirmed that to achieve similar crystallization char-

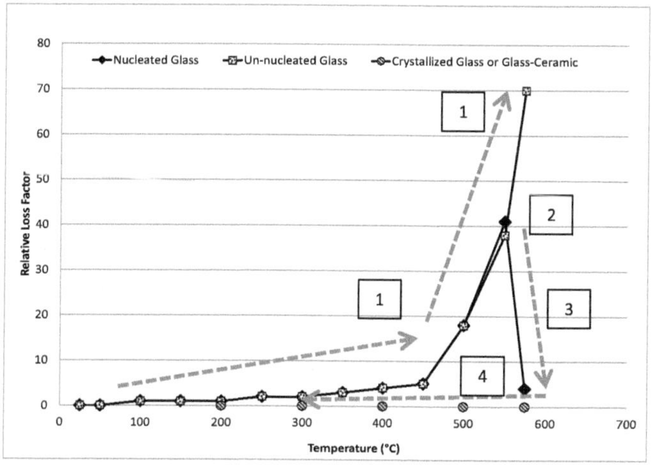

Figure 4.5 Relative dielectric loss for $Li_2Si_2O_5$ glass and glass-ceramic vs. temperature. Adapted from Ref. [44] with permission of the American Ceramic Society, www.ceramics.org All rights reserved.

acteristics, much greater times were required in the conventional process (hours vs. minutes).

Boonyapiwat et al. [44] also studied the $Li_2Si_2O_5$ glass and glass-ceramic system. In this work, a VFM was used at a central frequency of 9.5 GHz. It was found that once crystallization had taken place, the samples no longer absorbed microwave energy (Fig. 4.5). The dielectric loss factor for the unnucleated glass continued to rise (follows dashed line 1). However, at the onset of crystallization at about 550°C (point 2), the dielectric loss factor fell significantly after the crystals began forming (dashed line 3) and the fully crystallized samples stopped absorbing all together and cooled down to room temperature (followed line 4). This behavior opened up a new possibility that some compositions could exhibit phase changes during microwave processing that resulted in a self-limiting heating process. This self-limiting phenomenon is unique and not observed in all glass-ceramics.

The role of specific composition and the frequency of the microwave energy are quite important for these glass-ceramics. For further discussion of this topic, refer to Kolberg and Roemer

[45], who show that phase changes do not always bring about a nonabsorbing phase.

4.4 High-Temperature Applications: Sintering

While there is a wide variety of processing methods used to form ceramic products, a great many of the ceramic materials used commercially are processed at high temperatures to achieve densification through sintering [39]. Using conventional heating methods, sintering temperatures for ceramics usually range from 1000°C to 1700°C (depending on composition and particle size), with some exceeding even these temperatures. Depending on the material, a controlled environment (gases other than air) may be required.

High-temperature microwave processing appears to be a method where considerable cost savings can be realized compared to conventional processes. Studies by both the Ontario Ministry of Energy [46] shown in Table 4.4 and by EA Technologies in Europe (Fig. 4.4) [46, 47] indicated that higher temperatures and controlled environments could decrease the costs of producing densified ceramic products. These results show that it is in the high-temperature segment of heat treatments that microwave energy can provide a viable, cost-effective, and alternative manufacturing method.

A successful approach to drying and firing of whiteware products is discussed by Wroe [47]. Wroe illustrated the benefits of using the extrinsic hybrid approach during low- and high-temperature firing by combining gas heating and electric heating with microwave energy. Significant energy savings were observed in this approach, as seen in Fig. 4.6. It is very interesting to note that the most energy savings observed by Wroe occurred in the firing (high-temperature) step. While it is a simpler matter from an equipment standpoint to use microwaves in drying operations, the potential gain in efficiency is much higher for the firing step.

Research specifically into large-market materials, such as pure and composite alumina and zirconia products, has proven of great interest to microwave research groups [48–50]. In a study conducted

Table 4.4 Comparison of energy savings for conventional and microwave drying and firing of ceramics

	Energy Savings (× 10^6 kWh/year)				
	Conventional drying	**Microwave drying**	**Conventional firing**	**Microwave firing**	**Total energy savings**
Brick and tile	56.10	28.05	198.90	19.90	207.06
Electrical porcelain	3.52	1.76	12.48	1.25	12.99
Glazes	16.63	8.30	58.97	5.89	61.37
Pottery	1.96	0.98	6.94	0.69	7.23
Refractories	10.87	5.40	38.53	3.85	40.08
Sanitary ware	25.04	12.52	88.76	8.88	92.40
Advanced ceramics	1.30	0.65	4.60	0.46	4.79
Total (× 10^6 kWh/year)	115.42	57.66	409.18	40.92	425.92
(× 10^8 PJ/year)	0.42	0.21	1.47	0.15	1.5

Source: Adapted from Ref. [46] with permission of the American Ceramic Society, www.ceramics.org All rights reserved.

for the Department of Energy (DOE), Thridandapani et al. [51, 52] performed research on 8 mol% yttria-stabilized zirconia (8YZ) as a candidate inert matrix to replace uranium dioxide, a fissile matrix material. This new fuel pellet design would be used in next-generation reactors to expend the energy stored in current waste actinides, while not generating new wastes, as would be the case in uranium dioxide matrices. The main issue for the DOE was that the conventional sintering temperatures for these recycled waste pellets were too high to prevent vaporization of the actinides. This work proved ideal for the single-mode microwave environment due primarily to the size and geometry of the products. The research samples were the same general size of the actual fuels, allowing for a good understanding of the potential success possible with microwave sintering. The researchers determined that these pellets could be sintered to 99% theoretical density at 1300°C versus 1500°C–1600°C required with conventional sintering methods (advantage: lower processing temperatures). Figure 4.7 illustrates the difference in densification of the samples at 1100°C.

In this case, the microwave equipment could be used to concentrate the field in the samples and achieve rapid, controlled

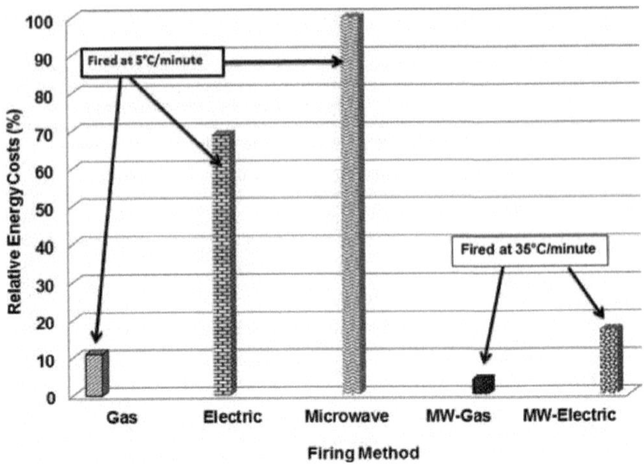

Figure 4.6 Relative energy costs for firing of alumina cylinders. Reprinted from Ref. [47] with permission of the American Ceramic Society, www.ceramics.org All rights reserved.

heating throughout the sample to the sintering temperature. At this temperature and heating rate, the potential was high for retaining the actinides in the matrix. Another benefit of using microwave energy for this particular application was the ability to configure the microwave equipment such that only the sample cavity would be exposed to a radioactive environment.

Two more recent studies, one by Alexeff and Meek [53] and the other by Santos et al. [54], indicate the ability to control the processing and properties of ceramics through manipulation of the microwave field and the application of specific microwave generating components. Alexeff and Meek showed that the intensity of the electric field applied to zirconia particles could increase the degree of sintering.

The work of Santos et al. took a different approach to apply homogeneous heating of various geometries for porcelain materials. They developed a microwave manufacturing system comprising a series of six microwave generators specifically placed and operated to input varying amounts of energy into a multimode cavity and create a uniform processing environment (advantage: volumetric heating). The samples produced (various geometries) showed

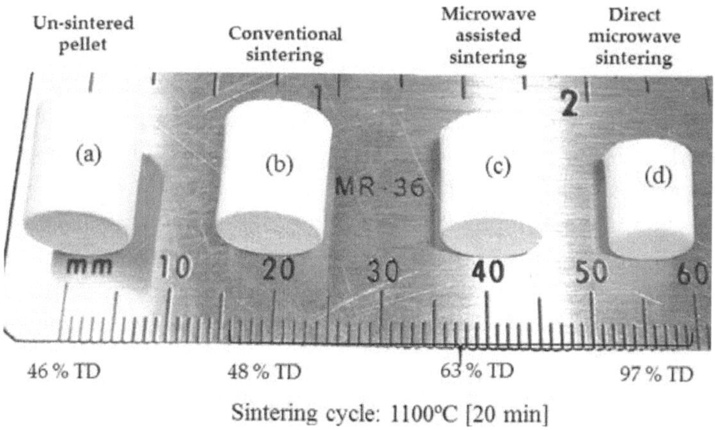

Figure 4.7 Samples of 8YZ samples: (a) isostatically pressed, (b) conventionally sintered at 1100°C, (c) single-mode, microwave-assisted (intrinsic hybrid) at 2.45 GHz, and (d) direct microwave sintered (same as (c) but with no suscepting casket) [52].

densities similar to those achieved with conventional gas-fired furnace heating. An evaluation of the mechanical characteristics of the products showed that the microwave-processed parts resulted in lower standard deviations than the gas-fired oven samples (rupture energy: 0.02–0.04 J versus 0.13 J, respectively; impact resistance: 0.06–0.14 J/cm^2 versus 0.34 J/cm^2, respectively) (advantage: uniform, improved properties) [54].

Santos et al. also noted that a significant advantage to this method was the rapid processing of these clay-based products that inhibited the emissions of fluorine gases, a well-known problem in conventional processing of clays. This large reduction in hazardous emissions of fluorine gas was first reported by Earl [55] when studying the effects of multimode microwave processing of floor tile compositions (advantage: rapid, volumetric heating). These observations support the claim that microwave energy can be used to create a greener processing environment for some ceramic materials.

De' et al. [49, 50] also presented data in the 1990s that showed the value of using microwave energy in sintering. Their work focused on alumina and showed that uniform thermal profiles could be achieved in pellets ranging from 6 to 20 g. Interestingly, the larger samples exhibited more uniform physical and mechanical properties throughout the cross section than the smaller samples (advantages: uniform thermal profile, uniform mechanical properties). The same-sized samples fired conventionally were not uniform at all in properties due primarily to the varied thermal profile, even in the small 6 g samples. De' et al. used the intrinsic microwave hybrid heating method to process the alumina—a ceramic that required initial heating (by external absorber elements) to the critical temperature where the alumina could become an absorber itself.

Strontium barium niobate (SBN) ceramic compositions are valuable in many applications, including electro-optics, piezoelectric, and photorefractive devices. Kumar et al. [56] found that the presence of strontium niobate and barium niobate in calcined powders resulted in enhanced microwave reactive sintering of SBN ceramic materials. By incorporating these relatively more lossy oxides into the sample materials, microwave energy was used to rapidly process high-density samples with uniform microstructures (advantage: uniform, volumetric heating). This method (microwave reactive sintering) allowed both chemical reaction and densification be accomplished in one heat treatment process. While the niobates in the calcined powders aided in sintering, the process still used a hybrid heating environment for sintering.

Apte and MacDonald [57] showed that large batches of nitrides could be sintered uniformly and that microwave sintering was not limited to small batches, such as those used in most of the research performed to date. This group's work showed that the net energy absorbed by the silicon nitride samples decreased the specific energy investment from 50.0 to 3.6 kWh/kg when the batch size was scaled up from 5 to 580 parts. These samples also were evaluated for their physical properties and showed that the density, hardness, and toughness were equal to or better than identical samples produced conventionally (advantages: rapid heating, improved mechanical properties). The crucial understanding required to achieve this level of efficiency or better was the dielectric loss for the material and the penetration depth of the microwave energy into the sample geometry's cross section.

The application of microwave energy for sintering ceramic materials can fill a vital niche in commercial applications. The papers discussed above comprise only a few of the candidate materials and processes that can serve to expand the use of microwave technology in manufacturing. Some additional applications of high-temperature sintering can be found in the following works. Wu et al. [58] and Veljovic et al. [59] provide examples of microwave sintering of hydroxyapatite bioceramics, Souto et al. [60] provide examples of the addition of an oxide additive to enhance microwave sintering of mullite, Lee et al. [61] and Geng et al. [62] studied the microstructure and properties of ceramics sintered using microwave energy, and Borrell et al. [63] examined mechanical and structural properties of microwave-sintered 3mol% yttria-stabilized zirconia nanoparticulates. The application of papers published in the past decade shows that the interest in microwave sintering continues to be strong among the research community.

4.5 Summary

The potential benefits of using microwave energy for processing ceramics are plentiful. Although only a few of the published papers can be provided in this overview, there is a wealth of information in global proceedings on this topic.

While the advantages associated with microwave energy are accepted by researchers ranging from academia and national laboratories to industry, widespread commercial application in microwave processing of ceramics has not been realized as yet. There are a number of reasons for this slow implementation, including the lack of available, off-the-shelf commercial microwave equipment; a general need for a significant database of dielectric properties for materials across a wide temperature and frequency range; and closer communication between researchers and potential end users of the technology on niche, modular scale-up of microwave processes into existing commercial production lines. There is much work that is being done independently on these various technical barriers; however, the need for a consolidated effort and sharing of data persists.

References

1. Sutton, W. H. (1987). Microwave processing of ceramics, *Am. Ceram. Soc. Bull.*, **68**(2), pp. 376–386.
2. Clark, D. E., Gac, F. D., and Sutton, W. H., eds. (1991). *Ceramic Transactions: Microwaves; Theory and Application in Materials Processing,* Vol. 21 (American Ceramic Society, OH).
3. Clark, D. E., Tinga, W. R., and Laia, J. R. Jr., eds. (1993). *Ceramic Transactions: Microwaves; Theory and Application in Materials Processing II,* Vol. 36 (American Ceramic Society, Westerville, OH).
4. Clark, D. E., Folz, D. C., and Silberglitt, R., eds. (1995). *Ceramic Transactions: Microwaves; Theory and Application in Materials Processing III,* Vol. 59 (American Ceramic Society, Westerville, OH).
5. Clark, D. E., Sutton, W., and Lewis, D. L., eds. (1997). *Ceramic Transactions: Microwaves; Theory and Application in Materials Processing IV,* Vol. 80 (American Ceramic Society, Westerville, OH).
6. Clark, D. E., Binner, J. G. P., and Lewis, D. A., eds. (2001). *Ceramic Transactions: Microwaves; Theory and Application in Materials Processing V,* Vol. 111 (American Ceramic Society, Westerville, OH).
7. Folz, D. C., Booske, J. H., Clark, D. E., and Gerling, J. F., eds. (2003). *Microwave and Radio Frequency Applications* (American Ceramic Society, Westerville, OH, USA).
8. Schulz, R. L., and Folz, D. C., eds. (2005). *Microwave and Radio Frequency Applications* (Microwave Working Group, Arnold, MD).
9. Willert-Porada, M., ed. (2006). *Advances in Microwave and Radio Frequency Processing* (Springer-Verlag, Berlin).
10. Microwave World Congress, Japan (2008). *1st Global Congress on Microwave Energy Applications* (JEMEA, Tokyo).
11. Schulz, R. L., and Folz, D. C., eds. (2013). *2nd Global Congress on Microwave Energy Applications* (Microwave Working Group, Arnold, MD).
12. Metaxas, A. C., and Meredith, R. J. (1983). *Industrial Microwave Heating* (Peter Peregrinus, London, UK).
13. Hutcheon, R. M., de Jong, M. S., Adams, F. P., Lucuta, P. G., McGregor, J. E., and Bahen, L., (1992). RF and microwave dielectric measurements to 1400°C and dielectric loss mechanisms, *MRS Proc.*, **269**, pp. 541–549.
14. Hutcheon, R. M., de Jong, M. S., and Adams, F. P. (1992). A system for rapid measurements of RF and microwave properties up to 1400°C part 1: theoretical development of the cavity frequency shift data

analysis equations, and part II: description of apparatus, data collection techniques and measurements on selected materials, *J. Microw. Power Electromagn. Energy*, **27**(2), pp. 87–100.

15. Von Hippel, A. R. (1995). *Dielectric Materials and Applications* (Artech House, Norwood, MA).
16. Von Hippel, A. R. (1954). *Dielectrics and Waves* (John Wiley and Sons, NY).
17. Nalwa, H. S., ed. (1999) *Handbook of Low and High Dielectric Constant Materials and Their Applications: Two Volume Set* (Academic Press, London, UK).
18. Frolich, H. (1958). *Theory of Dielectrics: Dielectric Constant and Dielectric Loss* (Clarendon Press, Oxford, UK).
19. Clark, D. E., Folz, D. C., Folgar, C. E., and Mahmoud, M. M., eds. (2005). *Microwave Solutions for Ceramic Engineers* (American Ceramic Society, Westerville, OH).
20. Schiffmann, F. R. (1997). *Principles of Industrial Microwave and RF Heating, Microwave: Theory and Application in Materials Processing IV*, Vol. 80 (American Ceramic Society, Westerville, OH).
21. Reddy, V. R., Upadhyay, S. K., Gupta, A., Awasthi, A. M., and Hussain, S. (2014). Enhanced dielectric and ferroelectric properties of $BaTi0_3$ ceramics prepared by microwave assisted radiant hybrid sintering, *Ceram. Int.*, **40**, pp. 8833–8339.
22. McInroy, A. B., Rowley, A. T., and Freer, R. (2010). Enhanced microwave-assisted sintering of X7R ceramic dielectrics for use in multilayer capacitors, *Ceram. Int.*, **36**(10), pp. 2007–2010.
23. Donai, P., Yang, W., Blasiak, W., and Forsgren, C. (2010). Recycling of automobile shredder residue with microwave pyrolysis combined with high temperature steam gasification, *J. Hazard. Mater.*, **182**(1–3), pp. 80–89.
24. Kashimura, K., Sato, M., Kishima, T., Mitani, T., Yoshikawa, N., Horioshi, S., and Shinohara, N. (2013). Applications of microwave energy for rubble processing from the Great East Japan earthquake – A test system for rendering asbestos harmless in building debris (slates). In *2nd Global Congress on Microwave Energy Applications*, Schulz, R. L., and Folz, D. C., eds. (Microwave Working Group, Ltd., Arnold, MD, USA).
25. Pfleging, W., Wesner, D. A., and Kreutz, E. W. (1996). $CaCl_4$ assisted-CF_4 etching of silicon in a microwave-assisted LDE (laser dry etching)-process, *Appl. Surf. Sci.*, **96–98**(*2*), pp. 496–500.

26. Wang, J., Binner, J., Pang, Y., and Vaidhyanathan, B. (2008). Microwave-enhanced densification of sol-gel alumina films, *Thin Solid Films*, **516**, pp. 5996–6001.

27. Boffa, L., Tagliapietra S., and Cravotto, G. (2013). Combined energy sources in the synthesis of nanomaterials. In *Microwaves in Nanoparticle Synthesis*, Horikoshi, S., and Serpone, N., eds. (Wiley-VCH, Verlag GmbH & Co., Germany), pp. 55–74.

28. Menezes, R. R., Souto, P. M., and Kiminami, R. H. G. A. (2012). Microwave fast sintering of ceramic materials. In *Sintering of Ceramics – New Emerging Techniques*, Lakshmanan, A., ed. (Intech, Rijeka, Croatia and Shanghai, China), pp. 1–23.

29. Wroe, F. C. R. (1994). Microwave-assisted firing of ceramics, Ceramics in Energy Applications, *Proc. Inst. Energy Conf., Lond.*, pp. 43–53.

30. Lasri, J., Ramesh, P. D., and Schächter, L. (2000). Energy conversion during microwave sintering of a multiphase ceramic surrounded by a susceptor, *J. Am. Ceram. Soc.*, **83**(6), pp. 1465–1468.

31. Leiser, K. (2001). *Microwave Behavior of Silicon Carbide/High-Alumina Cement Composites*, PhD Dissertation, University of Florida.

32. Atong, D-D. (2000). *Microwave Induced Combustion Synthesis of Al_2O_3-TiC Powder*, PhD Dissertation, University of Florida.

33. Clark, D. E., and Sutton, W. H. (1996). Microwave processing of materials, *Annu. Rev. Mater. Sci.*, **26**, pp. 299–331.

34. Goldblith, S. A. (1966). Basic principles of microwaves and recent developments, *Adv. Food Res.*, **15**, pp. 277–301.

35. Hemanthakumari, P. N., and Satapathy, L. N. (2008). A comparison of the effects of microwave versus conventional drying on the mechanical properties distribution of dried green porcelains, *Int. J. Appl. Ceram. Technol.*, **5**(1), pp. 94–100.

36. Breviary Technical Ceramics, *Informationszentrum Technical Ceramics part of Verband der Keramischen Industrie e.V.* http://www.nonmet.mat.ethz.ch/education/courses/Materialwissenschaft_2/brevier.pdf

37. Jones, D. A., Lelyveld, T. P., Mavrofidis, S. D., Kingman, S. W., and Niles, N. J. (2002). Microwave heating applications in environmental engineering – a review, *Resour., Conserv. Recy.*, **34**, pp. 75–90.

38. Tyagi, V. K., and Lo, S.-L. (2013). Microwave irradiation: a sustainable way for sludge treatment and resource recovery, *Renew. Sust. Energy Rev.*, **18**, pp. 288–305.

39. Carter, B.C., and Norton, M. G. (2013). *Ceramic Materials: Science and Engineering* (Springer-Verlag, NY).

40. Mahmoud. M. (2007). *Crystallization of Lithium Disilicate Glass using Variable Frequency Microwave Processing*, PhD Dissertation, Virginia Tech, Blacksburg, VA.

41. Saint-Jean, S. J. (2014). Dental glasses and glass ceramics. In *Advanced Ceramics for Dentistry* (Elsevier), pp. 255–277.

42. Mahmoud, M. M, Folz, D. C., Suchicital, C., and Clark, D. E. (2012). Crystallization of lithium disilicate glass using microwave processing, *J. Am. Ceram. Soc.*, **95**(2), pp. 579–585.

43. *Lambda Technologies, Microwave Fundamentals, VFM and Materials Heating,* http://www.microcure.com/pdf/microwave/ *%20FUNDAMENTALS,VFM&MATERIALSHEATING.pdf*

44. Boonyapiwat, A., Folz, D. C., and Clark, D. E. (2005). Microwave crystallization of glass. In *Microwave Solutions for Ceramic Engineers* (American Ceramic Society, Westerville, OH, USA), pp. 277–286.

45. Kolberg, U., and Roemer, H. (2005). Microwave heating of glass. In *Microwave Solutions for Ceramic Engineers* (American Ceramic Society, Westerville, OH), pp. 263–268.

46. Sheppard, L. M. (1988). Manufacturing ceramics with microwaves: the potential for economical production, *Ceram. Bull.*, **67**(10), pp. 1656–1661.

47. Wroe, F. C. R. (2005). Scaling up the microwave firing of ceramics. In *Microwave Solutions for Ceramic Engineers* (American Ceramic Society, Westerville, OH), pp. 447–456.

48. Janney, M. A., and Kimrey, H. (1988). *Ceramic Powder Science II* (American Ceramic Society, Westerville, OH), pp. 919–924.

49. De', A., Ahmad, I., Whitney, E., Dow, Clark, D. E. (1991). Microwave (hybrid) heating of alumina at 2.45GHz: I: microstructural uniformity and homogeneity. In *Ceramic Transactions: Microwaves; Theory and Application in Materials Processing,* Vol. 21 (American Ceramic Society, Westerville, OH), pp. 319–328.

50. De', A., Ahmad, I., Whitney, E., Dow, Clark, D.E. (1991). Microwave (hybrid) heating of alumina at 2.45GHz: II: effect of processing variables, heating rates and particle size. In *Ceramic Transactions: Microwaves; Theory and Application in Materials Processing,* Vol. 21 (American Ceramic Society, Westerville, OH), pp. 329–339.

51. Thridandapani, R. R., Folz, D. C., and Clark, D. E. (2011). Development of a microwave dilatometer for generating master sintering curves, *Meas. Sci. Technol.*, **22**(10), p. 105706.

52. Thridandapani, R. R. (2011). *The Effect of Microwave Energy on Sintering*, PhD Dissertation, Virginia Polytechnic Institute and State University.
53. Alexeff, I., and Meek, T. T. (2011). The effect of electric field intensity on the microwave sintering of zirconia, *Mater. Lett.*, **65**, pp. 2111–2113.
54. Santos, T., Costa, L. C., Hennetier, L., Valente, M. A., Monteiro, J., and Sousa, J. (2013). Microwave processing of porcelain tableware using a multiple generator configuration, *Appl. Therm. Eng.*, **50**, pp. 677–682.
55. Earl, D. (1998). *Glass Frit Composition Effects on Zircon Crystallization and Dissolution*, PhD Dissertation, University of Florida.
56. Kumar, S. N., Kumar, P., and Agrawal, D. K. (2012). Structural, dielectric and ferroelectric properties of SBN ceramics synthesized by microwave reactive sintering technique, *Ceram. Int.*, **38**, pp. 5243–5250.
57. Apte, P. A., and MacDonald, W. D. (2005). Microwave sintering kilogram batches of silicon nitride. In *Microwave Solutions for Ceramic Engineers* (American Ceramic Society, Westerville, OH), pp. 177–186.
58. Wu, Q., Zhang, X., Wu, B., and Huang, W. (2013). Effects of microwave sintering on the properties of porous hydroxyapatite scaffolds, *Ceram. Int.*, **39**, pp. 2389–2395.
59. Veljović, Dj., Zalite, I., Palcevskis, E., Smiciklas, I., Petrović, R., and Janaćković, Dj. (2010). Microwave sintering of fine grained HAP and HAP/TCP bioceramics, *Ceram. Int.*, **36**, pp. 595–603.
60. Souto, P. M., Menezes, R. R., and Kiminami, R. H. G. A. (2011). Effect of Y2O3 additive on conventional and microwave sintering of mullite, *Ceram. Int.*, **37**, pp. 241–248.
61. Lee, Y.-C., Yeh, Y.-Y., and Tsai, P.-R. (2012). Effect of microwave sintering on the microstructure and electric properties of (ZnMg)TiO_3-based multilayer ceramic capacitors, *J. Eur. Ceram. Soc.*, **32**, pp. 1725–1732.
62. Geng, H. X., Mei, A., Dong, C., Lin, Y. H., and Nan, C. W. (2009). Investigation of structure and electrical properties of Li0.5La0.5TiO3 ceramics via microwave sintering, *J. Alloys Compd.*, **481**, pp. 555–558.
63. Borrell, A., Salvador, M. D., Peńaranda-Foix, F. L., and Ćatala-Civera, J. M. (2013). Microwave sintering of zirconia materials: mechanical and microstructural properties, *Int. J. Appl. Ceram. Technol.*, **10**(2), pp. 313–320.

Chapter 5

Microwave Processing of Composites, Glass-Ceramic Coatings and Metallic Alloys: An Overview

Sumana Ghosh
CSIR-Central Glass and Ceramic Research Institute, Kolkata 700 032, India
sumana@cgcri.res.in

This chapter represents an overview of processing composites, glass-ceramic coatings, and metallic alloys by using the microwave heating method. In the past two decades, researchers have been utilizing the microwave heating technique to process a wide range of materials, including traditional and advanced ceramics, polymers, composites, coatings, glass, and even metals/metallic alloys.

5.1 Introduction

Microwave technology is being preferred for the processing of materials as it offers several advantages such as a unique microstructure and properties, improved product yield, energy savings, reduction in manufacturing cost, and synthesis of new materials compared to conventional processing methods [1]. In the case of conventional

Microwave Engineering of Materials: From Mesoscale to Nanoscale
Edited by Erwann Guénin

ISBN 978-981-4669-42-9 (Hardcover), 978-981-4669-43-6 (eBook)
www.panstanford.com

heating thermal energy is delivered to the surface of the material by radiant and/or convection heating, which is transferred to the bulk of the material via conduction, while microwave energy is delivered directly to the material through molecular interaction with the electromagnetic field. Thus, microwave heating is the energy conversion from electromagnetic energy to thermal energy rather than heat transfer. During microwave heating heat can be generated throughout the volume of the material, resulting in rapid, uniform volumetric heating. Therefore, the thermal gradient in the microwave-processed material is the reverse of that in the conventionally processed material. Microwave energy can selectively couple with the higher-loss-tangent material when it is in contact with materials having different dielectric properties. Therefore, the microwave heating technique can be used for the selective heating of materials [2].

Several researchers [1–6] have provided excellent reviews on the microwave processing of a variety of conventional as well as advanced materials. Here, an attempt has been made to focus on processing of composites, glass-ceramic coatings, and metallic alloys using microwave heating.

5.2 Microwave Processing of Composites

Minay et al. [7] developed borosilicate glass matrix composites containing molybdenum, tungsten, aluminum, titanium, nickel and iron metallic particles (particle size 2–50 μm) in a volume concentration of 10% by using the microwave processing method. Microwave heating was carried out in a self-constructed multimode microwave applicator operating at 2.45 GHz industrial, scientific, and medical (ISM) frequency. The microwave and conventionally processed samples were thoroughly characterized. The microwave-processed samples had densities up to 96% of those of conventionally processed samples and were obtained in significantly shorter times (7 min) while it was ~7 h in the case of the conventional processed ones [7]. A silicon-carbide-based susceptor and a boron-nitride-based insulating container was used to heat up the B_4C–Si system using a single-mode microwave cavity under an Ar–H_2

atmosphere to produce pore-free, reaction-bonded B_4C–SiC composites consisting of a mixture of B_4C and polygonal-shaped β-SiC within a residual silicon matrix. The microstructure and mechanical properties of the samples were compared to those obtained by conventionally processing. The average hardness ($H \approx 22$ GPa) value was quite constant along the sample thickness, which ensured good homogeneity of the samples [8].

The microwave sintering technique was used as an alternative to conventional polymer processing methods for the fabrication of microwave-transparent polypropylene (PP)-based composites. PP/multiwalled carbon nanotube (MWCNT)/hydroxyapatite (HA) biocomposites were produced by utilizing the microwave sintering method through incorporatation of 1 wt% of MWCNTs into PP as the microwave susceptors together with a different amount of HA for bone tissue engineering. The fabrication time was reduced, and the sintering time decreased with an increase in the HA content. The sintered composites showed homogeneous distribution of HA [9]. Crystallization and sintering of glass–phlogopite composites were investigated using microwave and conventional heating methods. The hydroxyl group in the phlogopite structure absorbs the microwave irradiation. It was found that the optimum sintering temperature was reduced with the increase in phlogopite content. A diopside phase was identified in the conventional and microwave-processed composites. The decomposition temperature of phlogopite was lowered by 100°C by microwave irradiation as compared to the conventional heating method. The easy machining of the microwave-sintered components was achieved due to the presence of the phlogopite phase. The morphology of the microwave-sintered diopside had almost a fiber-like structure whereas in conventionally processed composites crystals had a plate-like structure [10].

Jayalakshmi et al. [11] have shown that microwave sintering is an effective processing method to produce high performance composites containing amorphous/glassy alloys as better alternative reinforcements. They utilized microwave-assisted sintering and a subsequent hot extrusion technique to produce novel Al composites by reinforcing pure Al with different volume fractions of $Ni_{60}Nb_{40}$ amorphous alloy powders (V_p). Uniform distribution of reinforcement and absence of interfacial products were demonstrated by structural analyses. Microhardness and compressive strength increased significantly with V_p. The composites had good fracture resistance under compressive loading, even at about 50% strain. Tensile tests showed remarkable enhancement in a strength of ~60% at 25% V_p. Further, electrical resistivity of the composites increased with increasing V_p [11]. The synthesis of aluminum nitride–silicon carbide composites through a combustion reaction among aluminum, silicon nitride, and carbon powders was investigated by the microwave heating method [12]. The combustion reaction was initiated through the ignition of the powder mixture by microwave heating in a domestic oven leading to the synthesis of the desired composites. Processing parameters were altered during the synthesis to achieve a higher-quality product. However, the increase in the sample weight assisted the complete combustion and higher degree of conversion of the products [12].

Alumina matrix ceramics were prepared by microwave sintering using two kinds of different structured alumina–titania composite powders. In this study, the effect of the microstructures of the composite powders on rgw microstructure and properties of microwave-sintered alumina matrix ceramics was investigated. In one powder mixture fused and crushed micron-sized alumina particles were clad with submicron-sized titania while in the other powder mixture nanosized alumina and nanosized titania particles were homogeneously mixed by ball milling and spray-dried to prepare spherical alumina–titania composite powders. The nanosized composite (NC) powder showed enhanced sintering behavior in comparison to microsized composite (MC) powders. The as-prepared NC ceramic had a much denser, finer, and more homogenous microstructure than the MC ceramic. The flexural strength, Vickers hardness, and fracture toughness of the NC ceramic were higher than that of the MC ceramic. Improved mechanical properties of the NC ceramic were observed due to enhanced densification and a finer and more homogeneous microstructure [13].

Microwave processing led to glass matrix composites with well-defined spherical porosity (diameter 5–50 μm). High concentration of spherical pores in the central region of the sample and a relatively dense outer shell were obtained by using microwave heating. Pores in the central region were formed in the molten glass phase using gas evolution and entrapment. The outer region was sintered by viscous flow, which was kept at a lower temperature. The minimal distortion of the part was noted. In comparison to other methods such as hot pressing the present method is advantageous due to high heating rates resulting in saving of time and energy. Further, complex-shaped parts may be fabricated by this technique [14].

The carbon nanotube (CNT)-doped ferrite ceramics have good prospects to be used in an on-beam-line high-order-mode load in particle accelerators based on superconducting radio frequency due to their excellent low-temperature characteristics. Zhou et al. prepared CNT–$Ni_{0.5}Zn_{0.5}Fe_2O_4$ powders by in situ chemical precipitation and hydrothermal processing and thereafter sintering by the microwave heating technique. The CNT acted as a heating source and enhanced the consolidation of composites during microwave sintering. However, a higher amount of CNTs led to phase decomposition and reduction of ferrite materials due to an ultrahigh localized temperature increment at the interface of CNTs and ferrite grains. The electrical conductivity of the composites was increased by more than 7 orders of magnitude than that of pure $Ni_{0.5}Zn_{0.5}Fe_2O_4$. The developed composites had high electrical conductivity at a temperature of 70 K. The saturation magnetization was strongly dependent on the mass percentage of CNTs. By increasing the CNT content both the real and the imaginary permittivity were increased in the frequency region of 0.6–5 GHz [15].

Laser treatment is the most popular postprocessing technique. Recently, microwave heating is emerging as one of the potential postprocessing sources. Sharma and Krishnamurthy [16] utilized the microwave hybrid heating technique for postprocessing of plasma-deposited alumina–titania ceramic composite coatings on a steel substrate, and the resulting properties of the processed coatings were evaluated. Porosity reduction, microhardness, and wear resistance enhancement of the irradiated coatings were observed because of microstructural changes and densification [16]. An in situ composite coating of titania and HA on commercially pure titanium (CPTi) was obtained by a simple one-step process [17]. Microwave processing has been used as an alternative method of forming HA-based composite coatings on CPTi to enhance the bioactivity for orthopedic and dental implant applications. The coating was formed by processing CPTi metal packed in HA at 800 W microwave power for 22 min. The composition of the coating was found to be TiO_2 (rutile) as the major phase along with HA as the minor phase. The microwave absorption of the nonstoichiometric TiO_2 layer grown during the initial hybrid heating resulted in sintering of apatite particles interfacing them [17].

Tungsten-based carbide-reinforced metal–matrix composite cladding on austenitic stainless steel substrate was developed through microwave exposure of the preplaced clad materials at 2.45 GHz for 420 s. Microstructural studies of the clad section confirmed uniformly distributed reinforced particles in the metallic matrix. WC, W_2C, NiSi, NiW, and Co_3W_3C phases were detected in the clad. Clads exhibited good stiffness and good adhesion with the substrate [18]. Novel $Cu_{50}Zr_{45}Al_5$ metallic glass/polyphenylene sulfide (PPS) composites with high relative densities were prepared through the microwave processing of the constituents using a single-mode 915 MHz applicator in a separated **H** field with an applied pressure of about 5 MPa. The heating behaviors and structural changes of the composites were investigated. Good bonding between metallic glassy and PPS particles was found. The gradient structure was also induced by microwave heating of the composites with a high fraction of the PPS phase [19].

Recently, the development of cladding is being explored using microwave radiation. Ni-based emergency warning and caution (EWAC)/20% chromium carbide ($Cr_{23}C_6$) powder composite cladding has been developed on an austenitic stainless steel substrate (SS-316). A ~500 μm thick clad has been developed by microwave heating at a frequency of 2.45 GHz for 360 s in a domestic microwave oven. The X-ray diffraction (XRD) pattern of the developed clad identified the presence of $FeNi_3$, NiSi, and $Cr_{23}C_6$ phases. A typical composite clad showed good metallurgical bonding with the substrate by partial dilution. The developed clad was free from visible solidification cracking and has significantly less porosity (0.90%). The cross-sectional image of the clad showed that the reinforced $Cr_{23}C_6$ particles were uniformly distributed and well embedded in the nickle-based matrix. The average Vickers microhardness of the developed clad was observed to be 425 ± 140 HV [20].

Nanographite particles are being exploited as a reinforcement to the copper matrix due to their excellent properties. A pin-on-disc tribometer was used to study the effects of graphite particle size, spatial distribution, normal load, and sliding speed on the friction and wear performance of microwave-sintered copper metal matrix composites. Copper–nanographite composites showed a low coefficient of friction and higher wear resistance than copper–graphite composites. The high surface area of nanographite particles embedded in a copper matrix led to a high-adherent graphite tribolayer at the contact surface, thereby reducing the subsurface deformation of the composite due to reduced frictional force [21]. Ceramic composites can be rapidly cured by using microwave radiation and CNT additives. Bulk composite samples can attain temperatures above 500°C within 1 min by using less than 1 wt% CNT additives and 30–40 W of directed microwave power (2.45 GHz). MWCNTs, functionalized MWCNTs (f-MWCNTs), raw single-walled CNTs (r-SWCNTs), and purified SWCNTs (p-SWCNTs) were all used to produce composites in a Starfire® SMP-10 silicon carbide preceramic. MWCNTs loaded at 0.75 wt% in SMP-10 consistently displayed the fastest rate of heating (~500°C in 10 s) and highest temperatures (1150°C in 7 min) [22].

The powder metallurgy method was used for rapid in situ formation of a Ti alloy/TiC porous composite by mixing Ti_6Al_4V powders (particle size 5 μm) with 14.5 vol% of MWCNTs. The MWCNTs acted as microwave susceptors as well as reactants for the rapid sintering of Ti alloys. The green compact of the Ti_6Al_4V/MWCNTs powder was sintered in a 1.4 kW, 2.45 GHz bidirectional microwave furnace without using any protective gas environment. The microwave sintering was completed within 2 min at 1620°C and titanium carbide was formed. The composite had a porosity of ~25%, a compressive strength of 270.41 ± 24.97 MPa, a compressive strain of 3.75 ± 1.11%, a compressive yield stress of 145.48 ± 27.28 MPa, and a compressive Young's modulus of 10.87 ± 2.46 GPa. The Vickers hardness value in the TiC region (545.4 ± 13.9 HV) was significantly higher than that (435.9 ± 12.9 HV) in the Ti alloy region, whereas the overall Rockwell hardness was 47 HRA [23].

Al–Al_2O_3 core–shell composite powder was synthesized through microwave processing of commercial Al powder. XRD data confirmed the formation of Al–Al_2O_3 composite, whereas FTIR studies and scanning electron microscopy (SEM) observations indicated the core–shell type Al–Al_2O_3 composite formation. Transmission electron microscopy (TEM) images, corresponding selected-area electron diffraction (SAED) patterns, and energy-dispersive X-ray (EDX) analysis confirmed the Al–Al_2O_3 core–shell composite development. Zeta potential measurements also indicated that Al core particles were surrounded with an Al_2O_3 shell [24, 25]. Al–Al_2O_3 composites were prepared by microwave treatment of undoped and doped Al powder compacts. Mg, Mg–Si, and Al_2O_3 were used as dopants to study their effects on the oxidation of Al powder and resultant Al–Al_2O_3 composites. XRD and EDX data of the microwave-treated powder compacts confirmed the formation of Al–Al_2O_3 composites. Mg and Mg–Si doped Al powder compacts showed bulk composite growth, while in the undoped and Al_2O_3-doped Al powder compacts composites were formed on the surface and near surface regions only [26].

Successful production of basalt-fiber-reinforced furan composites was possible by utilizing microwave technology. The samples were tested by an impact tester and compared with conventionally cured ones. The microwave-cured composites presented higher values for interlaminar shear strength (ILSS) (13.1%), delamination threshold force (5.2%), maximum load (17.8%), and penetration threshold (14.5%). The better impact resistance of the microwave-cured composites is suggested to be due to its lower residual stresses and more homogeneous crosslinking induced by mass volumetric heating [27]. $(1-x)La_{0.67}Sr_{0.33}MnO_{3+x}$ $PbZr_{0.52}Ti_{0.48}O_3$ (x = 0, 5, 10, and 15 mol%) (LSMO–PZT) was synthesized using the wet chemical citrate sol–gel process and, finally, sintered by microwave heating. XRD identified the presence of two phases of LSMO and PZT in the composites. TEM of the pristine samples displayed particles sizes in the range of 22–58 nm. Electrical resistivity studies at a temperature range of 100–300 K showed that the metal–insulator transition temperature (T_P) decreased and the resistivity increased with increasing PZT concentration. Elastic moduli increased with increasing PZT concentration. Thus, it can be said that the addition of PZT improves the strength of the materials needed for practical applications [28].

To obtain fine-grained Y_2O_3–MgO composites for infrared application the kinetics of Y_2O_3–MgO nanopowders (~50 nm) during microwave sintering process was analyzed to track details of densification evolution. Finer structure and higher density were achieved by microwave sintering than those of conventional sintering. Y_2O_3–MgO composites with an average grain size of ~300 nm and Vickers hardness of 11.2±0.3 GPa were obtained by optimizing the microwave sintering parameters on the basis of the kinetic analysis data [29]. Microwave treatment was utilized to prepare an insoluble biocomposite consisting of potato starch and chitosan [30]. The effect of pH on swelling (liquid absorption) behavior of the composites was investigated at different pH values such as pH 2, pH 6.8, and pH 12. It was noted that the solution uptake ratio of composites depends both on the aqueous pH value and the content of the chitosan. Composites showed a relatively high swelling ratio at pH environments lower than the pK_a of amine groups on chitosan (about 6.3). There was little hydrolysis of such composites at low pH as determined by glucose analysis through ion exchange chromatography. Composites with higher chitosan content showed higher swelling ratios when they were immersed in the same pH solution [30].

5.2.1 *Microwave Processing of Glass-Ceramic Coatings*

A hard glass-ceramic coating from a glass based on MgO–Al_2O_3–TiO_2 system was applied on nickel-based superalloy substrate [31]. The coatings were characterized by XRD, SEM, image analysis, surface roughness measurement, and hardness evaluation by depth-sensitive indentation (DSI) technique. SEM studies of the polished samples revealed that the microwave-processed coating contained finer crystallites than the sizes of the crystallites obtained in the conventionally processed coating under identical conditions. The microwave-processed glass-ceramic coating exhibited much lower surface roughness (R_a) value than that of the conventionally processed ones. DSI results showed that the microwave-processed coating possessed much higher hardness (~6 GPa) compared to that of the conventionally processed coating [31].

Crystallization of $MgO-Al_2O_3-TiO_2$-based glass coating was achieved by microwave and conventional processing methods [32]. The resultant glass-ceramic coatings were evaluated by XRD, SEM, and image analyses. XRD analysis showed that the misfit strain in the lattice of conventionally processed coating was higher than that in the corresponding microwave-processed coating. SEM investigation showed that microwave heating generated finer crystallites in the glass coating than those obtained in the coating processed by the conventional heating method [32].

Mechanical properties of $MgO-Al_2O_3-TiO_2$-based glass-ceramic coatings formed on nimonic superalloy substrates by microwave and conventional heat treatment were evaluated by DSI technique [33]. Nanohardness and Young's modulus values of the microwave-processed glass-ceramic coatings were improved as compared to those of the conventionally treated glass-ceramic coatings due to presence of finer-sized crystallites in the microwave-processed coatings. A slight increase in the nanohardness and Young's modulus values with soaking time for the microwave-processed coatings occurred due to microstructural refinement and reinforcement of the parent glass matrix [33].

Glaze in the $CaO-MgO-Al_2O_3-SiO_2$ system was heated at 950°C–1190°C for 2 h and characterized. An XRD study showed that the glass-ceramic glaze heated at 950°C contained only trace amount of mullite. Both mullite and α-cordierite were formed as primary and secondary phases in the glass-ceramic glaze heated at 1050°C. Glass-ceramic glazes heated at 1120°C and 1190°C consisted of α-cordierite and mullite as major and minor phases. Rietveld analysis displayed the amount of α-cordierite increased and mullite decreased with the increase in heating temperature. Field-emission SEM showed the presence of mullite crystals dispersed within residual glassy phase in the glass-ceramic glazes heated at 950°C and 1050°C, whereas in the microstructures of glass-ceramic glazes heated at 1120°C and 1190°C α-cordierite crystals were mainly observed. EDX analysis confirmed the XRD results. Vickers microhardness measurement showed that the glass-ceramic glaze heated at 1190°C had the highest hardness (8.38 ± 0.07 GPa) [34].

Cordierite-based glass-ceramic glazes were applied on floor tiles by microwave and conventional processing methods [35]. The microwave and conventionally processed glazes had the surface roughness (R_a) value of ~3.73 and ~11.96 μm, respectively. Vickers hardness measurement showed that the microwave-processed glaze had almost five times higher hardness than that of conventionally processed glaze. Gradual transition of nanohardness and Young's modulus values was noted from the substrate to the glaze for both conventional and microwave-processed glazed tiles. Scratch and impact resistance of the microwave-processed glass-ceramic glaze was improved as compared to those of the conventionally processed one [35].

5.3 Microwave Processing of Metallic Alloys

Bulk metals act as reflectors. However, it is believed that multiple scattering coupled with eddy current loss plays a significant role in microwave absorption in finely divided metallic powder [36]. Sintering of metallic alloys was achieved by microwave processing wherein the sintered products showed better properties [37, 38]. Ni59.35Nb34.45Sn6.2 metallic glassy alloy powder was prepared by an argon gas atomization process. Microwave-induced heating and sintering of the glassy powders were carried out by a single-mode 2.45 GHz microwave applicator in the separated magnetic field or electric field. The structure and thermal stability of the sintered glassy alloy specimens were investigated [39].

Powder alloys of two compositions were sintered using both microwave and conventional vacuum sintering methods. The effects of sintering time, temperature, and weight percentage of copper and antimony on the mechanical and structural properties were examined for both sintering methods. Microwave-sintered samples had finer microstructures, higher densities, and higher hardness and tensile strength compared to the conventionally sintered samples and traditionally cast powder. Higher hardness was achieved by increasing the copper and antimony contents. Better mechanical properties were found for microwave-sintered products with shorter sintering times compared with conventional sintering. However, longer sintering times resulted in better diffusion for both sintering methods. Generally, the microwave-sintered samples achieved similar amounts of diffusion as obtained for conventionally sintered samples for the same time. The microwave sintering process is much faster than conventional sintering due to the rapid heating effect [40].

Aluminum alloys have a great demand for industrial applications such as aerospace and automotive industries. There is a lack of research on spark plasma and microwave processing of aluminum alloys. Nouari Saheb [41] showed the possibility for processing Al2124 and Al6061 alloys by spark plasma and microwave sintering techniques. The alloys were sintered at 400°C, 450°C, and 500°C for 20 min. It was found that spark plasma sintering is an effective way to obtain homogenous, dense, and hard alloys as compared to microwave sintering. Al6061 and Al2124 alloys with 100% density were obtained by spark plasma sintering for 20 min at 450°C and 500°C, respectively. Maximum relative densities achieved for microwave-sintered (500°C, 20 min) Al6061 and Al2124 alloys were 92.52% and 93.52%, respectively. The Vickers microhardness of spark-plasma-sintered samples increased with increasing the sintering temperature from 400°C to 500°C and reached the values of 70.16 HV and 117.10 HV for Al6061 and Al2124 alloys, respectively. For microwave-sintered samples, the microhardness initially increased with the sintering temperature from 400°C to 450°C and then decreased with the further increase of sintering temperature to 500°C [41].

The microwave plasma synthesis method was utilized by Sethi et al. [42] to prepare metal and metal alloy nanopowders. They studied the effect of various processing parameters such as plasma gas, carrier gas, cooling gas, precursor raw materials, and feeding rate on the synthesis of metallic nanopowders. The microwave plasma was operated in atmospheric pressure at a frequency of 2.45 GHz. The precursor was thermally decomposed in the plasma reaction region and the resultant products were then condensed in the heat exchanger, separated from the gas by the powder filter, and finally collected in the powder collector. Cu, Mo, W, Mo–Ni, and Fe–Co nanopowders were successfully prepared by using the microwave plasma synthesis method. The processing conditions can be tuned to manipulate the particle size of the nanopowders [42].

Fe, Fe–Cu, and Fe–Cu–C alloys were successfully consolidated through microwave sintering route. All samples showed nearly similar sintered densities. Iron copper (Fe–2Cu) showed less densification response than others because of compact swelling due to the copper growth phenomenon, whereas copper steel showed increasing response due to the role of carbon that restricted the growth of copper. All the compacts demonstrated shrinkage due to the sintering effect. Irrespective of the copper and carbon content, carbonyl powder microwave-sintered compacts had higher densification and a more uniform microstructure. Consequently, the mechanical properties of microwave-sintered carbonyl powder compacts were higher than their counterparts processed through water-atomized and sponge powders [43].

Both premixed and prealloyed Cu–Sn bronze were sintered in a microwave furnace for temperatures corresponding to transient, solid-state, and supersolidus sintering. Bronze was microwave-sintered in significantly less time than conventional sintering. Swelling did not occur during microwave sintering of both premixed as well as prealloyed compacts compacted at 150 MPa, whereas the premixed bronze compacts swelled at 775°C and 830°C during conventional sintering. Hardness of the premixed microwave samples was higher than that of the corresponding conventional premixed samples. The application of conventionally sintered premixed bronzes is restricted to making filters and bearings. However, higher hardness, lack of swelling, and uniform microstructure in premixed bronze during microwave sintering allow its use in structural applications. Microwave effects because of anisothermal heating caused high coalescence in microwave-premixed bronze. In the case of prealloyed bronze, the degree of densification remains unaffected by varying green density for both microwave as well as conventional sintering. The premixed bronze samples showed an increasing tendency for swelling as the initial porosity was reduced by increasing the compaction pressure. The swelling was expressed for microwave-sintered samples than that of the conventionally sintered ones. The microstructure was more uniform in the case of microwave-sintered samples. Therefore, the premixed bronze should be compacted at higher pressure followed by microwave sintering for fabricating bearing and filters with higher porosity [44].

92.5W–6.4Ni–1.1Fe alloys were sintered in a conventional furnace as well as in a 2.45 GHz microwave furnace at 1500°C [45]. Microwave sintering resulted in ~75% reduction in process time, leading to less W grain coarsening. Better mechanical properties were achieved in microwave-sintered compacts due to lower W concentration in the matrix and lack of intermetallic phase formation [45]. Si and Fe blend powders as well as Si-based alloy powders were heated in a single-mode microwave applicator [46]. The phase transformation and heating behavior of these powders were investigated in separate electric (**E**) and magnetic (**H**) fields. The increase of Fe content in the mixture of Si and Fe powders increased its microwave absorption capability in both the **E** and **H** fields. $Si_{90}Zr_{10}$ and $Si_{95}Y_5$ alloy powders were heated up more easily in the **H** field than the **E** field. Microwave heating of Si and Fe blend powders induced the formation of $FeSi_2$ and FeSi phases in the **E** and **H** fields under nitrogen gaseous atmosphere. A crystalline Fe_3O_4 phase was detected when $Si_{70}Fe_{30}$ blend powders were heated by microwaves in air. A SiY phase was formed by exposing $Si_{95}Y_5$ alloy powder to microwave radiation in the **H** field [46].

Padmavathi et al. [47] heated 6711 (Al–1Mg–0.8Si–0.25Cu) alloy at 570, 590°C, 610°C, and 630°C under vacuum (10^{-6} Torr) in conventional and microwave furnaces. Compared to conventional sintering microwaves coupled with the compacts and resulted in about 58% reduction in processing time along with higher heating rates. However, a rapid heating rate resulted in inhomogeneous microstructure with larger melt fraction at grain boundaries. XRD analysis showed absence of intermetallics as diffusion could not take place due to insufficient time. They showed that all compacts did undergo supersolidus liquid-phase sintering (SLPS), accordingly higher densification and shrinkage with increasing sintering temperature. Like sintered density the electrical conductivity and hardness followed similar trends in both cases. The alloy compact sintered at 630°C resulted in significant improvement in mechanical properties (transverse rupture strength [TRS]: 57%↑; ultimate tensile strength [UTS]: 27%↑; ductility: 41%↑) in the conventional method. In comparison to sintered compacts age-hardening treatment under T6 temper enhanced the tensile strength by 136% with drastic reduction in ductility [47].

Nitridation of metallic titanium is possible in air instead of the thermodynamically favorable oxidation in the presence of microwave radiation. Kashimura et al. [48] synthesized β-Ti–Cr alloy powders to these nitrides by microwaves. The alloy powders were heated at 1000°C or higher by microwaves at a high oxygen partial pressure and were analyzed by XRD and wavelength-dispersive X-ray spectroscopy. Nitridation was achieved for the β-Ti–Cr alloy but not for a Ti–Fe alloy. Furthermore, multilayer structures made of nitrides and oxides were observed [48]. The dynamic mechanical properties of microwave-sintered 93W–4.9Ni–2.1Fe alloy rods were investigated under dynamic compression using a split Hopkinson pressure bar. The microstructure and microhardness values of the sintered specimens after dynamic compression were analyzed. The results showed that the deformation and microhardness of specimens increased with increasing the strain rate. The deformation was increased to the maximum value of 59.8%, and the microhardness values of the tungsten grains and the matrix phase were 7.66 and 6.92 GPa, respectively, when the strain rate is 3000 S^{-1}. The formation of cracks during compressive deformation initiated before the appearance of the adiabatic shear bands. As the strain rate increased, cracks were initiated at the edge of specimens, gradually propagated to the bulk alloy, and the adiabatic shear band was observed at about 45°to the loading direction under the strain rate of 3000 S^{-1}. These findings suggested that tungsten-based alloys extruded rods sintered by microwave heating would be an ideal material with excellent self-sharpening and penetration performance for penetrators [49].

Porous nickel titanium (NiTi) shape memory alloy (SMA) was successfully fabricated by the microwave sintering method. This method allows the formation of porous structures without using any pore-forming agents. Moreover, microwave sintering of NiTi SMA can be successfully performed at a relatively low sintering temperature of 850°C and a short sintering time of 15 min. The pore characteristics, microstructure, phase transformation, and stress–strain behavior of the porous NiTi SMA were investigated. The porous NiTi SMA exhibited porosity ratios from 27% to 48% and pore sizes ranging from 50 to 200 μm when using different sintering temperatures and holding times. The predominant B2 (NiTi) and B19′ (NiTi) phases were identified in the porous NiTi SMA. A multistep phase transformation took place on heating, and a two-step phase transformation took place on cooling of the porous NiTi SMA. The irrecoverable strains decreased with increasing sintering temperature, but the holding time had little effect on the stress–strain behavior at 60°C [50].

Refractory metals and alloys are well known for their high mechanical properties, which make them useful for wide range of high-temperature applications. However, it is very difficult to consolidate them under moderate conditions owing to the refractoriness of these metals and alloys. Conventional powder metallurgy (P/M) processing is a viable sintering technique for these refractory metals. One of the constraints in conventional sintering is the long residence time resulting in undesirable microstructural coarsening. This problem becomes further aggravated by using smaller (submicron and nano) precursor powder sizes. Furthermore, conventional heating is mostly radiative that leads to nonuniform heating in large components. Refractory metals and alloys (W, Mo, Re, W–Cu, W–Ni–Cu, and W–Ni–Fe) have been successfully consolidated using microwave sintering. A comparative study with conventional data has been made. In most cases, microwave sintering resulted in an overall reduction of sintering time up to 80%. The reduction in sintering time prevents grain growth, substantially providing a finer microstructure; as a result, better mechanical properties have been observed [51].

5.4 Conclusion

In the past decades significant developments have taken place in the field of microwave processing of materials. However, more research must be conducted to obtain a detailed understanding about the nature and distribution of the electromagnetic field inside the microwave cavity, microwave–material interaction, material transformations, and heat transfer mechanisms for optimizing the process. Since microwave processing is a relatively new technology, considerable time may be required to implement this technology for the ceramic industry. It is hoped that significant advancements will take place in the science and technology of microwave processing so that one can envisage the introduction of batch-type microwave sintering furnaces in the production units of ceramic industries.

Acknowledgments

The authors are very grateful to Mr. K. Dasgupta, Director, CSIR-Central Glass and Ceramic Research Institute, for his kind permission to publish the chapter.

References

1. Sutton, W. H. (1987). Microwave processing of ceramics, *Am. Ceram. Soc. Bull.,* **68**(2), pp. 376–386.
2. Thostenson, E. T., and Chou, T.-W. (1999). Microwave processing: fundamentals and applications, *Compos. Part A: Appl. Sci.,* **30**, pp. 1055–1071.
3. Clark, D., and Sutton, W. H. (1996). Microwave processing of materials, *Annu. Rev. Mater. Sci.,* **26**, pp. 299–331.
4. Agrawal, D. K. (1998). Microwave processing of ceramics, *Curr. Opin. Solid State Mater. Sci.,* **3**, pp. 480–485.
5. Agrawal, D., Cheng, J., Peelamedu, R., and Roy, R. (2004). Microwave processing of ceramics and metallic materials: developments at Penn State, *Curr. Adv. Mater. Processes,* **17**, pp. 670–673.
6. Agrawal, D. (2005). Microwave processing of ceramics composites and metallic material, *High Tech. Ceram. News,* **16**, pp. 9–10.
7. Minay, E. J., Boccaccini, A. R., Veronesi, P., Cannillo, V., and Leonelli, C. (2004). Processing of novel glass matrix composites by microwave heating, *J. Mater. Process. Tech.,* **155–156** pp. 1749–1755.
8. Thuault, A., Marinel, S., Savary, E., Heuguet, R., Saunier, S., Goeuriot, D., and Agrawal, D. (2013). Processing of reaction-bonded B4C–SiC composites in a single-mode microwave cavity, *Ceram. Int.,* **39**, pp. 1215–1219.
9. Chen, L., Tang, C. Y., Ku, H. S., Tsui, C. P., and Chen, X. (2014). Microwave sintering and characterization of polypropylene/multi-walled carbon

nanotube/hydroxyapatite composites, *Compos. Part B: Eng.*, **56**, pp. 504–511.

10. Faeghi-Nia, A., and Ebadzadeh, T. (2012). Fabrication of machinable phlogopite–glass composite using microwave processing, *Ceram. Int.*, **38**, pp. 2653–2658.
11. Jayalakshmi, S., Sahu, S., Sankaranarayanan, S., Gupta, S., and Gupta, M., (2013). Structural and mechanical properties of $Ni_{60}Nb_{40}$ amorphous alloy particle reinforced Al-based composites produced by microwave-assisted rapid sintering, *Mater. Sci. Eng. A*, **581**, pp. 119–127.
12. Abbasi, Z., Shariat, M. H., and Javadpour, S. (2013). Microwave-assisted combustion synthesis of AlN–SiC composites using a solid source of nitrogen, *Powder Technol.*, **249**, pp. 181–185.
13. Bian, H., Yang, Y., Wang, Y., Tian, W., Jiang, H., Hu, Z., and Yu, W. (2013). Effect of microstructure of composite powders on microstructure and properties of microwave sintered alumina matrix ceramics, *J. Mater. Sci. Technol.*, **29**, pp. 429–433.
14. Boccaccini, A. R., Veronesi, P., and Leonelli, C. (2001). Microwave processing of glass matrix composites containing controlled isolated porosity, *J. Eur. Ceram. Soc.*, **21**, pp. 1073–1080.
15. Zhou, X., Shen, L., Li, L., Zhou, S., Huang, T., Hu, C., Pan, W., Jing, X., Sun, J., Gao, L., and Huang, Q. (2013). Microwave sintering carbon nanotube/$Ni_{0.5}Zn_{0.5}Fe_2O_4$ composites and their electromagnetic performance, *J. Eur. Ceram. Soc.*, **33**, pp. 2119–2126.
16. Sharma, A. K., and Krishnamurthy, R. (2002). Microwave processing of sprayed alumina composite for enhanced performance, *J. Eur. Ceram. Soc.*, **22**, pp. 2849–2860.
17. Siddharthan, A., Sampath Kumar, T. S., and Seshadri, S. K. (2010). In situ composite coating of titania-hydroxyapatite on commercially pure titanium by microwave processing, *Surf. Coat. Technol.*, **204**, pp. 1755–1763.
18. Sharma, A. K., and Gupta, D. (2012). On microstructure and flexural strength of metal–ceramic composite cladding developed through microwave heating, *Appl. Surf. Sci.*, **258**, pp. 5583–5592.
19. Li, S., Louzguine-Luzgin, D. V., Xie, G., Sato, M., and Inoue, A. (2010). Development of novel metallic glass/polymer composite materials by microwave heating in a separated H-field, *Mater. Lett.*, **64**, pp. 235–238.
20. Gupta, D., Bhovi, P. M., Sharma, A. K., and Dutta, S. (2012). Development and characterization of microwave composite cladding, *J. Manuf. Processes*, **14**, pp. 243–249.

21. Rajkumar, K., and Aravindan, S. (2013). Tribological behavior of microwave processed copper-nanographite composites, *Tribol. Int.*, **57**, pp. 282–296.

22. Higginbotham, A. L., Moloney, P. G., Waid, M. C., Duque, J. G., Kittrell, C., Schmidt, H. K., Stephenson, J. J., Arepalli, S., Yowell, L. L., and Tour, J. M. (2008). Carbon nanotube composite curing through absorption of microwave radiation, *Compos. Sci. Technol.*, **68**, pp. 3087–3092.

23. Tang, C. Y., Wong, C. T., Zhang, L. N., Choy, M. T., Chow, T. W., Chan, K. C., Yue, T. M., and Chen, Q. (2013). In situ formation of Ti alloy/TiC porous composites by rapid microwave sintering of Ti_6Al_4V/MWCNTs powder, *J. Alloys Compd.*, **557**, pp. 67–72.

24. Das, S., Datta, S., Mukhopadhyay, A. K., Pal, K. S., and Basu, D. (2010). Al–Al_2O_3 core–shell composite by microwave induced oxidation of aluminium powder, *Mater. Chem. Phys.*, **122**, pp. 574—581.

25. Ghosh, S., Pal, K. S., Dandapat, N., Mukhopadhyay, A. K., Datta, S., and Basu, D. (2011). Characterization of microwave processed aluminium powder, *Ceram. Int.*, **37**, pp. 1115–1119.

26. Ghosh, S., Mukhopadhyay, A. K., Datta, S., Dandapat, N., and Basu, D. (2010). Effect of dopants on the phase formation in microwave processed Al-Al_2O_3 composites, *J. Alloys Compd.*, **500**, pp. 231—236.

27. Vergara, U. L. de, Sarrionandia, M., Gondra, K., and Aurrekoetxea, J. (2014). Impact behaviour of basalt fibre reinforced furan composites cured under microwave and thermal conditions, *Compos. Part B: Eng.*, **66**, pp. 156–161

28. Dang-Hyok, Y., Raju, K., Bong-Ki, M., and Reddy, P. V. (2014). Synthesis and characterization of microwave sintered ferromagnetic–ferroelectric perovskite composites, *Ceram. Int.*, **40**, pp. 13497–13505.

29. Sun, H., Zhang, Y., Gong, H., Li, T., and Li, Q. (2014). Microwave sintering and kinetic analysis of Y2O3–MgO composites, *Ceram. Int*, **40**, pp. 10211–10215.

30. Deng, Y., and Catchmark, J. M. (2014). Insoluble starch composite foams produced through microwave expansion, *Carbohydr. Polym.*, **111**, pp. 864–869.

31. Das, S., Mukhopadhyay, A. K., Datta, S., Das, G. C., and Basu, D. (2008). Hard glass-ceramic coating by microwave processing, *J. Eur. Ceram. Soc.*, **28**, pp. 729–738.

32. Das, S., Basu, D., Datta, S., and Mukhopadhyay, A. K. (2008). Crystallization of glass coating by microwave heating, *Trans. Ind. Ceram. Soc.*, **67**, pp. 139—146.

33. Das, S., Mukhopadhyay, A. K., Datta, S., and Basu, D. (2010). Evaluation of microwave processed glass-ceramic coating on nimonic superalloy substrate, *Ceram. Int.*, **36**, pp. 1125–1130.
34. Ghosh, S., Pal, K. S., Dandapat, N., Ghosh, J., and Datta, S. (2013). Glass-ceramic glazes for future generation floor tiles, *J. Eur. Ceram. Soc.*, **33**, pp. 935–942.
35. Ghosh, S., Pal, K. S., Mandal, A., Biswas, K., N., Bhattacharya, M., and Bandyopadhyay, P. (2014). Cordierite based glass-ceramic glazed floor tile by microwave processing, *Mater. Charact.*, **95**, pp. 192–200.
36. Roy, R., Agrawal, D., Cheng, J., and Gedevanishvili, S. (1999). Full sintering of powdered-metal bodies in a microwave field, *Nature*, **399**, pp. 668–670.
37. Gedevanishvili, S., Agrawal, D., and Roy, R. (1999). Microwave combustion synthesis and sintering of intermetallics and alloys, *J. Mater. Sci. Lett.*, **18**, pp. 665–668.
38. Das, S., Mukhopadhyay, A. K., Datta, S., and Basu, D. (2008). Prospects of microwave processing: an overview, *Bull. Mater. Sci.*, **31**, pp. 943–956.
39. Xie, G., Li, S., Louzguine-Luzgin, D. V., Cao, Z., Yoshikawa, N., Sato, M., and Inoue, A. (2009). Fabrication of Ni-Nb-Sn metallic glassy alloy powder and its microwave-induced sintering behavior, *Int. Microw. Power Electromagn. Energy*, **43**, pp. 17–22.
40. Gabbitas, B. L., Ariff, T. F., and Cao, P. (2010). Synthesis of pewter alloy from tin–copper–antimony powder mixtures by microwave and conventional sintering, *Powder Metall.*, **54**, pp. 488–496.
41. Saheb, N. (2013). Spark plasma and microwave sintering of Al6061 and Al2124 alloys, *Int. J. Miner., Metall. Mater.*, **20**, pp. 152–159.
42. Chau, J. L. H., Yang, C.-C., and Shih, H.-H. (2014). Microwave plasma production of metal nanopowders, *Inorganics*, **2**, pp. 278–290.
43. Annamalai, R., Upadhyaya, A., and Agrawal, D. (2013). An investigation on microwave sintering of Fe, Fe-Cu, Fe-Cu-C alloys, *Bull. Mater. Sci.*, **36**, pp. 447–456.
44. Sethi, G., Upadhyaya, A., and Agrawal, D. (2003). Microwave and conventional sintering of premixed and prealloyed Cu-12Sn bronze, *Sci. Sinter.*, **35**, pp. 49–65.
45. Upadhyaya, A., Tiwari, S. K., and Mishra, P. (2007). Microwave sintering of W–Ni–Fe alloy, *Scripta Mater.*, **56**, pp. 5–8.
46. Li, S., Xie, G., Louzguine-Luzgin, D. V., Cao, Z., Yoshikawa, N., Sato, M., and Inoue, A. (2010). Phase transformations in Si-based alloy powder

mixtures induced by microwave heating in a 2.45 GHz single-mode applicator, *Intermetallics,* **18**, pp. 2030–2033.

47. Padmavathi, C., Upadhyaya, A., and Agrawal, D. (2011). Effect of microwave and conventional heating on sintering behavior and properties of Al-Mg-Si-Cu alloy, *Mater. Chem. Phys.*, **130**, pp. 449–457.
48. Kashimura, K., Fukushima, J., Mitani, T., Sato, M., and Shinohara, N. (2013). Metal Ti-Cr alloy powders nitriding under atmospheric pressure by microwave heating, *J. Alloys Compd.*, **550**, pp. 239–244.
49. Ma, Y., Zhang, J., Liu, W., Yue, P., and Huang, B. (2014). Microstructure and dynamic mechanical properties of tungsten-based alloys in the form of extruded rods via microwave heating, *Int. J. Refract. Metals Hard Mater.*, **42**, pp. 71–76.
50. Tang, C. Y., Zhang, L. N., Wong, C. T., Chan, K. C., and Yue, T. M. (2011). Fabrication and characteristics of porous NiTi shape memory alloy synthesized by microwave sintering, *Mater. Sci. Eng. A*, **528**, pp. 6006–6011.
51. Mondal, A., Agrawal, D., and Upadhyaya, A. (2010). Microwave sintering of refractory metals/alloys: W, Mo, Re, W-Cu, W-Ni-Cu and W-Ni-Fe alloys, *J. Microw. Power Electromagn. Energy*, **44**, pp. 28–44.

Chapter 6

Microwave Engineering for Synthesizing Clays and Modifying Properties in Zeolites

Olga Bergadà, Fiseha B. Gebretsadik, M. Dolores González, Judith Granados-Reyes, Elena Pérez, Tatiana Sánchez, Isabel Vicente, Pilar Salagre, and Yolanda Cesteros

Departament de Química Física i Inorgànica, Universitat Rovira i Virgili, C/Marcel·lí Domingo s/n, Tarragona 43007, Spain

pilar.salagre@urv.cat, yolanda.cesteros@urv.cat

In this chapter, we report the main results obtained in our research group during the past five years about the effect of microwaves on the synthesis of cationic clays (saponites and hectorites) and anionic clays (hydrotalcites and hydrocalumites) and on the modification of the surface and acidic properties of several pentasil-type zeolites (beta, mordenite, ZSM-5) by cation exchange, dealumination, and sulfonic acid functionalization. Microwaves reduced the synthesis time and modified the surface and acid–base properties of these materials.

Microwave Engineering of Materials: From Mesoscale to Nanoscale

Edited by Erwann Guénin

ISBN 978-981-4669-42-9 (Hardcover), 978-981-4669-43-6 (eBook)

www.panstanford.com

6.1 Introduction

Microwave technology has been used since the 1980s as a heating source in different research fields, such as solid dehydration, and in promotion of both organic and inorganic reactions [1, 2]. The heating of samples using microwaves is mainly produced through two mechanisms, ionic conduction and dipolar rotation. The heating by ionic conduction is due to the resistance to the ionic flux in the medium, while in the dipolar rotation mechanism, the polar molecules rotate to get oriented with the electromagnetic field, produced by the microwaves radiation, resulting in frictions among them that generate an increase of temperature in the sample. For a typical frequency of 2450 MHz the alignment of molecules followed by their relaxation occur 4.9×10^9 times/s, leading to fast heating. Because of the high rate of heating, hot spots may happen.

In the past few years, there has been an increasing interest in using microwaves to synthesize inorganic materials such as hydrotalcites [3–6], zeolites [7, 8], and mesoporous materials [9] and, more recently, to synthesize saponites [10], not only due to the particular characteristics that the synthesized solids can achieve, but also due to a decrease in the synthesis time with subsequent energy saving. Therefore, microwave reactions constitute valuable processes in green chemistry.

Clay minerals are widespread materials on the earth. There are signs that people already employed them many thousands of years ago to create pottery and decorative items. Nowadays, millions of tons of these minerals are used because of their applications, not only as ceramics in building and pharmaceuticals and beauty treatments, but also as catalysts, adsorbents, or catalytic supports, depending on their properties. Clays are phyllosilicates, the layers of which consist of $M(O, OH)_6$ octahedral sheets (where $M = Al^{3+}$, Mg^{2+}, Fe^{3+}, or Fe^{2+}) and tetrahedral sheets, mainly composed of Si(O, OH). In both sheets, isomorphic substitution by less charged cations is possible. In the hectorite, Mg^{2+} is replaced by Li^+ in the octahedral positions ($(Si_8)^{Td}[Mg_{6-x}Li_x]^{Oh}(OH)_4O_{20}M^{n+}{}_{x/n}{\cdot}mH_2O$), whereas in saponite, Si^{4+} in the tetrahedral sheet is replaced by Al^{3+}($(Si_{8-x}Al_x)^{Td}[Mg_6]^{Oh}(OH)_4O_{20}M^{n+}{}_{x/n} \cdot mH_2O$). These substitutions provoke negative charge of layers, which is compensated by cations located in the interlayer space [11]. On the whole, the properties of these clays depend on their composition and specific surface area. Interesting properties are the interlayer swelling [12] and cation exchange, which is expressed as cationic exchange capacity (CEC) [13, 14]. Additionally, hectorites and saponites can have Brønsted and Lewis acidity. Brønsted acid sites are related to the presence of H^+, which are neutralizing the negative charge of the layers, whereas Lewis acidity corresponds to interlayer cations and cations from deficiently coordinated layers. It is important to remark the importance of obtaining saponites with a high occupation of Al^{3+} in the tetrahedral position in order to increase the amount and the strength of their Brønsted acid sites. This has an associated difficulty, which is the tendency of Al^{3+} to have octahedral coordination. Then, the control of physical and chemical properties of saponites becomes very important [15]. Hectorites and saponites have been extensively applied in adsorption processes from the early years of the 20th century for waste treatment or in the pharmaceutical industry [16, 17]. Afterward, they have been also applied in catalytic processes for organic synthesis as active phases or as supports [18–21]. Another interesting application of these materials is the preparation of nanocomposites [22–24]. There are no references in the literature about the use of microwaves for the synthesis of hectorites and only some for the synthesis of saponites [10].

Layered double hydroxides (LDHs) minerals have similar properties to clays minerals, but considering their anion exchange capacity, LDHs are called anionic clays. These layered materials have the general formula $[M(II)_{1-x}M(III)_x(OH)_2][A_{x/n}]^{n-} \cdot mH_2O$, where M(II) and M(III) are the divalent and trivalent cations and A is the interlayer anion. If M(II) is Mg^{2+} or Ca^{2+}, the compound is named hydrotalcite or hydrocalumite, respectively. M(III) is Al^{3+} for both materials. Structurally, LDHs have brucite-like ($Mg(OH)_2$) sheets where isomorphous substitution of M^{2+} by a trivalent cation, such as Al^{3+}, occurs. The resulting excess positive charge of the layered network is compensated by anions, which occupy the interlayer space along with water molecules. On the whole, they have basic characteristics, while the mixture of the oxides obtained after their calcination presents interesting acid–base properties. Microwaves have been widely used for the synthesis of hydrotalcites [3–6, 25–

28]. The use of microwaves resulted very interesting for improving textural and basic properties [25–28]. Hydrocalumite materials have been receiving increasing attention in recent years, owing to their potential applications as basic heterogeneous catalysts [29–31], ion exchangers [32], adsorbents [33], and precursors for composite materials [34]. There are no references in the literature about the use of microwaves for the synthesis of hydrocalumites.

Zeolites, with the general formula $(M^{n+})_{x/n}\{(AlO_2)_x(SiO_2)_y\}\cdot mH_2O$, are well-known microporous materials widely used as solid acid catalysts in many industrial processes due to their large surface areas, shape selectivity, and tunable acidity [35, 36]. These properties can be modified with different treatments such as by introducing other extra framework cations (ion exchange), by dealumination, or by incorporating organosulfonic acid groups (functionalization).

The type of cation present not only influences the ion exchange properties of the zeolite but also is a factor in its adsorptive and catalytic properties [37]. For instance, introducing transition metal cations into zeolite modifies its Brønsted and Lewis acidity, affecting its catalytic behavior. In the most part of studies, cations have been exchanged in the liquid phase, putting zeolite in contact with aqueous solutions of the exchanging cation. When the size of the exchanging cation is higher, due to hydration or hydrolysis processes [38, 39], it is necessary to apply temperature during ion exchange or to make an intermediate calcination step between several ion exchange steps to increase the cation exchange [40, 41]. Additionally, ion exchange can be also achieved in the solid-state phase by keeping in contact a zeolite and a metal salt at elevated temperatures with prolonged heating [42].

Dealumination of zeolites is useful to reduce the acid site concentration, improve thermal stability, and modify pore structure [36, 43]. There are several factors that influence the dealumination of zeolites, such as zeolite synthesis conditions, their structure type, and the dealumination method used [44–48]. A considerable number of zeolite dealumination techniques have been developed: treatment of zeolites with steam or $SiCl_4$ vapor at elevated temperatures or treatment with $(NH_4)SiF_6$, mineral acids (i.e., HCl, HNO_3), organic acids (i.e., acetic acid, oxalic acid), F_2, chelating agents (i.e., ethylenediaminetetraacetic acid [EDTA]) etc. [44–51]. Conventional heating is mostly used when applying temperature during dealumination.

The organic tailoring of the surface of materials has received great attention in the last decade in terms of their applications in the fields of catalysis, sensing, and adsorption. Specifically, the incorporation of organosulfonic acid groups in zeolites can generate effective solid Brønsted acid catalysts with enhanced catalytic properties. It is difficult to introduce bulky organic species in zeolites due to their microporosity. There are only few studies about the preparation of zeolites containing sulfonic acid groups within their pores by direct synthesis through the incorporation of phenethyl groups (phenethyltrimethoxysilane [PETMS]) covalently tethered to silicon atoms and subsequent sulfonation of the phenyl rings [52, 53]. The incorporation of sulfonic acid groups in zeolites by postsynthesis is more difficult due to the lack of reactant silanol groups (≡Si–OH), which are required to react with the sulfonating agents [54].

In the field of zeolites, microwaves were initially mostly applied for their drying or synthesis [55, 56]. Later, microwaves were used for the cation exchange of Na^+ by La^{3+}, Co^{2+}, Cu^{2+}, and Zn^{2+} in NaX and Na–ZSM-5 zeolites or to study modifications in Brønsted and Lewis acid centers in the X zeolite by ion exchange of Na^+ by K^+ [57–61]. Nowadays, there are a considerable number of papers about the use of microwaves for ion exchange in liquid and solid phases [62–65]. However, much less literature can be found about dealumination or sulfonation processes of zeolites with microwaves [66].

In this chapter, we report our recent work referring to the synthesis with microwaves of anionic and cationic clays and with respect to the effect of microwaves on the surface and acidic properties of several pentasil-type zeolites, which were modified by Ni exchange, dealumination, and sulfonic acid functionalization by postsynthesis. The microwave furnace used for these studies was Milestone Ethos Touch Control (2.45 GHz).

6.2 Synthesis of Clays Using Microwaves

To have reproducible clays (cationic or anionic) for their applications, they must be synthesized. However, their synthesis often involves long hydrothermal treatments. The use of microwaves for clay synthesis could considerably decrease the synthesis time and confer special features to the final materials. For the characterization of the synthesized clays we used the conventional solid characterization techniques (X-ray fluorescence [XRF], X-ray diffraction [XRD], nitrogen physisorption, transmission electron microscopy [TEM], and CEC determination). For the characterization of saponites ^{27}Almagic-angle spinningnuclear magnetic resonance (Al-MAS-NMR) was also used.

6.2.1 *Cationic Clays: Hectorites and Saponites*

6.2.1.1 Hectorites

Hectorite was prepared according to the method reported by Granquist and Pollack [67] and modified in our research group using microwaves [68, 69]. In this method, a slurry of SiO_2, $Mg(OH)_2$, and LiF in the molar ratio SiO_2:$Mg(OH)_2$:LiF=4:3:1 was first prepared. The resulting slurries were aged by autoclaving in a microwave oven at 373 and 393 K for 4, 8, and 16 h. One sample was aged in an autoclave using conventional heating at 393 K for 8 days for comparison. Several effects were observed. The most significant is that at 393 K, the hydrothermal microwave aging treatment provided 24 times faster synthesis of hectorite than that obtained by conventional heating (8 h in front to 8 days). Higher crystallinity of brucite, longer aging time, and higher aging temperature also favored a faster synthesis [68, 69]. The highest surface area was obtained at 393 K after 16 h of hydrothermal treatment with microwaves (170 m^2/g). To increase the surface area of these hectorites (reference samples $MwH_{(8h)}$ and $CH_{(8d)}$, Table 6.1), two strategies were used, in situ construction of silicate pillars (silicate bearing) and delamination.

Silicate-bearing hectorites were prepared in our research group following the method reported by Torii et al. [70] but modifying several parameters and using microwaves for the hydrothermal aging [71]. The molar ratio of reagents was Si:Mg:Li = 8.0:5.4:0.6, and several steps were followed to obtain a slurry that was aged by autoclaving in the microwave oven at 453 K for 1 h or by conventional heating at 453 K for 2 h. The resulting solids

Table 6.1 Microwaves effect in the synthesis of hectorites

Hectorite	Mw/Conv	Crystallite size (060) (XRD) (nm)	CEC (meq/100 g)	BET (m^2/g)	Mg vacancies (mmol/100 g)
H_{393}	$MwH_{(8h)}$	9.3	68	156	–
	$CH_{(8d)}$	7.9	53	165	–
$7HD_{20}$	$Mw7HD_{20}$	6.5	76	603	33.21
	$C7HD_{20}$	6.0	62	563	26.23
$7HD_{40}$	$Mw7HD_{40}$	7.3	76	601	29.35
	$C7HD_{40}$	6.5	53	561	10.58
$10HD_{20}$	$Mw10HD_{20}$	7.1	84	599	34.80
	$C10HD_{20}$	6.3	73	590	32.13
10HB	Mw10HB	11.9	39	262	–
	C10HB	8.7	34	356	–

were suspended in water, mixed with trimethyldodecylammonium chloride (AQ) in a molar ratio Li:AQ= 1:1 and refluxed. Finally, the products were calcined. The sample obtained by microwaves was designated as Mw10HB, whereas that prepared with conventional heating was named as C10HB (Table 6.1).

We prepared delaminated hectorites following the method reported by Iwasaki et al. [72] but introducing ultrasounds for mixing the starting reagents and microwave irradiation during the hydrothermal treatment [71]. The slurry of Si and Mg reagents with LiOH (molar ratio Si:Mg:Li = 8.0:5.2:0.8) was maintained for different times (20 or 40 min) in a Selecta ultrasound water bath and later filtered and washed until pH 10 or 7. The precipitates were suspended in water and mixed with the AQ salt in a molar ratio Li:AQ = 1:1. The hydrothermal treatment was performed by autoclaving with microwaves at 453 K for 1 h or by conventional heating at 453 K for 2 h. Finally, the products were calcined. The delaminated samples, as indicated in Table 6.1, were named, including the type of heating (Mw or C), the slurry pH (7 or 10), and the ultrasounds time (20 or 40), as follows: $Mw7HD_{20}$, $Mw7HD_{40}$, $C7HD_{20}$, $C7HD_{40}$, $Mw10HD_{20}$, and $C10HD_{20}$.

The use of microwaves favored the synthesis of more crystalline materials at shorter time for the three types of hectorites (H, HD, or HB) (Table 6.1, Figs. 6.1–6.3). The silicate-bearing and delaminated

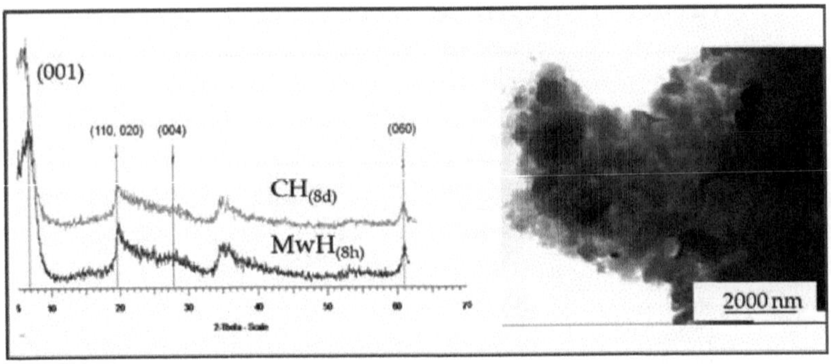

Figure 6.1 XRD patterns of samples $MwH_{(8h)}$ and $CH_{(8d)}$. TEM image of sample $MwH_{(8h)}$.

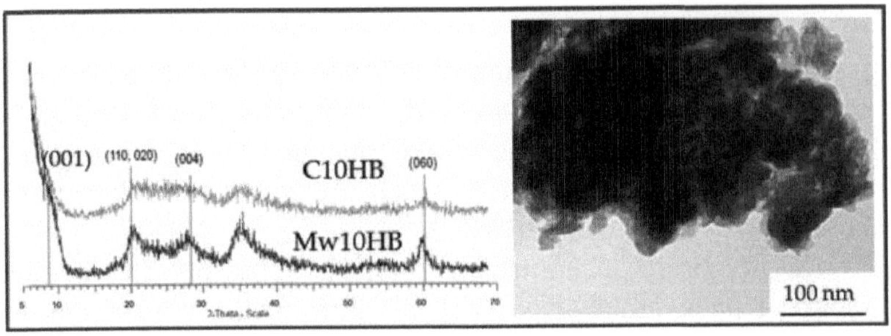

Figure 6.2 XRD patterns of samples Mw10HB and C10HB. TEM image of sample Mw10HB.

hectorites presented the expected clay pattern, except the (001) reflection, related to the layer stacking, which was not observed in HD samples (Fig. 6.3) and was a shoulder in HB samples (Fig. 6.2). For silicate-bearing hectorites, this can be related to some delamination in a limited order in the *c* axis.

TEM images (Figs. 6.1–6.3) and N_2 adsorption–desorption isotherms were in agreement with the XRD results. For HD and HB hectorites, the Brunauer–Emmett–Teller (BET) areas were higher than those of their corresponding H hectorites (Table 6.1). This increase in area was more significant for delaminated samples and was higher for the HD hectorites prepared with microwaves (around 600 m^2/g) (Table 6.1). The BET area and CEC values were related to

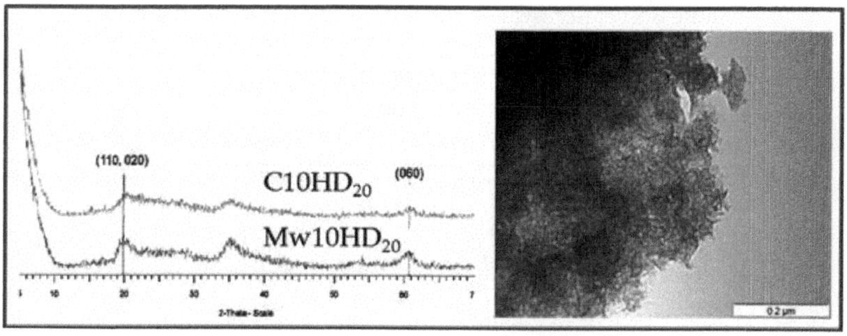

Figure 6.3 XRD patterns of samples $Mw10HD_{20}$ and $C10HD_{20}$. TEM image of sample $Mw10HD_{20}$.

the presence of octahedral vacancies in the delaminated hectorites. Microwaves enhance vacancy formation, whereas ultrasound decreases this effect.

6.2.1.2 Saponites

Saponites were prepared in our research group following the method reported by Vogels et al. [73], but modifying several parameters, and using microwaves for the hydrothermal aging. An appropriate molar ratio of Si^{4+}:Al^{3+}:Mg^{2+}:NH_4^+ was used to obtain a saponite with the theoretical formula $NH_{4(1.2)}[Mg_6Al_{1.2}Si_{6.8}O_{20}(OH)_4]$. NH_4^+ was in excess. Fumed SiO_2 was the silica source, and different Mg and Al compounds were used in order to achieve different pH values in the initial slurry (7 and 8), obtaining samples S7 and S8, respectively. Then, they were aged by autoclaving at 453 K using a microwave oven for 6 h (samples $MwS7_{(6h)}$ and $MwS8_{(6h)}$) or by conventional heating for 3 days (samples $CS7_{(3d)}$ and $CS8_{(3d)}$). The influence of the presence of F^- (sample $MwS7F_{(6h)}$) and the modification of the aging time (12 h) were also studied [74]. These saponites are included in group 1S of Table 6.2.

To improve the saponite properties we performed several modifications in the preparation of saponites at pH = 8 with microwaves at 453 K [75]: (1) the use of ultrasounds for reagent mixing, (2) the use of different dilution slurries (H_2O/Si molar ratios = 250, 125, 50), and (3) the presence of polyvinylpyrrolidones (PVPs), with different molecular masses (40,000 g/mol [P1] and

Table 6.2 Microwaves effect in the synthesis of saponites

Saponite	Mw/Conv	Crystallite size (060) (XRD) (nm)	CEC (meq/100g)	BET (m^2/g)
1S	$MwS7_{(6h)}$	[1pt] 6.0	52	226
	$MwS7F_{(6h)}$	6.2	65	279
	$CS7_{(3d)}$	6.3	50	346
	$MwS8_{(6h)}$	6.2	55	270
	$CS8_{(3d)}$	7.8	53	165
2S*	$MwS8_{250}$	7.7	58	438
	$MwS8_{125}$	8.4	87	461
	$MwS8_{50}$	8.9	55	394
	$CS8_{50}$	10.8	37	254
3S*	$MwS8_{125}P1$	7.5	63	396
	$MwS8_{125}P2$	6.5	59	354
	$MwS8_{250}P1$	6.9	61	353
	$CS8_{250}P1$	7.3	55	259

*Use of ultrasound in the step of reagent mixing.

360,000 g/mol [P2]), as a template in the aged slurry. The same modifications were made in the preparation of saponites using conventional heating during the aging treatment (453 K for 3 days) for comparison. For the samples prepared with the polymer, the template was removed by calcination [75]. For better comparison, the saponites used to study the dilution effect are included in group 2S of Table 6.2 ($MwS8_{250}$, $MwS8_{125}$, $MwS8_{50}$, and $CS8_{50}$), whereas the saponites used to study the effect of the presence of the polymer during their synthesis ($MwS8_{125}P1$, $MwS8_{125}P2$, $MwS8_{250}P1$, and $CS8_{250}P1$) are reported in group 3S of the same table.

XRD patterns of all the prepared saponites presented the expected reflections of a trioctahedral clay. Table 6.2 summarizes the main characterization results corresponding to the influence of microwaves in the synthesis of saponites considering the modification of several parameters. In the first group, that is, 1S, we analyze the effect of the pH (7 or 8) and the presence of fluoride during the synthesis. The saponites aged with microwaves at pH 7 with F^- for 6 h and at pH 8 for 12 h presented the same crystallinity than the corresponding saponites prepared by conventional heating

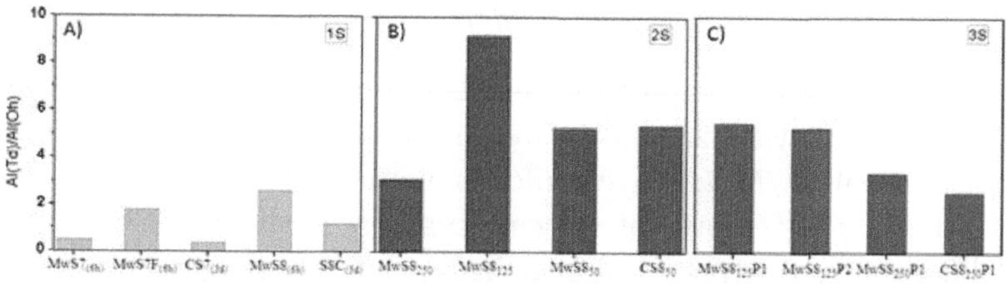

Figure 6.4 The Al(Td)/Al(Oh) ratio for (a) 1S, (b) 2S, and (c) 3S saponites.

for 3 days [74]. However, when the pH used was 8 for the saponite aged with microwaves for 6 h, the Al(Td)/Al(Oh) ratio was the highest (2.6) (Fig. 6.4).

All the saponites of group 2S presented an important degree of delamination. The decrease of slurry dilution supposed the expected increase of the stacking order proposed in the literature [76]; the increase of the tetrahedral sheet order, measured by the shift to lower cm^{-1} values of the O–Si–O stretch; and the decrease of the amorphous silica presence. On the whole, the order in the stacking and in the layer was slightly higher for the saponites prepared by conventional heating, which were aged at a longer time. However, the results obtained at medium dilution ($H_2O/Si = 125$) for the sample prepared with microwaves ($MwS8_{125}$) were very interesting because the CEC, the BET area, and the Al(Td)/Al(Oh) ratio were higher (Table 6.2, Fig. 6.4). The action of microwaves and a sufficient mobility of the ions favor the incorporation of the cations in the expected positions (Si and Al in the tetrahedral positions and Mg in the octahedral sheets).

The effect of the presence of a polymer is summarized in group 3S of saponites. The presence of the polymer favored stacking disorder at medium and lower slurry dilutions when microwaves were used. However, the BET areas and the Al(Td)/Al(Oh) ratios were lower than or similar to those of the corresponding sample prepared without the polymer (Table 6.2, Fig. 6.5). The molecular mass of the polymer had an influence on the average mesoporous pore radius, which was 2.5 nm for sample $MwS8_{125}P1$ and 3.2 nm for sample $MwS8_{125}P2$. For the rest of the dilutions, similar differences were observed [75].

6.2.2 Anionic Clays: Hydrotalcites and Hydrocalumites

In this section we report the results of our research group for the synthesis of hydrotalcites and hydrocalumites using microwaves.

Several Ni/Mg/Al hydrotalcites with a M^{2+}/Al^{3+} ratio of 4 and Ni^{2+}/Mg^{2+} ratio of 6 were synthesized by modification of different parameters: (1) type of heating during the aging treatment (microwave or conventional), (2) type of reactors (refluxing or autoclaving), (3) temperature of aging (343 K for refluxing and 403, 423, and 453 K for autoclaving), and (4) time of aging (60–1080 min for refluxing, 1–360 min for autoclaving) [77]. The study was performed with the idea to obtain materials with different textural and basic properties.

The hydrotalcite phase was the only crystalline phase observed for all samples. We confirmed that the use of microwaves during aging, and the increase of the aging temperature or time led to more crystalline hydrotalcites. These effects were observed using both types of reactors, refluxing and autoclaving. The hydrotalcite aged at 453 K for 120 min was the most crystalline. It is interesting to remark that this hydrotalcite presented the highest BET area (87 m^2/g) and the highest amount or basic sites (453 μmol OH/g hydrotalcite). These properties were related to the defects generated in the lamellae as a consequence of the aluminium loss with the generation of a new porosity [77]. Figure 6.5 shows the nitrogen adsorption–desorption isotherm and the pore size distribution of

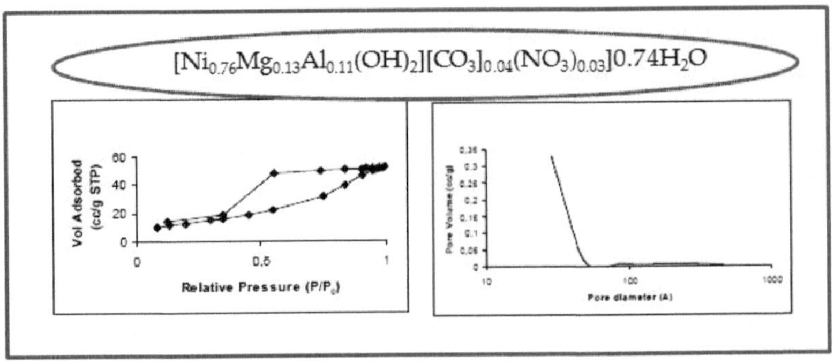

Figure 6.5 Composition, N_2 adsorption–desorption isotherm, and pore size distribution of the hydrotalcite MwHT120 (453).

this hydrotalcite (MwHT120 [453]). As we can observe, this is a mesoporous material with small mesopores of diameter <3 nm. In the figure, we have also written its formula, determined by atomic absorption. We obtained a Ni+Mg/Al ratio of 8.1, which was higher than the theoretical expected ratio (6.0). Both results were in agreement with the loss of aluminium observed because of the effect of microwaves at higher temperature and longer time, as commented above.

In Section 6.1, we commented that hydrocalumites are LDH compounds as hydrotalcites. However, there were no references in the literature about the use of microwaves for their synthesis. To study the influence of the microwaves on the synthesis of hydrocalumites, several samples were synthesized using microwaves for the hydrothermal treatment. The method selected was the coprecipitation of the corresponding chloride salts at constant pH (11.5) with a Ca/Al molar ratio of 2. This method has been described in the literature by Vieille et al. [34] using conventional heating. The parameters studied were the type of aging heating (microwave or conventional), the type of reactor (refluxing or autoclave), the aging temperature, and the aging time [78]. A faster synthesis of crystalline Cl-hydrocalumite was possible after 1 h at 453 K with microwaves in an autoclave (Mw1-HC [453]). Figure 6.6 shows the TEM image and the XRD pattern of this sample.

The defined large hexagonal particles of 4000 nm observed, the low BET area (2 m^2/g), and the narrow XRD reflections of the characteristic pattern of the hydrocalumite phase confirm its very high crystallinity.

6.3 Microwave-Assisted Modification of Zeolites

The modification of the surface and acidic properties in zeolites using different treatments, such as ion exchange, dealumination, or functionalization, led to the obtention of modified zeolites, which can have interesting applications for adsorption, ion exchange, or catalysis [35, 37, 44]. Microwaves can have significant effects in the variation of these properties when used during the modification treatments. For the characterization of the zeolites we used XRF,

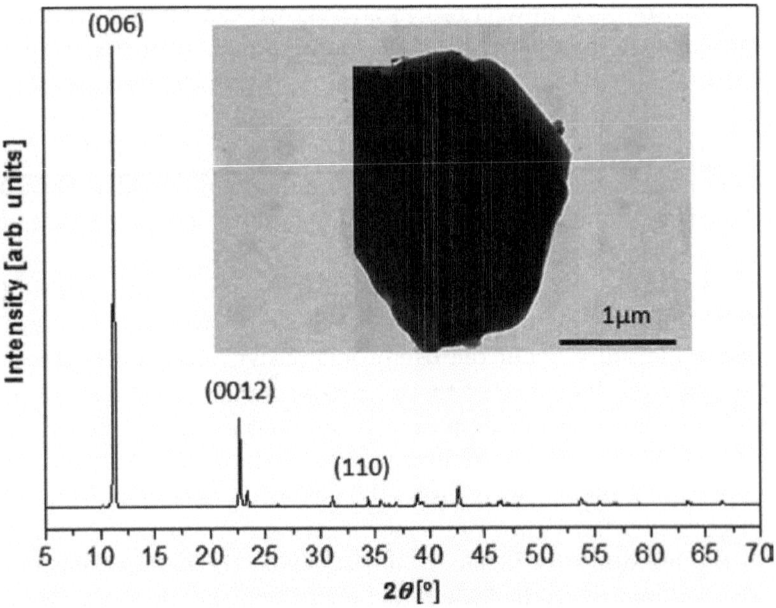

Figure 6.6 TEM and XRD images of the sample Mw1-HC (453).

XRD, N2 physisorption, X-ray photoelectron spectroscopy (XPS), Fourier transform infrared spectroscopy (FTIR), ^{1}H NMR, and thermogravimetric analysis (TGA) techniques.

6.3.1 *Effect of Microwaves on the Acidic Properties of Ni-Exchanged Mordenites*

We prepared several Ni,Na-mordenites by ion exchange of commercial Na-mordenite (Si/Al = 6.5) in the liquid phase under microwave irradiation varying the volume and concentration of $Ni(NO_3)_2 \cdot 6H_2O$ solution, the exchange temperature (333–393 K), and the exchange time (15–45 min) [79]. All samples showed similar Ni^{2+} exchange (40%–47%), as determined by atomic absorption. These moderate values of Ni^{2+} exchange can be explained by the increase in the effective size of the nickel cations in aqueous solutions due to hydrolysis and/or hydration. Several authors concluded that exchangeable sites are mainly situated in the main channels, whereas the rest of sites, which are hardly exchangeable by the hydrated nickel ions, are located in the side pockets and compressed channels [80].

Another group of Ni,Na-mordenites was prepared by conventional heating for comparison [79]. After one exchange at 333 K, using the same volume and concentration of nickel nitrate solution, similar amounts of exchanged Ni^{2+} (around 47%) were achieved with and without microwaves but much faster under microwave irradiation (15 min vs. 24 h) [79]. Making an intermediate thermal treatment under microwaves at 453 K for 6 h between two exchanges, also performed with microwaves, the exchanged Ni^{2+} increased up to 84%, whereas when the intermediate thermal treatment was made by conventional heating at 673 K for 12 h, the exchanged Ni^{2+} only increased until 52%. Therefore, it seems that microwave irradiation favors the migration of Ni^{2+} from the main channels to the rest of the sites of the mordenite. Some other studies pointed out that microwaves might enhance diffusion, for example, the diffusion of various cations in Pyrex glasses [81].

Interestingly, microwaves strongly affected the acidity of the exchanged samples since the samples exchanged with microwaves showed the appearance of new weak acid sites, the amounts of which increased as longer the microwaves treatment was (peak 1, NH_3-temperature-programmed desorption [TPD], Fig. 6.7a). From these results, we concluded that microwaves could interact with nickel cations, provoking the formation of hydrolyzed nickel species $[Ni(OH)]^+$, which were not present in the conventional exchanged sample (Fig. 6.7a). This species should have weakly Lewis acidic character, as suggested by other authors [8]. This agrees with the higher amounts of Lewis acid centers observed by adsorbed pyridine FTIR for the microwaved samples (Fig. 6.7b).

With respect to the rest of peaks observed in NH_3-TPD (Fig. 6.7a), peak 2 can be associated to Lewis acid sites due to nonhydrated Ni^{2+} cations and remaining Na^+, whereas peak 3, which appeared at higher desorption temperatures with lower intensity, could be related to some proton exchange that occurred during Ni exchange due to the acidic medium ($pH = 4$), as suggested by Tsipis and

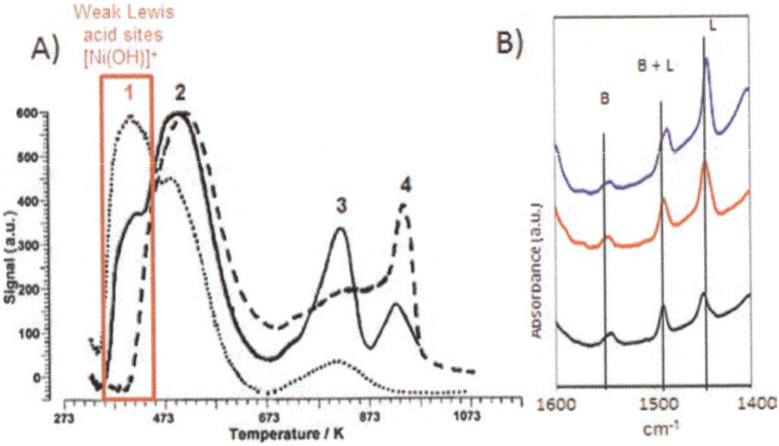

Figure 6.7 (a) NH_3-TPD curves of the samples: (—) conventionally exchanged sample, (microwave-exchanged sample, and (···) sample exchanged twice with microwaves with an intermediate microwave thermal treatment. (b) Adsorbed pyridine FTIR spectra for (black line) conventionally exchanged sample, (red line) microwave-exchanged sample, and (blue line) sample exchanged twice with microwaves with an intermediate microwave thermal treatment.

coworkers [82] and/or the hydrolysis of Ni^{2+} cations ($Ni^{2+} + H_2O \rightarrow [Ni(OH)]^{+} + H^{+}$ [83]). Additionally, the acidity of these protons can be enhanced as a result of a synergy between protons and Ni species, as reported elsewhere [84], explaining the appearance of peak 4 with higher intensity for the samples with higher amounts of nonhydrated Ni^{2+} (peak 2).

These acid characterization results were confirmed by testing these Ni,Na-mordenites in several reactions catalyzed by different acid sites: the isomerization of styrene oxide to β-phenylacetaldehyde (mainly catalyzed by Brønsted acid sites), the styrene oxide ring opening to 2-ethoxy-2-phenylethanol (catalyzed by both Brønsted and Lewis acid sites) [79], and the isomerization of linoleic acid to conjugated linoleic acids (CLAs), where the presence of these new weak Lewis acid centers yielded higher conversion and higher selectivity to the desired isomer *cis*-9, *trans*-11-CLA [85].

Table 6.3 Characterization results of commercial and partially dealuminated mordenite samples

Sample	Si/Al (XRF)	Crystallinity (%) (XRD)	BET area (m^2/g)	NH_3-TPD results Peak 1	Peak 2
M	6.5	100	303	475 (l)	–
R_{15min}	14.9	76	384	433 (m)	793 (m)
R_{2h}	18.2	75	412	433 (m)	803 (h)
A_{15min}	11.2	73	376	435 (h)	813 (m)
A_{2h}	17.1	75	397	423 (h)	835 (h)
MwR_{15min}	15.5	67	388	413 (m)	788 (m)
MwR_{2h}	20.2	68	409	413 (m)	783 (h)
MwA_{15min}	15.8	70	419	513 (h)	843 (l)
MwA_{2h}	23.6	68	430	418 (h)	778 (m)

l: low intensity; m: medium intensity; h: high intensity.

6.3.2 *Modification of Surface and Acidic Properties during Dealumination of Zeolites by Acid Treatment With Microwaves*

In a first work, we studied the effect of microwaves on the surface and acidic properties of one commercial mordenite (M) (Si/Al = 6.5) partially dealuminated in HCl medium by refluxing (R) or autoclaving (A) at 373 K for 15 min or 2 h, which were compared with the use of conventional heating during dealumination under the same conditions [86]. Table 6.3 summarizes the main characterization results. The samples, the name of which starts with Mw, are those dealuminated under microwaves.

We observed dealumination for all the acid-treated zeolites since they showed higher Si/Al ratio (XRF), lower cell volumes (XRD), higher surface area (N_2 physisorption), and a shift to higher values of the infrared (IR) bands assigned to symmetric and asymmetric stretching of the T–O bond (T = Si, Al) than commercial mordenites. The increase in the strength of the T–O bond when the Al content decreases was explained by the fact that the Si–O bond is shorter than the Al–O bond and that Al has lower electronegativity than Si [87]. Microwaves led to faster dealumination than conventional heating by autoclaving as well as by refluxing. In addition to the enhancement of dealumination observed, microwaves favored the later elimination of the Al extracted during washing and affected the acidic characteristics of the resulting samples.

NH_3–TPD thermograms presented two nonsymmetrical desorption peaks for all the partially dealuminated samples (Table 6.3). The first desorption peak was related to the presence of weak–medium acid centers, which were a contribution of silanol groups (Brønsted acidity) together with $Al(OH)_2^+$ or $Al(OH)^{2+}$ Lewis species generated during dealumination. The second peak was associated to strong Lewis acid centers due to extra-framework insoluble aluminium species also formed during dealumination. Samples dealuminated under refluxing have less acidity than the autoclaved ones due to the best elimination, during washing, of the Al extracted at atmospheric pressure. All samples showed similar low amounts of Brønsted acid sites. However, the catalyst treated under microwaves by autoclaving for a shorter time (15 min) presented acid centers with medium strength and, interestingly, lower amounts of strong Lewis acid sites than the rest of the catalysts (Table 6.3). These results were confirmed by testing these samples as catalysts for the two acid-catalyzed reactions previously commented: the isomerization of styrene oxide to β-phenylacetaldehyde and the styrene oxide ring opening to 2-ethoxy-2-phenylethanol [86].

From these results, we developed a second work with the aim to extend the investigation of the use of microwaves during dealumination in HCl medium to other two zeolite structures: ZSM-5 ($Si/Al = 20$) and beta ($Si/Al = 10$) [88]. Dealumination experiments were performed at a short treatment time (15 min) in autoclave to compare with the previous mordenite dealumination results. Zeolites were also dealuminated in an acidic medium by conventional heating under identical conditions for comparison. Table 6.4 shows some characterization results of the samples. Commercial mordenites, beta, and ZSM-5 were designated as M, B, and Z, respectively. The nomenclature of the samples partially dealuminated by conventional heating starts with C, while that of the samples partially dealuminated by microwaves starts with Mw.

The extent of dealumination was a function of the zeolite structure and the heating method used. Beta zeolite was easier to dealuminate than the mordenite, whereas dealumination of ZSM-5 was very low. This can be related to the flexibility of each zeolite

Table 6.4 Characterization of commercial and partially dealuminated samples by XRF, XRD, nitrogen physisorption, and ^{1}H NMR techniques

Samples	Si/Al	Crystallinity	BET area	^{1}H NMR
	(XRF)	(%) (XRD)	(m^2/g)	(ppm)
M	6.5	100	303	4.58
CM	11.2	73	376	5.07
MwM	15.8	70	419	5.11
B	10.0	100	573	4.24
CB	110.8	62	554	4.17
MwB	121.9	69	451	3.98, 2.22
Z	20.0	100	300	3.79
CZ	21.3	99	306	4.98
MwZ	22.4	100	300	5.47, 2.27, 1.36

framework and the accessibility of the aluminium atoms, depending on the pore arrangement and sizes. Zeolite beta, for example, has a three-dimensional 12-ring pore system (straight channels of diameter 6.6 × 6.7 Å and sinusoidal channels of diameter 5.6 × 5.6 Å), and because of this property, its framework is very flexible. The zeolite mordenite has a onc-dimensional pore system with main channels of diameter 6.7 × 7.0 Å and compressed channels of diameter 2.6 × 5.7 Å, whereas ZSM-5 has a three-dimensional 10-ring pore system with channels of diameter 5.1 × 5.5 Å. Microwaves led to faster dealumination than conventional heating for the three zeolites [88].

The mordenite treated with HCl under microwave irradiation showed higher mesoporosity, higher surface area, and Brønsted acid sites with higher strength, as confirmed from ^{1}H NMR, than the mordenite treated by conventional heating. After acid treatment with microwaves, beta zeolite exhibited a lower surface area accompanied with some variations in micromesoporosity and similar Brønsted acidity than the conventionally acid-treated beta. However, some silanols, with very weak acidity, were only detected for the microwaved sample (Table 6.4). Lastly, ZSM-5 treated in an acidic medium with microwaves had similar surface and porosity characteristics than the conventionally heated sample, probably due to the very low dealumination achieved, but higher Brønsted acidity, which can be explained by some proton exchange that occurred during dealumination due to the acid medium used. Interestingly, for the microwaved ZSM-5 sample, new signals appeared between 1.2 and 2.3 ppm, which can be associated to the presence of some external silanol groups that appeared after acid treatment. Additionally, a slightly higher amount of extra-framework Al and, therefore, slightly higher Lewis acidity were observed for this sample [88].

The effect of microwaves on dealumination, when compared to conventional heating, can be mainly explained by taking into account that chemicals do not interact equally with the commonly used microwave frequencies for dielectric heating, and consequently, selective heating may be achieved [89, 90]. This may lead to a significantly different temperature regime, caused by microwave dielectric heating, being the main contributing factor to the acceleration observed with respect to conventional heating.

Partially dealuminated samples were also tested as catalysts for the two acid-catalyzed reactions that we use to check Brønsted acidity and Lewis acidity: the isomerization of styrene oxide to β-phenylacetaldehyde and the styrene oxide ring opening to 2-ethoxy-2-phenylethanol [88]. Catalytic results were correlated with the number and strength of acid sites, together with the accessibility of the reagents to the acid sites, depending on the zeolite structure [88].

6.3.3 *Microwave-Assisted Sulfonic Acid Functionalization of Zeolites by Postsynthesis*

Commercial beta (Si/Al = 10), ZSM-5 (Si/Al = 20), and mordenite (Si/Al = 6.5) zeolites were successfully sulfonated with 2-(4-chlorosulfonylphenyl) ethyltrimethoxysilane (CSPTMS) in one step, for the first time, in our research group, by a postsynthesis method with microwaves [91].

The XPS technique is useful for evaluating qualitatively the type of sulfur species and measuring quantitatively the sulfonic acid groups near the surface region [92, 93]. The S2p X-ray photoelectron spectra of the most representative sulfonated samples (Fig. 6.8a) only showed one peak at ca. 168–169 eV associated with sulfate

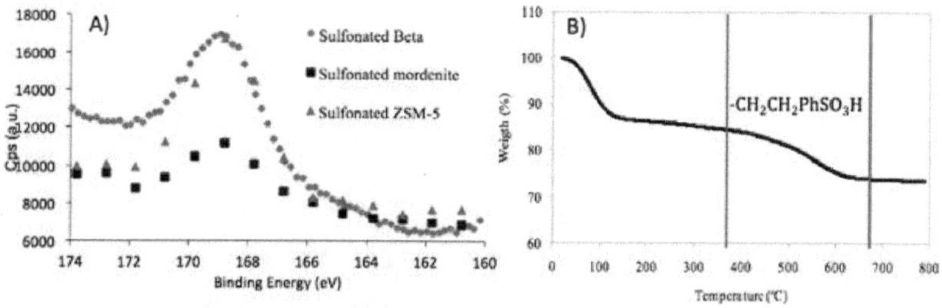

Figure 6.8 (a) XP spectra of the zeolites sulfonated with microwaves in the S2p core level region; (b) TGA curve for sulfonated beta.

(S^{6+}) species due to sulfonic ($-SO_3H$) acid groups [92, 93]. Additionally, the TGA technique was used to confirm and quantify the introduction of the sulfonic acid groups since the weight loss observed between 360°C and 660°C for the organosulfonated samples can be related to the loss of organosulfonic acid groups (Fig. 6.8b).

Taking into account that the main difficulty to introduce bulky organic species in zeolites is their microporous structure together with the lack of reactant silanol groups (≡Si–OH) [54], the zeolite dealumination occurred due to the acidic medium used during acid functionalization appears to be a key factor in the generation of new silanol groups that can react with the organosulfonating agent (Fig. 6.9).

Dealumination during sulfonation was confirmed by FTIR, XRF, and XPS techniques. The highest incorporation of sulfonic acid groups was achieved with different amounts of sulfonating agent for each zeolite structure. This can be explained by the different dealumination degree suffered by the starting samples at the acidic conditions used during sulfonation. Additionally, we observed that the arrangement and size of the pores also affected the accessibility of the silanol groups to the sulfonating agent. The presence of the sulfonic acid groups, which resulted in a higher number and strength of Brønsted acid sites than those of the starting commercial zeolites, resulted in high active and high selective catalysts to di- and triether of glycerol (h-GTBE), in spite of their microporosity, for the etherification of glycerol with isobutene [91].

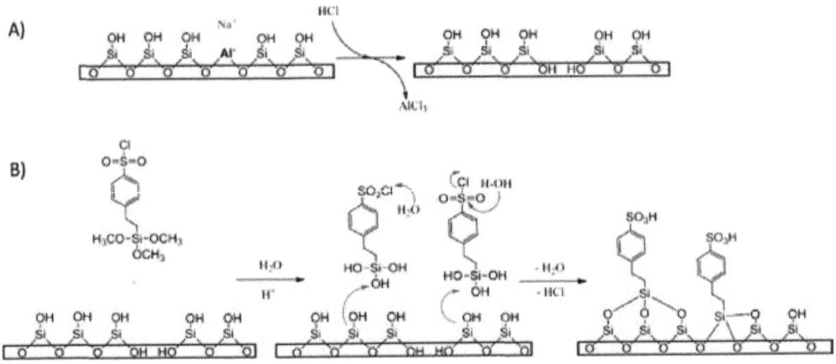

Figure 6.9 (a) Formation of silanols during dealumination in acidic medium and (b) reaction of silanols with CSPTMS to form the sulfonating groups.

Table 6.5 Characterization of the acidity of commercial and sulfonated beta samples

Samples	S/Si surface ratio (XPS)	Sulfur content (mmol organic sulfonic acid group/g sample) (TGA)	Acid capacity (meq H^+/g) (potentiometric titration)
Beta	-	-	-
CS-beta	0.06	0.74	0.75
MwS-beta	0.06	0.70	0.72

For commercial beta zeolite, we also performed this sulfonation process by conventional heating for comparison [94]. Although the two samples had similar acidity (Table 6.5), the sample sulfonated with microwaves (MwS-Beta) presented a higher surface area (micropore and external), a slightly higher pore volume, slightly higher mesoporosity, and a less decrease in zeolite crystallinity than the conventional sulfonated sample (CS-Beta) (Table 6.6).

Interestingly, the sample sulfonated with microwaves showed a remarkably higher selectivity to h-GTBE (DTBG + TTBG) than that functionalized by conventional heating after 4 h of reaction (Fig. 6.10).

Taking into account that both catalysts had similar acidity (Table 6.5), the higher selectivity to h-GTBE of catalyst beta-MwS could be attributed to a higher accessibility of its acid sites, since this catalyst showed a slightly higher pore volume (Table 6.6),

Table 6.6 Surface characterization of commercial and sulfonated beta samples

Samples	Crystallinity (%) (XRD)	BET area (m^2/g)	Micropore area (m^2/g)	External surface area (m^2/g)	Pore volume (cm^3/g)
Beta	100	584	462	122	0.23
CS-beta	27	330	243	87	0.14
MwS-beta	45	505	394	111	0.18

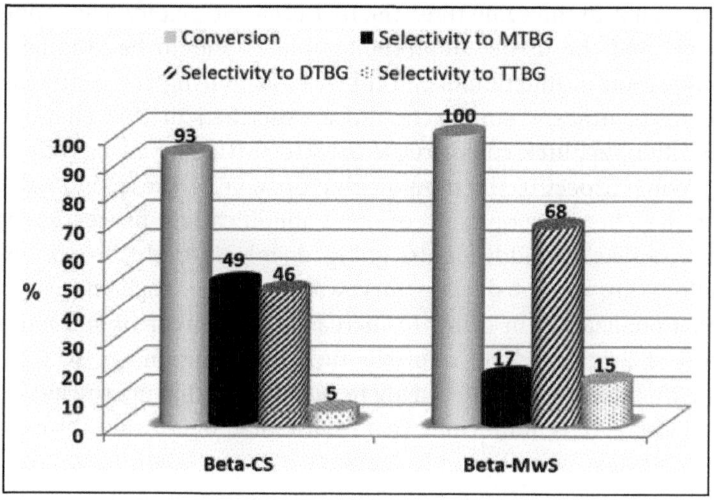

Figure 6.10 Catalytic activity of CS-beta and MwS-beta for the etherification of glycerol with isobutene. MTBG: glycerol monoether; DTBG: glycerol diether; TTBG: glycerol triether.

slightly higher mesoporosity, and a higher external surface area after sulfonation than the beta catalyst obtained by sulfonation using conventional heating (Table 6.6). The effect of microwaves on sulfonic acid functionalization, and therefore on the catalytic activity results, when compared to conventional heating, can be mainly related to the most homogeneous heating achieved with microwaves [2, 95], which favors crystallinity of the synthesized samples [2, 77]. In addition, for modified beta samples, microwaves in acidic medium favored dealumination, and therefore the formation of slight higher mesoporosity, as previously reported [88].

6.4 Future Prospects

From the results shown in previous sections we can realize the importance of introducing the use of microwaves in the synthesis of clays (cationic and anionic) and in the modification of zeolites. In the field of clays, the following steps should be used to optimize the synthesis parameters of delaminated saponites using microwaves during hydrothermal aging in order to increase delamination and favor the incorporation of aluminium in the tetrahedral positions, improving, at the same time, the BET area and acidity. The influence of pH and the use of different templates should be studied. The application of functionalization processes during clay synthesis or by postsynthesis, similar to those described in this chapter for modifying zeolites, can increase the strength acidity in cationic clays.

With respect to the modification of zeolites, the main challenge can be the development of new modification processes using microwaves, in addition to going deeply into the use of other sulfonating agents during microwave-assisted postsynthesis acid functionalization in order to increase the number of sulfonic acid sites incorporated or improve their distribution, as well as to functionalize the zeolite during its synthesis using microwaves.

Finally, it is important to remark that microwave processing of materials is still mainly restricted to the laboratory, although in the past years, the development of industrial microwave ovens has increased considerably, and actually, they are used in food processing and drying applications.

Acknowledgments

The authors are grateful for the financial support of Ministerio de Ciencia e Innovación and FEDER funds (CTQ2005-02384/PPQ, CTQ2008-04433/PPQ, and CTQ2011-24610).

References

1. Komarneni, S., Roy, R., and Li, Q. H. (1992). Microwave–hydrothermal synthesis of ceramic powders, *Mater. Res. Bull.*, **27**, pp. 1393–1405.

2. Kingston, H. M., and Haswell, S. J. (1997). *Microwave-Enhanced Chemistry Fundamentals, Sample Preparation and Applications* (American Chemical Society, Washington DC).
3. Komarneni, S., Li, Q. H., and Roy, R. (1996). Microwave–hydrothermal processing of layered anion exchangers, *J. Mater. Res.*, **11**, pp. 1866–1869.
4. Tichit, D., Rolland, A., Prinetto, F., Fetter, G., Martínez-Ortíz, M. J., Valenzuela, M. A., and Bosch, P. (2002). Comparison of the structural and acid-base properties of Ga- and Al-containing layered double hydroxides obtained by microwave irradiation and conventional ageing of synthesis gel, *J. Mater. Chem.*, **12**, pp. 3832–3838.
5. Climent, M. J., Corma, A., Iborra, S., Epping, K., and Velty, A. (2004). Increasing the basicity and catalytic activity of hydrotalcites by different synthesis procedures, *J. Catal.*, **225**, pp. 316–326.
6. Rivera, A., Fetter, G., and Bosch, P. (2006). Microwave power effect on hydrotalcite synthesis, *Microporous Mesoporous Mater.*, **89**, pp. 306–314.
7. Pilter, Z., Szabó, S., Hasznos-Nezdei, M., and Pallai-Varsányi, E., (2000). X-ray diffraction study of the effect of microwave treatment of zeolite Na-A, *Microporous Mesoporous Mater.*, **40**, pp. 257–262.
8. Romero, M. D., Gómez, J. M., Ovejero, G., and Rodríguez, A. (2004). Synthesis of LSX zeolite by microwave heating, *Mater. Res. Bull.*, **39**, pp. 389–400.
9. Newalkar, B. L., Olanrewaju, J., and Komarneni, S. (2001). Microwave-hydrothermal synthesis and characterization of zirconium-substituted SBA-15 mesoporous silica, *J. Phys. Chem. B*, **105**, pp. 8356–8360.
10. Trujillano, R., Rico, E., Vicente, M. A., Rives, V., Ciuffi, K. J., Cestari, A., Gil, A., and Korili S. A. (2011). Rapid microwave-assisted synthesis of saponites and their use as oxidation catalysts, *Appl. Clay Sci.*, **53**, pp. 326–330.
11. Vaccari, A. (1999). Clays and catalysis: a promising future, *Appl. Clay Sci.*, **14**, pp. 161–198.
12. Stepkowska, E. T., Pérez-Rodríguez, J. L., Maqueda, C., and Starnawska, E. (2004). Variability in water sorption and in particle thickness of standard smectites, *Appl. Clay Sci.*, **24**, pp. 185–199.
13. Bergaya, F., and Vayer, M. (1997). CEC of clays: measurement by adsorption of a copper ethylenediamine complex, *Appl. Clay Sci.*, **12**, pp. 275–280.

14. Czimerová, A., Bujdák, J., and Dohrmann, R. (2006). Traditional and novel methods for estimating the layer charge of smectites, *Appl. Clay Sci.*, **34**, pp. 2–13.
15. Vogels, R. J. M. J., Kloprogge, J. T., and Geus, J. W. (2005). Catalytic activity of synthetic saponite clays: effects of tetrahedral and octahedral composition, *J. Catal.*, **231**, pp. 443–452.
16. Zielke, R. C., and Pinnavaia, T. J. (1988). Modified clays for the adsorption of environmental toxicants: binding of chlorophenols to pillared, delaminated, and hydroxyinterlayeredsmectites, *Clays Clay Miner.*, **36**, pp. 403–408.
17. Vico, L., and Acebal, S. (2006). Some aspects about the adsorption of quinoline on fibrous silicates and Patagonian saponite, *Appl. Clay Sci.*, **33**, pp. 142–148.
18. Moreno, S., Kou, R. S., and Poncelet, G. (1996). Hydroconversion of heptanes over Pt/Al-pillared montmorillonites and saponites. A comparative study, *J. Catal.*, **162**, pp. 198–208.
19. Vaccari, A. (1998). Preparation and catalytic properties of cationic and anionic clays, *Catal. Today*, **41**, pp. 53–71.
20. Varma, R. S. (2002). Clay and clay-supported reagents in organic synthesis, *Tetrahedron*, **58**, pp. 1235–1255.
21. Aggarwal, V., Li, H., Boyd, S. A., and Teppen, B. J. (2006). Enhanced sorption of trichloroethane by smectite clay exchanged with Cs^+, *Environ. Sci. Technol.*, **40**, pp. 894–899.
22. LeBaron, P. C., Wang, Z., and Pinnavaia, T. J. (1999). Polymer-layered silicate nanocomposites: an overview, *Appl. Clay Sci.*, **15**, pp. 11–29.
23. Liu, X., and Breen, C. (2005). High-temperature crystalline phases in nylon-6/clay nanocomposites, *Macromol. Rapid Commun.*, **26**, pp. 1081–1086.
24. Chen, B., Evans, J. R. G., Greenwell, H. C., Boulet, P., Coveney, P. V., Bowden, A. A., and Whiting, A. (2008). A critical appraisal of polymer-clay nanocomposites, *Chem. Soc. Rev.*, **37**, pp. 568–594.
25. Benito, P., Labajos, F. M., Rocha, J., and Rives, V. (2006). Influence of microwave radiation on the textural properties of layered double hydroxides, *Microporous Mesoporous Mater.*, **94**, pp. 148–158.
26. Benito, P., Herrero, M., Barriga, C., Labajos, F. M., and Rives, V. (2008). Microwave-assisted homogeneous precipitation of hydrotalcites by urea hydrolysis, *Inorg. Chem.*, **47**, pp. 5453–5463.
27. Ayala, A., Fetter, G., Palomares, E., and Bosch, P. (2011). CuNi/Al hydrotalcites synthesized in presence of microwave irradiation, *Mater. Lett.*, **65**, pp. 1663–1665.

28. Alvarez, A., Trujillano, R., and Rives, V. (2013). Gallium-containing layered double hydroxides, *Appl. Clay Sci.*, **80–81**, pp. 326–333.

29. Campos-Molina, M. J., Santamaría-González, J., Mérida-Robles, J., Moreno-Tost, R., Albuquerque, M. C. G., Bruque-Gámez, S., Rodríguez-Castellón, E., Jiménez-López, A., and Maireles-Torres, P. (2010). Base catalysts derived from hydrocalumite for the transesterification of sunflower oil, *Energy Fuels*, **24**, pp. 979–984.

30. Cota, I., Ramírez, E., Medina, F., Sueiras, J. E., Layrac, G., and Tichit, D. (2010). New synthesis route of hydrocalumite-type materials and their application as basic catalysts for aldol condensation, *Appl. Clay Sci.*, **50**, pp. 498–502.

31. Mora, M., López, M. I., Jiménez-Sanchidrián, C., and Ruiz, J. R. (2010). Ca/Al mixed oxides as catalysts for the Meerwein–Ponndorf–Verley reaction, *Catal. Lett.*, **136**, pp. 192–198.

32. Grover, K., Komarneni, S., and Katsuki, H. (2010). Synthetic hydrotalcite-type and hydrocalumite-type layered double hydroxides for arsenate uptake, *Appl. Clay Sci.*, **48**, pp. 631–637.

33. Kuwahara, Y., Ohmichi, T., Kamegawa, T., Mori, K., and Yamashita, H. (2010). A novel conversion process for waste slag: synthesis of a hydrotalcite-like compound and zeolite from blast furnace slag and evaluation of adsorption capacities, *Mater. Chem.*, **20**, pp. 5052–5062.

34. Vieille, L., Rousselot, I., Leroux, F., Besse, J. P., and Taviot-Guého, C. (2003). Hydrocalumite and its polymer derivatives. 1. Reversible thermal behavior of Friedel's salt: a direct observation by means of high-temperature in situ powder X-ray diffraction, *Chem. Mater.*, **15**, pp. 4361–4368.

35. Marxwell, I. E., and Stork, W. H. J. (2001). Hydrocarbon processing with zeolites. In *Introduction to Zeolite Science and Practice, Studies in Surface Science and Catalysis,* Vol. 137, Van Bekkum, H., Flanigen, E. M., Jacobs, P. A., and Jansen, J. C., eds. (Elsevier Science, Amsterdam), pp. 747–820.

36. Corma, A. (1995). Inorganic solid acids and their use in acid-catalyzed hydrocarbon reactions, *Chem. Rev.*, **95**, pp. 559–614.

37. Townsend, R. P., and Coker, E. N. (2001). Ion exchange in zeolites. In *Introduction to Zeolite Science and Practice, Studies in Surface Science and Catalysis*, Vol. 137, Van Bekkum, H., Flanigen, E. M., Jacobs, P. A., and Jansen J. C., eds. (Elsevier Science, Amsterdam), pp. 467–524.

38. Barrer, R. M. (1978). *Zeolites and Clay Materials as Sorbents and Molecular Sieves* (Academic Press, London, UK).

39. De Lucas, A., Valverde, J. L., Dorado, F., Romero, A., and Asencio, I. (2005). Influence of the ion exchanged metal (Cu, Co, Ni and Mn) on the selective

catalytic reduction of NO_X over mordenite and ZSM-5, *J. Mol. Catal. A: Chem.*, **225**, pp. 47–58.

40. Dedecek, J., and Wichterlova, B. (1999). Co^{2+} ion siting in pentasil-containing zeolites. I. Co^{2+} ion sites and their occupation in mordenite, a vis–NIR diffuse reflectance spectroscopy study, *J. Phys. Chem. B*, **103**, pp. 1462–1476.
41. Kieger, S., Delahay, G., Coq, B., and Neveu, B. (1999). Selective catalytic reduction of nitric oxide by ammonia over Cu-FAU catalysts in oxygen-rich atmosphere, *J. Catal.*, **183**, pp. 267–280.
42. Karge, H. G. (1997). Post-synthesis modification of microporous materials by solid-state reactions, *Stud. Surf. Sci. Catal.*, **105**, pp. 1901–1948.
43. Whyte, T. E., Betta, R. D., Derouane, E. G., and Baker, R. T. K. (1984). *Catalytic Materials: Relationship between Structure and Reactivity* (American Chemical Society, Washington DC).
44. Szostak, R. (2001). Chapter 16, Secondary synthesis methods. In *Introduction to Zeolite Science and Practice, Studies in Surface Science and Catalysis,* Vol. 137, Van Bekkum, H., Flanigen, E. M., Jacobs, P. A., and Jansen J. C., eds. (Elsevier Science, Amsterdam), pp. 261–297.
45. Müller, M., Harvey, G., and Prins, R. (2000). Comparison of the dealumination of zeolites beta, mordenite, ZSM-5 and ferrierite by thermal treatment, leaching with oxalic acid and treatment with $SiCl_4$ by ^{1}H, ^{29}Si and ^{27}Al MAS NMR, *Microporous Mesoporous Mater.*, **34**, pp. 135–147.
46. Triantafillidis, C. S., Vlessidis, A. G., and Evmiridis, N. P. (2000). Dealuminated H–Y zeolites: influence of the degree and the type of dealumination method on the structural and acidic characteristics of H–Y zeolites, *Ind. Eng. Chem. Res.*, **39**, pp. 307–319.
47. Shu, S., Husain, S., and Koros, J. (2007). Sonication-assisted dealumination of zeolite a with thionyl chloride, *Ind. Eng. Chem. Res.*, **46**, pp. 767–772.
48. Moreno, S., and Poncelet, G. (1997). Dealumination of small- and large-port mordenites: a comparative study, *Microporous Mesoporous Mater.*, **12**, pp. 197–222.
49. Nesterenko, N. S., Thibault-Starzyk, F., Montouillot, V., Yuschenko, V. V., Fernandez, C., Gilson, J. P., Fajula, F., and Ivanova, I. I. (2004). Accessibility of the acid sites in dealuminated small-port mordenites studied by FTIR of co-adsorbed alkylpyridines and CO, *Microporous Mesoporous Mater.*, **71**, pp. 157–166.

50. Chung, K. H. (2008). Dealumination of mordenites with acetic acid and their catalytic activity in the alkylation of cumene, *Microporous Mesoporous Mater.*, **111**, pp. 544–550.

51. Viswanadham, N., and Kumar, M. (2006). Effect of dealumination severity on the pore size distribution of mordenite, *Microporous Mesoporous Mater.*, **92**, pp. 31–37.

52. Jones, C. W., Tsuji, K., and Davis, M. E. (1998). Organic-functionalized molecular sieves as shape-selective catalysts, *Nature*, **393**, pp. 52–54.

53. Holmberg, B. A., Hwang, S. J., Davis, M. E., and Yan, Y. (2005). Synthesis and proton conductivity of sulfonic acid functionalized zeolite BEA nanocrystals, *Microporous Mesoporous Mater.*, **80**, pp. 347–356.

54. Lee, D. H., Choi, M., Yu, B. W., and Ryoo, R. (2009). Organic functionalization of mesopore walls in hierarchically porous zeolites, *Chem. Commun.*, pp. 74–76.

55. Chanaa, M. B., Lallemant, M., Simonot-Grange, M. H., and Beretrand, G. (1987). Sur quelques aspects thermodynamiques et cinetiques de la deshydratation de la zeolithe 13X sous champ microonde, *Thermochim. Acta*, **115**, pp. 317–333.

56. Whittington, B. I., and Milestone, N. B. (1992). The microwave heating of zeolites, *Zeolites*, **12**, pp. 815–818.

57. Arafat, A., Jansen, J. C., Ebaid, A. R., and van Bekkum, H. (1993). Microwave preparation of zeolite Y and ZSM-5, *Zeolites*, **13**, pp. 162–165.

58. Chang, C., Zhang, W. F., Zhang, L. H., Li, R. S., Tian, M. W., Yang, X. W., and Hou, L. (1996). Studies on preparation of LaHZSM-5 by microwave technique and its cracking properties, *Chem. J. Chin. Univ.*, **17**, pp. 1914–1917.

59. Kuroda, Y., Okamoto, T., Kumashiro, R., Yoshikawa, Y., and Nagao, M. (2002). Microwave-assisted simple ion-exchange of ZSM-5-type zeolites with copper ions and their specific adsorption properties for N_2 molecules at room temperature, *Chem. Commun.*, pp. 1758–1759.

60. Xiao, F. S., Xu, W. G., Qiu, S. L., and Xu, R. R. (1994). The high dispersionof $CuCl_2$ in NaZSM-5 by using microwave technique, *Catal. Lett.*, **26**, pp. 209–215.

61. Xiao, F. S., Xu, W. G., Qiu, S. L., and Xu, R. R. (1994). New route for dispersion of inorganic salts onto the channel surfaces of microporous crystals: high dispersion of $CuCl_2$ in zeolites using a microwave technique, *J. Mater. Chem.*, **4**, pp. 735–739.

62. Romero, M. D., Ovejero, G., Uguina, M. A., Rodriguez, A., and Gómez, J. M. (2004). Fast removal of the acid properties in the NaX zeolite by ion-exchange under microwave heating, *Catal. Commun.*, **5**, pp. 157–160.
63. Li, X., Zhang, X., and Lei, L. (2009). Preparation of CuNaY zeolites with microwave irradiation and their application for removing thiophene from model fuel, *Sep. Purif. Technol.*, **64**, pp. 326–331.
64. Zendehdel, M., Cruciani, G., and Dondi, M. (2012). Appraisal of microwave-assisted ion-exchange in mordenite by crystal structure analysis, *J. Porous Mater.*, **19**, pp. 361–368.
65. Song, H., Wan, X., and Sun, X. (2013). Preparation of AgY zeolites using microwave irradiation and study on their adsorptive desulphurisation performance, *Can. J. Chem. Eng.*, **91**, pp. 915–923.
66. Chandra Shekara, B. M., Jai Prakash, B. S., and Bhat, Y. S. (2011). Dealumination of zeolite BEA under microwave irradiation, *ACS Catal.*, **1**, pp. 193–199.
67. Granquist, W. T., and Pollack, S. S. (1959). A study of the synthesis of hectorite, *Clays Clay Miner.*, **8**, pp. 150–169.
68. Vicente, I., Salagre, P., Cesteros, Y., Guirado, F., Medina, F., and Sueiras, J. E., (2009). Fast microwave synthesis of hectorite, *Appl. Clay Sci.*, **43**, pp. 103–107.
69. Vicente, I, Salagre, P., and Cesteros, Y. (2010). Preparation of pure hectorite using microwaves, *Phys. Procedia*, **8**, pp. 88–93.
70. Torii, K., Iwasaki, T., Onodera, Y., and Hatakeda, K. (1991). Mesoporous-materials produced from hydrothermally synthesized hectorites, *Stud. Surf. Sci. Catal.*,**60**, pp. 81–88.
71. Sánchez, T., Salagre, P, and Cesteros, Y. (2013). Ultrasounds and microwave-assisted synthesis of mesoporoushectorites, *Microporous Mesoporous Mater.*, **171**, pp. 24–34.
72. Iwasaki, T., Reinikainen, M., Onodera, Y., Hayashi, H., Ebina, T., Nagase, T., Torii,K., Kataja, K., and Chatterjee, A. (1998). Use of silicate crystallite mesoporous material as catalyst support for Fischer-Tropsch reaction, *Appl. Surf. Sci.*, **130–132**, pp. 845–850.
73. Vogels, R. J. M. J., Breukelaar, J., Kloprogge, J. T., Jansen, J. B. H., and Geus, J. W. (1997). Hydrothermal crystallisation of ammonium-saponite at 200°C and autogeneous water pressure, *Clays Clay Miner.*, **45**, pp. 1–7.
74. Vicente, I., Salagre, P., Cesteros, Y., Medina, F., and Sueiras, J. E. (2010). Microwave-assisted synthesis of saponite, *Appl. Clay Sci.*, **48**, pp. 26–31.

75. Gebretsadik, F. B., Salagre, P., and Cesteros, Y. (2014). Use of polymer as template in microwave synthesis of saponite. Study of several factors of influence, *Appl. Clay Sci.*, **87**, pp. 170–178.

76. Bisio, C., Gatti, G., Boccaleri, E., Marchese, L., Superti, G. B., Pastore, H. O., and Thommes, M. (2008). Understanding physic-chemical properties of saponite synthetic clays, *Microporous Mesoporous Mater.*, **107**, pp. 90–101.

77. Bergadà, O., Vicente, I., Salagre, P., Cesteros, Y., Medina, F., and Sueiras, J. E. (2007). Microwave effect during aging on the porosity and basic properties of hydrotalcites, *Microporous Mesoporous Mater.*, **101**, pp. 363–373.

78. Pérez-Barrado, E., Pujol M. C., Aguiló, M., Cesteros, Y., Díaz, F., Pallarés, J., Marsal, L. F., and Salagre, P. (2013). Fast aging treatment for the synthesis of hydrocalumites using microwaves, *Appl. Clay Sci.*, **80–81**, pp. 313–319.

79. Bergadà, O., Boix, E., Salagre, P., Cesteros, Y., Medina, F., and Sueiras, J. E. (2009). Acidity properties of Ni-exchanged mordenites prepared with and without microwaves, *Appl. Catal. A: Gen.*, **368**, pp. 163–169.

80. Suzuki, M., Tsutsumi, K., and Takahashi, H. (1982). Characterization and catalytic activity of nickel-zeolite catalysts. I. Reduction properties of nickel ions in zeolites, *Zeolites*, **2**, pp. 51– 58.

81. Meek, T. T., Blake, R. D., Katzs, J. D., Bradbury, J. R., and Brooks, M. H. (1988). Action diffusion in glass heated in a microwave field, *J. Mater. Sci. Lett.*, **7**, pp. 928–931.

82. Godelitsas, A., Charistos, D., Tsipis, A., Tsipis, C., Filippidis, C., Triantafyllidis, C., Manos, G., and Siapkas, D. (2001). Characterisation of zeolitic materials with a HEU-type structure modified by transition metal elements: definition of acid sites in nickel-loaded crystals in the light of experimental and quantum-chemical results, *Chem. Eur. J.*, **17**, pp. 3705–3721.

83. Mosqueda-Jiménez, B. I., Jentys, A., Seshan, K., and Lercher, J. A. (2003). Reduction of nitric oxide by propene and propane on Ni-exchanged mordenite, *Appl. Catal. B: Environ.*, **43**, pp. 105–115.

84. Medina, Y., Hernández, M., and Alcaraz, J. (2006). Dimerization of isobutene over nickel modified zeolites to obtain isooctane, *Catal. Lett.*, **110**, pp. 107–113.

85. Cardó, X., Bergadà, O., Cesteros, Y., and Salagre, P. (2012). Effect of catalyst acidity and porosity on the catalytic isomerization of linoleic acid to obtain conjugated linoleic acids (CLAs), *Chem. Eng. J.*, **183**, pp. 459–465.

86. González, M. D., Cesteros, Y., Salagre, P., Medina, F., and Sueiras, J. E. (2009). Effect of microwaves in the dealumination of mordenite on its surface and acidic properties, *Microporous Mesoporous Mater.*, **118**, pp. 341–347.

87. Imelik, B., and Vedrine, J. C. (1994). *Catalyst Characterization, Physical Techniques for Solid Materials* (Plenum Press, New York).

88. González, M. D., Cesteros, Y., and Salagre, P. (2011). Comparison of dealumination of zeolites beta, mordenite and ZSM-5 by treatment with acid under microwave irradiation, *Microporous Mesoporous Mater.*, **144**, pp. 162–170.

89. Sonnemans, M. H. W., den Heijer, C., and Crocker, M. (1993). Studies on the acidity of mordenite and ZSM-5. 2. Loss of Broensted acidity by dehydroxylation and dealumination, *J. Phys. Chem.*, **97**, pp. 440–445.

90. Magrioti Noronha, Z. M., Fontes Monteiro, J. L., and Gélin, P. (1998). The role of matrix embedding on the properties of steamed mordenites, *Microporous Mesoporous Mater.*, **23**, pp. 331–344.

91. González, M. D., Salagre, P., Taboada, E., Llorca, J., and Cesteros, Y. (2013). Microwave-assisted synthesis of sulfonic acid-functionalized microporous materials for the catalytic etherification of glycerol with isobutene, *Green Chem.*, **15**, pp. 2230–2239.

92. Cano-Serrano, E., Blanco-Brieva, G., Campos-Martin, J. M., and Fierro, J. L. G. (2003). Acid-functionalized amorphous silica by chemical grafting–quantitative oxidation of thiol groups, *Langmuir*, **19**, pp. 7621–7627.

93. Kim, Y., Choi, Y., Kim, H. K., and Lee, J. S. (2010). New sulfonic acid moiety grafted on montmorillonite as filler of organic–inorganic composite membrane for non-humidified proton-exchange membrane fuel cells, *J. Power Sources*, **195**, pp. 4653–4659.

94. González, M. D., Cesteros, Y., Llorca, J., and Salagre, P. (2012). Boosted selectivity toward high glycerol tertiary butyl ethers by microwave-assisted sulfonic acid-funcionalization of SBA-15 and beta zeolite, *J. Catal.*, **290**, pp. 202–209.

95. Gabriel, C., Gabriel, S., Grant, E. H., Halstead, B. S. J., and Mingos, D. M. P. (1998). Dielectric parameters relevant to microwave dielectric heating, *Chem. Soc. Rev.*, **27**, pp. 213–224.

Chapter 7

Microwave Engineering of Carbon Nanotubes

Madhulika Sinha and Yong-Chien Ling

Department of Chemistry, National Tsing Hua University, Hsinchu 30013, Taiwan
ycling@mx.nthu.edu.tw

This chapter portrays a brief literature on microwave (MW)-based methods for synthesis of carbon nanotubes (CNTs). It lays a future reference on the recent development in MW engineering for synthesis, purification, and functionalization of carbon nanomaterials. The topic describes the use of MWs in different aspects of CNT chemistry and its development over other conventional methods of synthesis. Here we discuss various advantages of using MW-assisted synthesis, purification, and functionalization techniques for CNTs on the basis of the time taken, convenience of the method, a quicker process, high purity of products, and cost effectiveness over other traditional processes.

7.1 Introduction

Carbon nanotubes (CNTs) are hollow cylinders formed by rolled graphite sheets, fundamentally made of an sp^2 bonded hexagonal

Microwave Engineering of Materials: From Mesoscale to Nanoscale
Edited by Erwann Guénin

ISBN 978-981-4669-42-9 (Hardcover), 978-981-4669-43-6 (eBook)
www.panstanford.com

lattice. CNTs were a fortunate accidental discovery made by Japanese physicist Iijima [1] in 1991, during the synthesis of fullerenes. Since their discovery, CNTs have attracted enormous attention in the field of nanotechnology worldwide, owing to their unique physical, chemical, mechanical, and thermal properties originating from the nanometer size, cylindrical structure, and high aspect ratio of length to diameter [2]. Great progress has been made toward the synthesis and application of CNTs. Upscale production of CNTs has been subjected to exploration. The present methods include the arc discharge method [3], the laser vaporization method [4], chemical vapor deposition [5], the solar energy route [6], and the fluidized bed method [7].

CNTs synthesized by conventional methods are found with structural defects, which prevents their much promised applications. Conventionally, a high-temperature thermal annealing process is needed for CNTs' defect reduction [8]. However, this usually leads to the degradation of their mechanical properties [9]. The major drawback lies in the poor solubility of CNTs in most of the common solvents due to their hydrophobic nature and their tendency to agglomerate (due to π–π stacking) within the matrix or in the processing solvent due to strong tube–tube interactions. Covalent functionalization of the side walls and tips has been proved to increase the solubility of CNTs [10]. A broad range of reactions have been used so far, like refluxing and/or sonication using solvents [11], biopolymers [12], and surfactants [13], which quite frequently make use of high temperature and/or pressure [14], long reaction time [15], or highly reactive species [16]. A notable downside of covalent functionalization is the disruption of the extended π-conjugation in CNTs, which may significantly alter their electrical or mechanical properties [17].

The microwave (MW) engineering technique such as heating, termed as the "Bunsen burner of the 21st century," has emerged as a valuable alternative in the synthesis of inorganic nanomaterials (NMs) [18]. The MW-based method has shown several benefits over conventional methods in the preparation of CNTs, such as being noninvasive, simple, instantaneous, rapid, and environmentally benign. MW-assisted preparation of inorganic nanostructured materials in the liquid phase is of great importance and significance

[19]. MW-assisted functionalization of CNTs reduces the reaction time and offers higher degrees of functionalization as compared to conventional thermal methods [20]. The energy from MWs can assist in assembling atoms into thin films directly on a substrate at significantly lower temperatures than conventional processes, potentially enabling plastic-based electronics [21]. In many cases, MW-assisted syntheses have yielded new nanoporous structures. By reduction of reaction time over an order of magnitude, it would be possible for continuous production to replace batch synthesis [22]. The competitive or time-bound effect of MW irradiation on functionalization and removal of initial functional groups on CNTs has been reported [23]. The MW-assisted, reagent-free method is fast and gives clean CNTs with significantly less purification-induced defects [24]. MW-assisted cycloaddition of pristine CNTs in solvent-free conditions under 1 h MW irradiation has been reported, implying large-scale synthesis of NMs is promising [16].

7.2 Microwave Chemistry

MW heating has been known as a faster, simpler, and more cost-effective process, often affording high-yield and high-purity products [25]. MW energy is a form of electromagnetic energy, which falls between infrared and radio waves in the electromagnetic spectrum. Its wavelength ranges between 0.01 and 1 m with corresponding frequencies between 0.3 and 300 GHz. The first use of MWs in chemical research was reported in 1971 [26]. For chemical synthesis purposes, MW reactors were operated at a frequency of 2.45 GHz, corresponding to a wavelength of 12.25 cm, in order to avoid interference with telecommunication and cellular phone frequencies. Figure 7.1 shows the difference between conventional and MW heating mechanisms.

The basis of MW-assisted synthesis lies in specific parameters such as temperature, rampant temperature, pressure, homogeneity, and choice of solvents or solvent-free system. The control of aforementioned parameters can lead to future advancement and modernization of MW methods in chemical research. Figure 7.2 illustrates the heating curves of MW heating and conventional

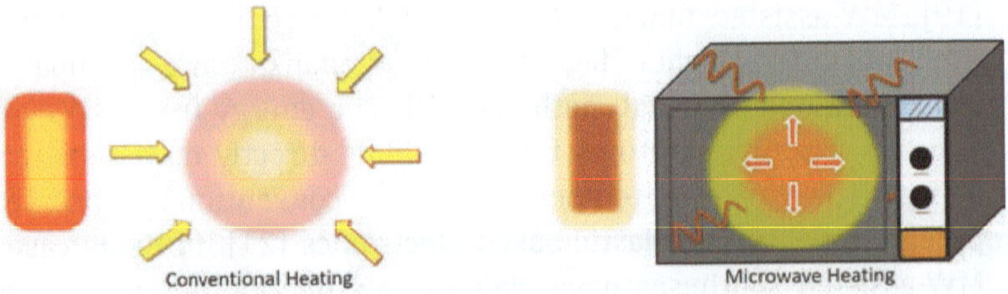

Figure 7.1 A comparative difference between conventional and MW heating mechanisms.

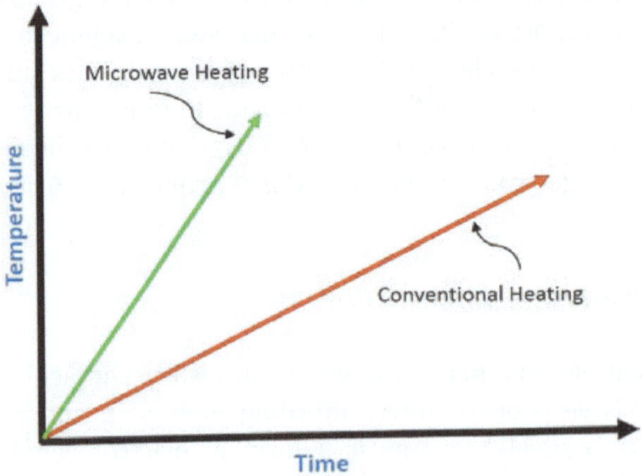

Figure 7.2 Schematic of heating curves of MW heating and conventional heating.

heating, showing that MW heating rapidly and simultaneously raises the temperature of the whole volume and leads to bulk heating, whereas conventional heating occurs when the reaction mixture is in contact with the vessel wall and heat transfer is relatively slow [27]. The rapid heating phenomenon by MW heating is more prominent for smaller objects. Different properties of MW heating, like direct, volumetric, instantaneous, and selective heating, have been largely discussed [28]. Direct heating involves direct interaction of MW irradiation with reactants [29]. Volumetric heating refers to uniform heating throughout the reactants in the case of homogeneous samples [30]. Direct and volumetric heating

Table 7.1 Experimental parameters for MW-assisted synthesis of CNTs, graphene, fullerene, and carbon dots

	Power (W)	Time (min)	Solvent	Starting material	Reference
CNT	300	60	Ag acetate	MWCNTs powder	Lin et al. [33]
Graphene	700	~1	No solvent	GO	Zhu and coworkers [34]
Fullerene	200–300	2–4	No solvent	Graphite powder	Hetzel et al. [35]
Carbon dot	700	10	Phosphate aq. solution	Glycerol	Liu et al. [36]

offers extremely fast heating and results in rapid temperature rise with power on and rapid temperature reduction with power off, yielding instantaneous heating [29, 31]. Selective heating involves direct heating of reactants having strong MW absorption [29, 32].

Different experimental parameters necessary for the MW-assisted synthesis of carbon-based NMs such as CNTs, graphene, fullerene, and carbon dots are listed in Table 7.1. The synthesis was carried out using a conventional domestic MW oven operating at 2.45 GHz frequency. The power setting is used to indicate the heating temperature [33–36].

Recent development in MW-assisted synthesis of NMs has found a huge base in the scientific community. Figure 7.3 shows the richness of MW-assisted synthesis, including reduction, condensation, cyclization, isomerization, and coupling reactions [18]. A simple and facile approach for the synthesis of palladium nanobelts, nanoplates, and nanotrees has been developed using vitamin B_1 in water without the use of any capping agent [37]. MW heating required lower amount of heat transfer via thermal conduction compared to conventional heating, implying that relatively larger samples can be efficiently heated and in far less time by MW heating.

7.3 General Synthesis of CNTs

7.3.1 *Properties of CNTs*

CNTs grabbed much attention by scientists around the world owing to their unique physical, chemical, mechanical, and thermal

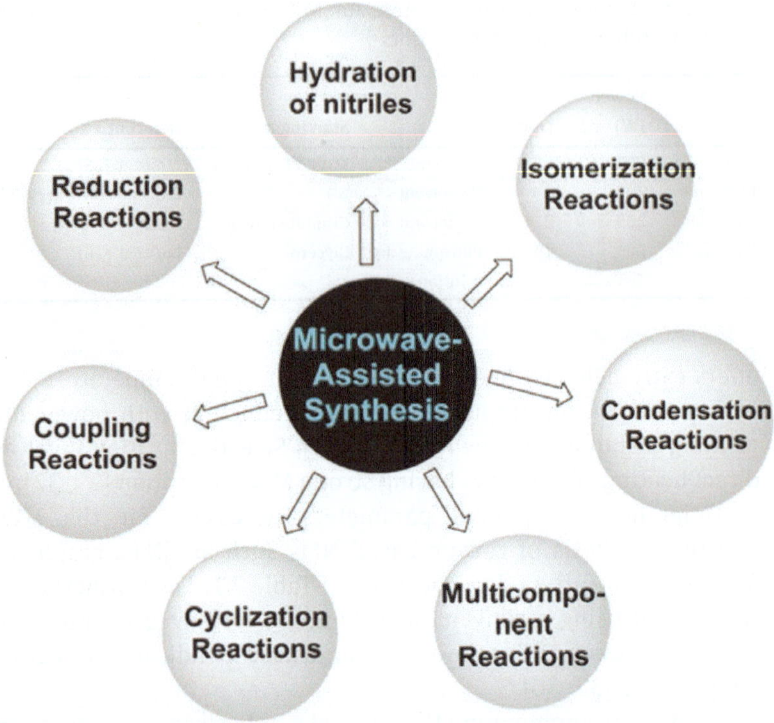

Figure 7.3 The richness of MA-assisted synthesis. Reprinted with permission from Ref. [18], Copyright 2014 American Chemical Society.

properties partially originated from their nanosize, cylindrical structure and high aspect ratio of length to diameter. CNTs are mostly found in two distinct types, single-walled carbon nanotubes (SWCNTs) and multiwalled carbon nanotubes (MWCNTs). SWCNTs are made of a single graphene sheet wrapped around to form a cylinder, whereas MWCNTs consists of concentrically nested cylinders with an interlayer spacing of 3.4 Å and a diameter in the range of 10–20 nm [2].

7.3.2 *Arc Discharge Method*

The arc discharge method was first described and used for the synthesis of CNTs by Iijima [1]. Two graphite electrodes are placed about 1 mm apart from each other in the presence of an inert gas atmosphere, like helium, at a pressure of 500 Torr [38]. An arc appeared between the two electrodes on applying a voltage of 20–25 V and a current of 50–120 A at very high temperatures, following carbon evaporation from the electrodes. The evaporated material recondensed on the cathode, which consists of CNTs and other carbon materials [39]. The method involves the use of electrodes with metals such as Ni, Fe, Co, Gd and Y for preparation of SWCNTs [40]. MWCNTs can be synthesized using arc discharge under liquid N_2 in an open vessel [41]. It is an easy method for the synthesis of CNTs at the cost of few limitations mostly due to the impurities formed during the synthesis process. Additionally, synthesis of MWCNTs and polyaromatic carbon shells by arc discharge in deionized water environment other than liquid nitrogen has also been reported [42, 43]. Bulk production of CNTs include the use of helium at a pressure of 500 or 700 Torr to offer high-yield SWCNTs, equivalent to 10 g/day [44, 45], confirming the finding that the vapor pressure of the metal catalyst is an important factor to grow SWCNTs [46].

7.3.3 *Laser Ablation Method*

Using similar operating conditions as the arc discharge method, the laser ablation method also involves condensing of the vapor formed from the graphitic carbon. The laser-ablation-assisted synthesis of CNTs is carried out at a temperature of 800°C–1500°C, where the target graphite electrodes are placed in a quartz tube and heated under an argon gas atmosphere at a pressure of 500 Torr. The carbon soot formed on a water-cooled Cu collector is carried by the argon gas [47]. In case the source graphite is doped with Ni, Co, or Pt, it results in the synthesis of SWCNTs. The process has resulted in 70%–90% conversion of graphite into CNTs. A "CVD-like" pulse laser deposition method has been used to deposit catalyst (nanoparticles) NPs on SiO_2/Si substrates, which are subsequently exposed to the carbon vapor produced by KrF laser ablation of a pure graphite target [48]. Laser ablation suffers several disadvantages, including very high cost of production due to the use of expensive laser and high power consumption. Carbonaceous soot contains other carbon and metal impurities, thus involving extensive purification processes adding to the already elevated production cost.

7.3.4 *Chemical Vapor Deposition Method*

Chemical vapor deposition uses metal catalysts to crack down the molecules of carbon sources to synthesize CNTs [49–51]. A temperature of 600°C–1000°C is used to heat the supported catalysts in a furnace in the presence of hydrocarbon gas for a period of time [50]. Following the heat treatment, the sample is cooled in an inert gas environment, avoiding the etching away of CNTs on reaction with oxygen. A variety of reagents such as methane, benzene, camphor, ethanol, ethane, alcohol, carbon monoxide, hexane, cyclohexane, naphthalene, and anthracene have been used as carbon source [52]. For the synthesis of SWCNTs, carbon monoxide and methane have been found to be effective. The most popular metals used to produce CNTs are iron, nickel, cobalt, and molybdenum [50]. MWCNTs are mainly formed at lower temperatures (300°C–800°C), whereas SWCNTs require higher temperatures (600°C–1000°C). Very recently, metal-free catalysts such as diamond, Al_2O_3, SiO_2, ZrO_2, and graphite have been used for the production of CNTs [53–61].

7.4 Microwave-Assisted Synthesis of CNTs

7.4.1 *CNT Applications*

CNTs reveal greater potential in applications such as electrical interconnects [62, 63], thermal interface materials [64, 65], high-performance fibers [66, 67], environmental monitoring [68, 69], tissue engineering [70], and health and biotechnology [71, 72]. In reality, synthesized CNTs aren't defect-free. An ultrafast MW annealing process to reduce the defect density in vertically aligned CNTs has been developed [73]. The as-annealed CNTs showed dramatic improvement in thermal stability, mechanical properties, and electrical conductivity. This can be the basis of defect-free synthesis, modification, and functionalization of CNTs for various

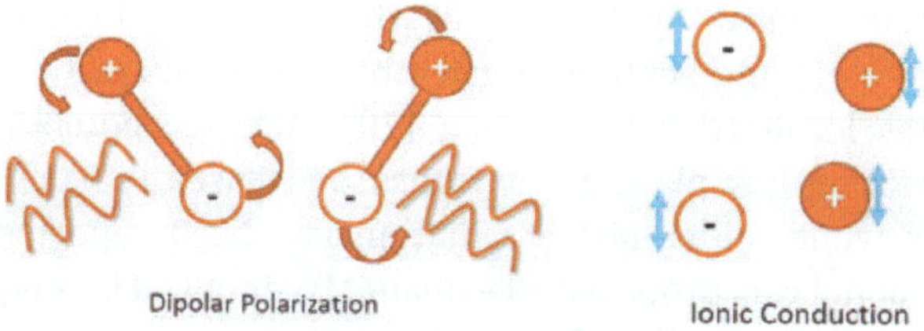

Figure 7.4 Two main heating mechanisms under MW irradiation: dipolar polarization and ionic conduction [78].

applications, lightening the future prospects for MW technology in material chemistry.

7.4.2 *Interaction Mechanisms between CNTs and MW*

The interactions between CNTs and MWs are of great interest from both a fundamental and a practical point of view [20]. Under MW irradiation, intense heat release, outgassing, and light emission are observed in CNTs due to their ability to absorb strong MW radiation [74–76]. CNTs temperature is rapidly promoted to above 1500°C due to intense release of heat, leading to reconstruction and higher crystallization of CNTs [73, 75]. Heating by MW irradiation takes place mainly by two main mechanisms, that is, dipolar polarization and ionic conduction (Joule heating) [77].

MW heating, as shown in Fig. 7.4, demonstrates two different heating mechanisms: Dipolar polarization involves charge carriers bound as dipoles and the electric field induces motion, whereas ionic conduction involves mobile charge carriers and the MW-induced current traveling in phase with the field, leading to resistive heating [28]. Dipole polarization results from the rotation of permanent molecular dipoles being hindered, presumably due to intermolecular forces. The frequency dependence of loss exhibits dispersions at particular frequencies. In contrast, conduction losses are attributed to the motion of free-charge-carrying currents induced by the electric field [74].

Metal and carbon impurities present in the electrically conductive CNTs support the conduction heating mechanism. No immediate and direct heating of material by MWs takes place; rather sample heating is caused by motion of the electrons induced by an electric field. Joule heating induced by the metal catalyst particles generates localized superheating [79]. Another potential source of generating gas plasma from absorbed gases (particularly H_2) in CNTs while under MW irradiation has been reported [75]. A simple model was proposed to explain MW-induced heating of CNTs through transformation of electromagnetic energy into mechanical vibrations [80]. Within this model, CNTs subjected to MW irradiation undergo ultraheating due to transverse parametric resonance, which arises from the polarization of CNTs in the MW field. The residual metals in the system appear to play the role of donors of sufficient free charge to expedite the MW coupling [80]. In a dense and viscous local environment, CNTs will not vibrate. Therefore, no heating occurs due to the absence of transverse resonance or the Joule effect. Perhaps this will help to explain the different results of absorption in the presence and absence of solvents [20].

7.4.3 *Effect of Modifiers*

In the case of reagents undergoing MW-assisted synthesis, the basic requirement is that the material be MW absorbing. Often conventional reagents are poor MW absorbers and therefore undergo a slow or no reaction in an MW field. In such an event, substances that can absorb MWs and act as a heat source are required. These materials are commonly referred to as susceptors, which are defined simply as substances that have the ability to absorb electromagnetic energy and convert it to heat, or any material with a high dielectric loss tangent [28]. Examples of such materials include carbon (graphite, amorphous carbon), silicon carbide, and copper (II) oxide.

Susceptors can be either used in direct contact with the sample (by mixing it with other reagents or used to surround a pellet) or kept separated by surrounding the reaction vessel. The use of carbon-based materials might create a problem due to the difficulty encountered in their removal from the required sample. The problem of precursor coupling and contamination due to susceptors can be avoided by using microwave-induced plasma (MIP) [81]. The MIP shown in Fig. 7.5 is used to replace susceptors in such reactions.

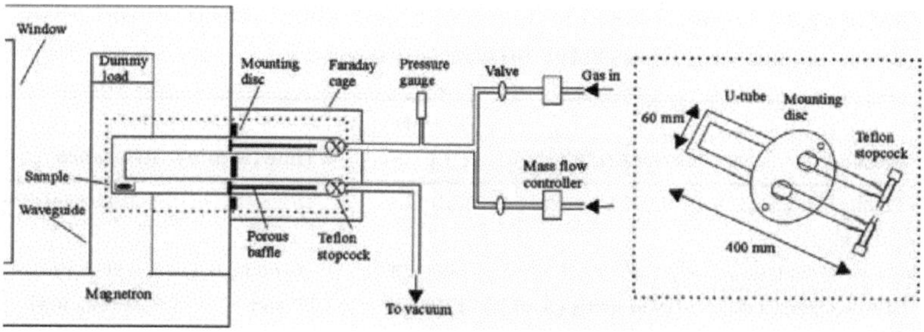

Figure 7.5 Schematic of an MIP reactor. The portion of the apparatus shown in the dotted box can be isolated to allow manipulation of air-sensitive products and precursors. Reprinted with permission from Ref. [82]. Copyright 2004, AIP Publishing LLC.

The plasma helps transfer energy between the MW radiation and the reactants, causing heating of the solids [82].

7.4.4 *MW-Assisted Purification of CNTs*

For successful applications, high-purity CNTs are a major requirement [83]. Many techniques have been employed for purification of CNTs. Most common techniques involve chemical oxidation, thermal oxidation, filtration [84], and chromatography. It usually employs liquid-phase refluxing in acids or gas-phase oxidation in oxidants (such as water vapor or ozone) [85]. However, these techniques have led to introduction of more defects in CNTs due to higher thermal budgets, long durations, and high sample loss under a nonoptimized purification condition, which fails to provide selectivity toward CNTs of different types [86, 87].

MW-assisted purification has been realized as an efficient method for purification of CNTs, offering CNTs with lesser defects [73] and minimal damage, demonstrated under optimized conditions as compared to conventional acid reflux methods [77].

Table 7.2 presents a comparison between traditional solution-based and MW-assisted purification methods. The poor dispersion of CNTs leading to their insolubility even in most common solvents is a matter of concern. The MW method has shown successful dispersion of CNTs at 100°C and 140°C for 10 and 30 min, showing better dispersion at 140°C for 10 and 30 min [88].

Table 7.2 Comparison of the traditional solution-based purification methods and the MW-assisted purification methods

Method	Solvent (mL)	Temperature (°C)	Time (h/min)	Reference
Chemical method (refluxing)	3 M HNO_3 (60 mL)	120	16 h	Lobach et al. [89]
Thermal oxidation	O_2/Air	500–800	40–60 min	Liu et al. [90]
MW assisted (with solvent)	15.8 M HNO_3 (5 mL)	120–180	30 min	Lee et al. [69]
MW assisted (solvent-free)	N/A	100–150	5 min	Chajara et al. [24]
Ultrasonic-assisted acid digestion	0.5 M HNO_3 (100 mL)	120	24 h	Tsai et al. [91]

Source: Modified from Ref. [88].

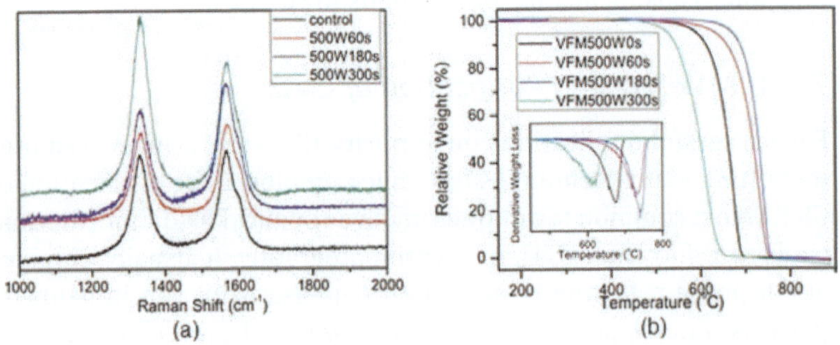

Figure 7.6 (a) Raman spectra of control VACNTs and VFM VACNTs with 500 W power for different durations and (b) TGA results of the control VACNTs and VFM VACNTs; (inset) the derivative weight loss plot against temperature. Reprinted with permission from Ref. [73], Copyright 2011 American Chemical Society.

An ultrafast MW-assisted annealing process has been developed to reduce the defect density in vertically aligned CNTs (VACNTs). Raman and thermogravimetric analyses (TGA) have shown a distinct defect reduction in MW-annealed variable-frequency microwave ([VFM]) CNTs (Fig. 7.6a,b) [73]. Within a short duration (i.e., $<$180 s) of MW irradiation, I_D/I_G and $I_{D'}/I_{G'}$ decrease with increase in time, indicating an improvement in the structural order of the VFM CNTs. TGA results indicate an effective improvement in oxidative stability under VFM (500 W) treatments and a shift in the peak temperature up to ~730°C and ~740°C, at 1 and 3 min, respectively. TGA has shown no detectable amorphous carbon. The high purity of VACNTs proves that the oxidative stability is a mark of pure crystalline CNTs. This result is helpful in successfully postulating the theory of reconstruction into the graphitic structure of defect sites on CNTs, explaining the decrease in I_D/I_G and $I_{D'}/I_{G'}$ [73]. An easy way for improving the purity of raw high-pressure carbon monoxide (HIPCO) nanotubes by reduction of their iron content using a domestic multimode MW oven has been reported, in which selective burning of metal particles helps to remove most of the iron content [92]. Another novel method of preparing narrow-diameter-distributed SWCNTs is achieved by combining MW irradiation (500 W, 2.45 GHz) with acid mixture (1:1 H_2SO_4:HNO_3)-assisted dispersion at 100°C for 0–25 min, making this method much more viable for achieving uniformly distributed CNTs required in scaled-up future applications [93].

A closed-vessel MW-assisted purification of MWCNTs at 180°C for 30 min was carried out in the presence of 20% H_2O_2. Better efficiency is obtained compared to HNO_3. H_2O_2 functions as a more favorable purification agent than HNO_3, indicating the effect of oxidation by the reagents, especially at higher temperature on defects decomposition. The TGA results show that prior to purification two significant weight loss temperatures occur at 195°C and 550°C, whereas after purification with H_2O_2, only one significant weight loss temperature is observed [94].

Figure 7.7a–c shows an optical image of purified CNT fibers sealed in a quartz tube before and after MW irradiation and scanning electron microscopy (SEM) images of purified CNT bundles before and after MW irradiation, respectively. Some carbon evaporates and deposits on the inner wall of the tube. Figure 7.7d exhibits a higher-magnification SEM image of the carbon flakes, showing that the carbon flakes stand separately to each other on the CNT bundles, which can be due to the electric repulsion generated by the electromagnetic wave [95].

Figure 7.8a demonstrates the presence of amorphous carbon and metal catalysts seen in the transmission electron microscopy (TEM) images. Figure 7.8b is the TEM image of purified SWCNTs

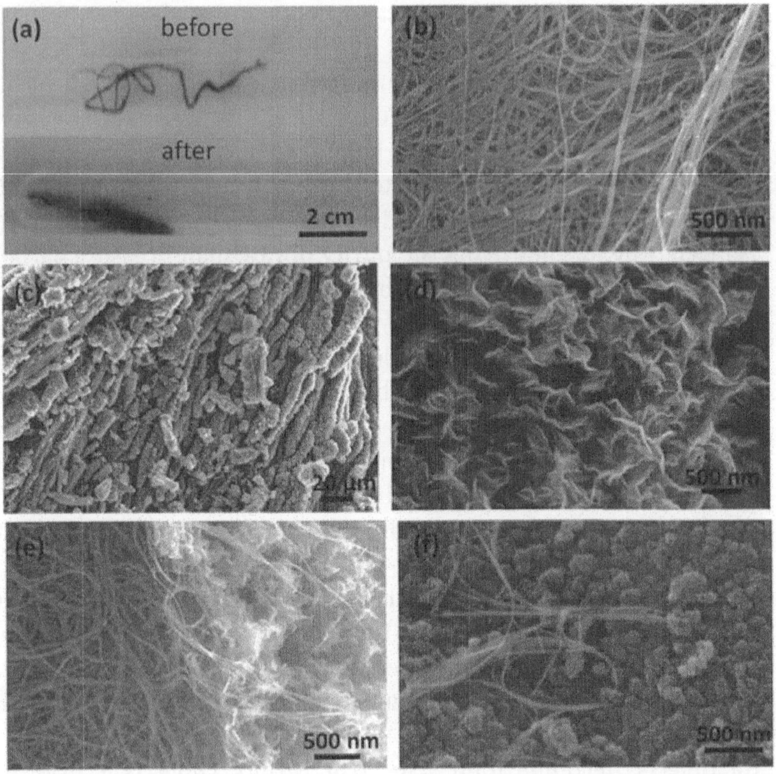

Figure 7.7 (a) Optical image of CNT fibers sealed in a quartz tube (up) before and (down) after MW irradiation. SEM images of purified CNT fibers (b) before and (c) after MW irradiation. High-magnification SEM images of (d) carbon flakes on the CNT bundles and (e) the interface between the CNT bundles and the carbon flakes. (f) SEM image of the carbon flakes collected from the inner wall of a quartz tube. Reproduced from Ref. [95], Copyright (2014), with permission of the Royal Society of Chemistry.

and double-walled carbon nanotubes (DWCNTs), whereas Fig. 8c,d shows the structural changes, mainly reconstruction, breakdown, and improved crystallinity of the sample [95]. The reagent-free MW-assisted purification process is fast and does not require time-consuming isolation and waste disposal. It gives cleaner CNTs, producing light emission under MW irradiation (Fig. 7.8e,f) with lower purification-induced defects as compared to CNTs purified by

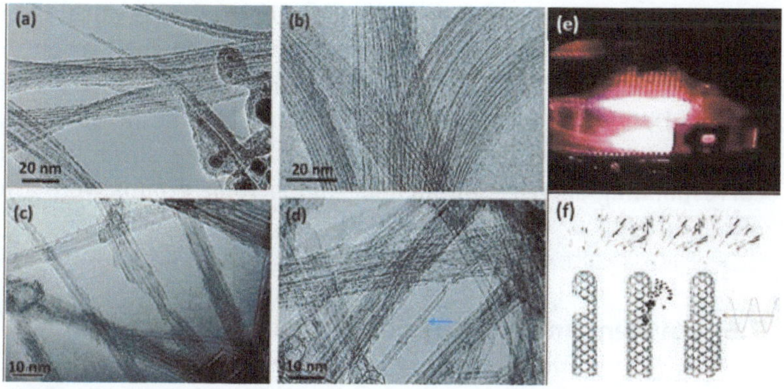

Figure 7.8 TEM images of (a) as-grown and (b) MW-assisted purified SWCNTs and DWCNTs before MW irradiation. (c) CNTs after MW excitation, showing that some bundles merged into large-diameter CNTs. (d) SWCNTs and DWCNTs with diameter varied from 3 to 10 nm, which are derived from melting and reconstruction of bundles. (e) Optical image of light emission from a purified CNT fiber under MW irradiation and (f) schematic of carbon exfoliation and light emission from CNTs under MW irradiation. Reproduced from Ref. [95], Copyright (2014), with permission of the Royal Society of Chemistry.

treatments involving oxidative and/ or acidic reagents. The results are obtained with the help of Raman, TGA, SEM, TEM, and infrared spectroscopies [95].

The morphology of MW-assisted CNTs compared to CNTs treated under conventional purification conditions suggests that thermal delamination of non-CNT carbon from CNTs is the key step. It is attenuated by solubilization of amorphous carbon, preventing reaggregation. The residual metal impurities form overheated aggregates and evaporate from carbon structures, forming tree-like precipitates, which can be handpicked [24].

Raman spectra show a significant difference between untreated and MW-purified MWCNTs with a decreased intensity of the D band, with the D/G ratio being ~0.8 for untreated and ~0.7 for MW-treated ones. The purification protocol can be repeated more than once. However, it leads to some damage to MWCNTs and increases the D band intensity [24]. Dilute base washing has been shown to remove oxidation debris by converting the acidic groups to their dissolvable conjugate salts. The effect of base washing on MW-induced carboxylated CNTs induces change in the carbon-to-oxygen ratio, the Brunauer–Emmett–Teller (BET) surface area, the ratio of D/G bands based on Raman measurements, and solubility. A dramatic increase in carboxylic groups on the surface and a corresponding decrease in phenolic and lactonic groups were also observed [96].

7.5 Covalent and Noncovalent Functionalization of CNTs

CNTs tend to form aggregates in solvents and become harder to purify or separate. MW-assisted functionalization and solubilization are a unique way to make dispersible and soluble CNTs, as they tend to integrate easily with organic, inorganic, and biological systems. Many studies have focused on covalent functionalization of the sidewalls and tips of CNTs. Some possible advantages include increased solubility in classical solvents, lower reaction time, cleaner products, and better conversion over conventional methods. Covalent functionalization is required to allow the unique properties of CNTs to be combined with those of other types of materials [16]. Covalent functionalization is based on the formation of a covalent linkage between functional entities and the carbon skeleton of CNTs, whereas noncovalent functionalization is mainly based on supramolecular complexation using various adsorption forces such as van der Waals forces, hydrogen bonds, electrostatic force, and π-stacking interactions [97].

7.5.1 *MW-Assisted Synthesis of CNTs-Based Nanocomposites*

A new approach that uses MWs to rapidly functionalize CNTs by using 1,3-dipolar cycloaddition of aziridines has been reported. An indirect proof of the covalent sidewall functionalization of the tubes provided through Raman and TGA studies. Raman spectra display a gradual increase in the intensity, with the number of groups added covalently to the surface of the CNTs. TGA thermograms show a weight loss of 12%, corresponding to one functional group, which

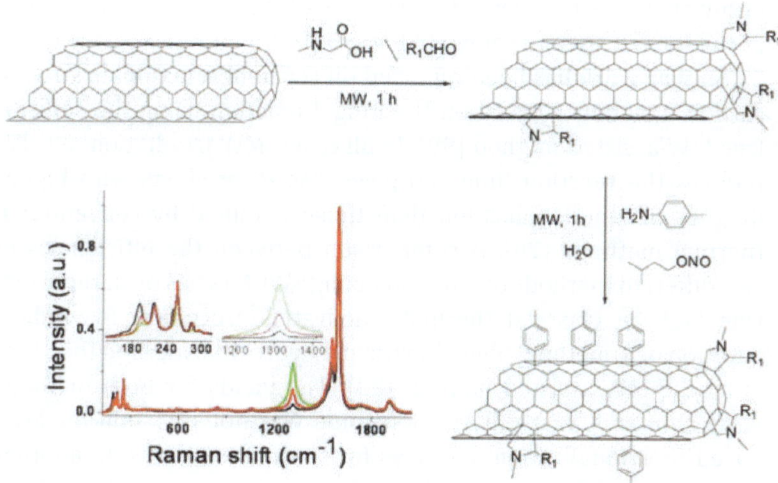

Figure 7.9 Schematic of MW-assisted double functionalization of SWCNTs and Raman spectra (λ_{exc} 633 nm) of pristine SWCNTs (black line), SWCNTs functionalized by 1,3-dipolar cycloaddition (red line), and doubly functionalized SWCNTs (green line). Reprinted with permission from Ref. [98], Copyright 2009 American Chemical Society.

explains that the MW-assisted functionalization preserves all types of CNTs [16].

Multifunctionalization of SWCNTs is demonstrated by the same research group. The MW-induced reactivity of SWCNTs is accomplished by using two different well-established addition reactions via a simple, fast, and environmentally friendly method. Figure 7.9 shows the scheme for 1,3-dipolar cycloaddition of azomethine that can be achieved in only 1 h using MWs. This is much shorter than the 5 days necessary for the conventional thermal conditions [98].

In solvent-free conditions, SWCNTs absorb the MW radiation directly and possibly take full advantage of the strong MW absorption typical of pristine CNTs. Moreover, solvent-free systems pave the way to green protocols and large-scale functionalization. The functionalization of arene radicals on SWCNTs shows a significant increase of the D band (the D/G ratio), explaining arene additions are more spread out, whereas 1,3-dipolar cycloaddition produces a smaller increase in the D band, implying that the cycloadditions occur at close sites, leaving large parts of the π-system untouched, as the Raman spectra shown in Fig. 7.9 [98].

Similar work has been reported on covalent functionalization of carbon nanohorns based on the Bingel reaction using the solvent-free MW-assisted method [99]. In all cases, MW irradiation of CNTs reduces the reaction time and gives rise to products with higher degrees of functionalization than those obtained by conventional thermal methods [20]. A comparison between the efficiencies of two different methods of functionalizing SWCNTs has been reported: One is by a classical thermally activated procedure for radical addition of 4-methoxyphenylhydrazine hydrochloride, and the other is by the MW-assisted method [23]. The yields for both methods were similar. A vast difference in time was noted: 5 min for MW-based functionalization and days for thermal methods. In another report, pyrrolidine units with gold-targeting groups have been added in a fast and convenient reaction. By controlling the amount of addends at the outermost layer of the MWCNTs, the amount of added groups determined by using TGA, after 2 h of MW heating at 200°C, was found to be in the same range as that obtained after 100–120 h of conventional heating of soluble and insoluble MWCNTs [100]. Usually MW-assisted heating is 50–60 times faster than thermal method and causes no damage, presumably due to heating to the structure of the added groups on MWCNTs [100].

7.5.2 *MW-Assisted Synthesis of CNTs-Based Hybrids*

Noncovalent linkages tend to form hybrids with CNTs mostly due to interactions such as van der Waals forces, electrostatic forces, π–π stacking, and hydrogen bonding. Modification of the surface functional groups to create CNT-based hybrid structures has been used to assemble and separate CNTs with different structures [101]. An ecofriendly approach of using MWs without the use of any corrosive oxidants (e.g., HNO_3) or ultrasonication for functionalization of MWCNTs with carbonyl, carboxyl, and hydroxyl or allyl termini groups was reported.

MW irradiation is an effective tool for solvent-free rapid decomposition of organic metal salts (such as silver acetate) in a solid mixture with various carbon and noncarbon substrates

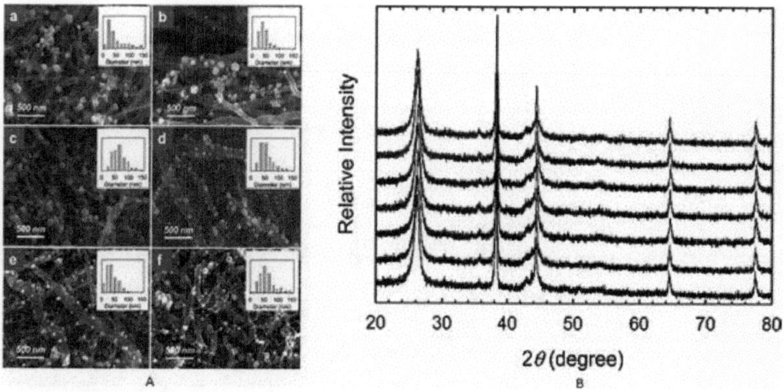

Figure 7.10 SEM images and XRD patterns of various samples from modulated MW irradiation (maximum power = 50 W, set temperature = 250°C) of the same silver acetate/MWCNTs starting material (1 mol% Ag-to-C content) for (a) 1, (b) 2, (c) 5, (d) 10, (e) 30, and (f) 60 min, respectively, after ramping to the set temperature. Reprinted with permission from Ref. [33], Copyright 2011 American Chemical Society.

[33]. Low-dimensional sp^2 carbon NMs such as CNTs and graphene are excellent MW absorbers due to their electrical conductivity. Researchers demonstrate that the direct solid-state Joule heating of CNTs or graphene with organic metal salts using MW irradiation is a highly effective and versatile approach for the preparation of a wide variety of metal and metal oxide NP-decorated substrates. Figure 7.10 shows the SEM images and X-ray diffraction (XRD) patterns of the Ag/MWCNTs prepared under MW irradiation (maximum power = 50 W, set temperature = 250°C) with different time durations after ramping to a set temperature. XRD patterns show that the Ag signals of all samples appeared with similar intensity, indicating complete salt-to-metal conversions within the hold time (the amount of time for which the reaction mixture was held at a set temperature after ramping to that temperature). The MW method time is 1 min as compared to 1–2 h for conventional thermal methods [33]. Samples heated with no hold time offered similar XRD patterns as those with hold times. Local heating of the carbon substrates decomposed the nearby salt particles to afford the near-instantaneous formation of metal or metal oxide NPs,

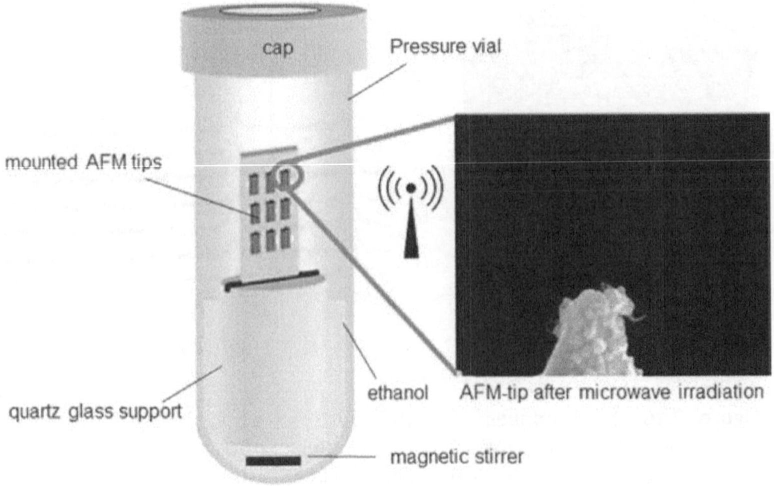

Figure 7.11 Schematic of a quartz glass pressure vial with mounted AFM cantilevers, support, and liquid ethanol reservoir. The pressure vial is transferred to the cavity, and MW irradiation is applied. Morphology of CNTs after MW application on the AFM tip. Reprinted with permission from Ref. [102], Copyright 2010 American Chemical Society.

decorating the substrate surfaces within seconds of MW exposure. Each metal salt has its own thermal decomposition process. The reaction profiles like ramp times and conversion yields were not significantly affected by the salt used. Instead, they were directly related to the MW absorptivity of the substrate [33].

CNT atomic force microscope (AFM) tips fabricated under MW irradiation show another interesting and unique application of MW chemistry. In this method, CNTs are grown on the apex of an AFM tip by MW irradiation and selective heating of catalysts. Deposition of catalysts can be improved further by the selective particle-picking approach [102]. Figure 7.11 shows the schematic of the method in which the reactions were performed in capped pressure vials first. The samples were placed onto a quartz glass support loaded with 5 mL ethanol. The allowed reaction time was 5 min. This is a quick and promising method for the preparation of CNT AFM tips with great efficiency.

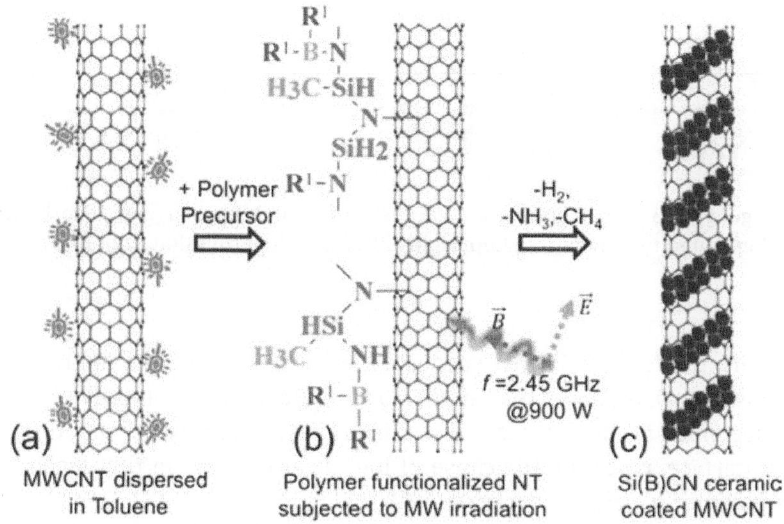

Figure 7.12 Schematic illustrating the mechanism of noncovalent exohedral sequential sidewall functionalization of MWCNTs with (a) a solvent (toluene), (b) a boron-modified polysilazane polymeric precursor, and (c) a pyrolyzed Si(B)CN ceramic following MW irradiation. Reprinted with permission from Ref. [103], Copyright 2012 American Chemical Society.

Synthesis of a polymer-derived ceramic (PDC)–MWCNTs composite under MW irradiation (in a domestic MW oven) at 2.45 GHz was reported for polymer-to-ceramic conversion [103]. Figure 7.12 shows the schematic for preparation of PDC–MWCNTs by MW irradiation (power 900 W) of polymer-functionalized CNTs for 10 min, resulting in the formation of Si(B)CN-coated MWCNTs. X-ray photoelectron spectroscopy (XPS) and Fourier transform infrared (FTIR) analysis reveal the dominance of Si—O, Si—C, B—O, and B—N bonds in MW specimens, analogous to that found in Si(B)CN ceramics processed at 800°C following conventional routes. TGA results depict the high-temperature oxidation resistance of MW specimens comparable to or better than that of Si(B)CN-MWCNTs by the conventional method [103].

Following dielectric measurements at room temperature, an exponentially decaying dielectric response was recorded, suggesting a β-relaxation phase due to interfacial polarization. The low

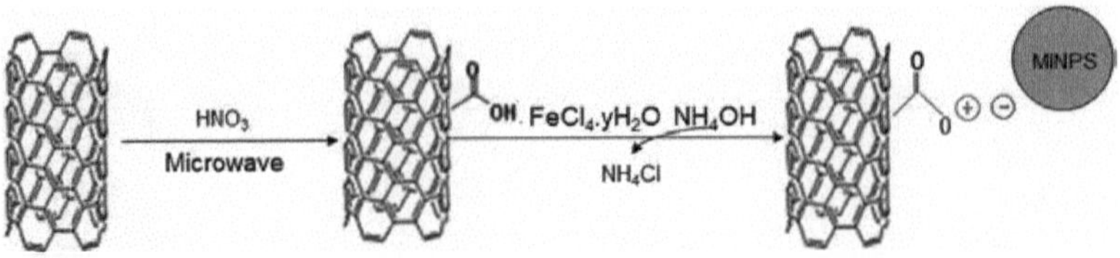

Figure 7.13 A diagram for synthesis of MWCNT-based gas adsorbent. Reprinted from Ref. [69], Copyright (2010), with permission from Elsevier.

permittivity at 2.45 GHz suggests low energy retention. The return loss analysis shows that about 70% of incident MW energy on the specimen is effectively converted into heat at the polymer–nanotube interface, revealing the cause of high heat generation that leads to polymer to ceramic conversion [103].

We have successfully carried out the MW-assisted acidification of MWCNTs as the first step toward the synthesis of CNT-based hybrid materials. Experimental parameters such as heating temperature, time period, types of oxidants, and the amounts used were considered in the preparation of MWCNT-hybrid materials used as gas adsorbents [69]. Figure 7.13 shows the electrostatic interaction between magnetic iron oxide nanoparticles (MINPs) and MW-assisted acidified MWCNTs, resulting in the formation of these hybrid materials (MCNTs) with promising lower detection limit and larger breakthrough volume, with respect to Tenax GR (Table 7.3).

Table 7.3 The adsorption efficiency of MCNTs and commercially available Tenax GR

Compound	**Breakthrough volume(L/g)**		**Detection limit(ppbv)**	
	MCNT[a]	Tenax GR[a]	MCNT[a]	Tenax GR[a]
Benzene	100	20	0.7	7.7
Toluene	>100	40	0.9	9.1
Ethyl benzene	>100	40	0.8	8.0
Xylene	>100	40	0.7	7.9
Styrene	>100	40	0.8	8.5

[a]50 mg amount

Source: Reproduced from Ref. [69], Copyright (2010), with permission from Elsevier.

Three-dimensional vertically aligned CNTs patterns were utilized as templates for fabricating mesoporous hybrid architectures composed of CNTs and various crystalline metal oxide (MO; M = Co, Zn, Mn) NPs by an MW-assisted chemical approach. Vertically aligned CNTs substrates were immersed into a metallic precursor solution under mild vacuum conditions and MW-irradiated for 30–60 s, leading to rapid localized heating and formation of MO NPs [104]. The allowed decoration of SWCNT tubes with MO NPs in this MW-assisted chemical process was due to their porous nature and high specific surface area (548 m^2/g). The CNTs/MO hybrid structures did not damage the alignment, 3D morphology, and porous nature of the original CNTs patterns. The high surface area (350 m^2/g for CNTs/Co_3O_4) is preserved with a slightly larger pore size of 9–18 nm than the pristine aligned CNTs [104].

A Pt/CNT-based nanocomposite electrocatalyst for a proton-exchange membrane in fuel cells was synthesized by using MW-assisted heating of functionalized SWCNTs and demonstrated superior catalytic performance as compared to Pt NPs supported by raw CNTs or by carbon black prepared under identical conditions [105]. MW-assisted solid-state grafting of MWCNTs into a polyurethane composite was reported and exhibited excellent optical limiting properties [106]. An MW-assisted improved impregnation method was developed by heating an aqueous Pt solution and formaldehyde under MW irradiation as a simple time- and energy-saving technique for the preparation of a Pt/MWCNT as a catalyst for use in the oxygen reduction reaction. Results show good electrocatalytic activity for oxygen reduction following a four-electron path to produce water molecules [107].

7.6 Summary and Future Scope of Microwave-Assisted Synthesis of CNTs

Upscale production of CNTs is highly desirable. MW-assisted synthesis and purification of CNTs provide a faster, benign, and greener method, resulting in high purity, lower defects, and higher sample yields as compared to other conventional methods. The MW-based method can be considered as one of the promising methods for large-scale production of CNTs.

MW chemistry still shows plenty of room at the bottom for exploration. Whether or not it will parallel the market share of traditional methods is still debatable. Although properties like lesser reaction time can be one of the important factors, they are not the only ones. MW heating methods can address problems related to inhomogeneous heating found in conventional thermal techniques. Selective (volumetric) heating is another important feature that can be considered favorable over the rest. MWs have attracted tremendous research in the last decade, but there still are many aspects left that need to be explored. Key ones are the mechanism of interactions with CNTs, MW absorption properties of CNTs, and interaction with a wider range of materials and compounds.

MW chemistry is a noninvasive, simple, fast, and cleaner technique that offers excellent reproducibility. As discussed in the examples here, it sure is a greener method for the purification of CNTs and graphitic materials and provides increased reaction kinetics and rapid initial heating, leading to enhanced reaction rates, and can work in solvent-free media, thereby reducing the use of toxic and corrosive solvents. Rapid consumption of starting materials gives higher yields by reduction in unwanted by-product formation during the reaction. With every passing year, there has been enormous development in MW chemistry, leading to further sophistication of the method and researchers opening new opportunities for its extensive use. Improved crystallinity due to high purity and electrical conductivity of CNTs due to MW-based purification methods opens the ground for further research as optical sensors for various applications in the field of electronics. Selective absorption of MW irradiation of different types of CNTs can also prove to be a critical parameter in learning about the semiconducting properties of graphitic materials.

In light of these developments, it is apparent that the synergy between MW reaction conditions and carbon materials commands further discussion, keeping in mind an environmentally friendly large-scale production of CNTs in the future.

Acknowledgments

We would like to express our sincere gratitude to all those who provided their support during the writing of this chapter. We would also like to thank the editor, Dr. Erwann Guénin, for his patience and prompt response to our queries. Last but not at the least, we would like to express our gratitude toward the Ministry of Science and Technology as well as National Tsing Hua University, Taiwan, for their continuous support of our research.

References

1. Iijima, S. (1991). Helical microtubules of graphitic carbon, *Nature*, **354**(6348), pp. 56–58.
2. Dasgupta, K., Banerjee, S., and Joshi, J. B. (2011). Fluidized bed synthesis of carbon nanotubes: a review, *Chem. Eng. J.*, **171**, pp. 841–869.
3. Journet, C., Bernier, P., de la Chapelle, M. L., Deniard, P., Fischer, J. E., Lee, R., Lefrant, S., Loiseau, A., and Maser, W. K. (1997). Large-scale production of single-walled carbon nanotubes by the electric-arc technique, *Nature*, **388**(6644), pp. 756–758.
4. Guo, T., Nikolaev, P., Colbert, D. T., Smalley, R. E., and Thess, A. (1995). Catalytic growth of single-walled nanotubes by laser vaporization, *Chem. Phys. Lett.*, **243**, pp. 49–54.
5. Campos-Delgado, J., Cullen, D. A., Jorio, A., Maciel, I. O., Pimenta, M. A., Smith, D. J., Terrones, H., and Terrones, M. (2010). Chemical vapor deposition synthesis of n-, p-, and Si-doped single-walled carbon nanotubes, *ACS Nano*, **4**(3), pp. 1696–1702.
6. Laplaze, D., Bernier, P., Flamant, G., Guillard, T., Loiseau, A., and Maser, W. K. (1998). Carbon nanotubes: the solar approach, *Carbon*, **36**(5–6), pp. 685–688.
7. Li, Y.-L., Geng, J., Johnson, B., Kinloch, I. A., Shaffer, M. S. P., and Windle, A. H. (2004). Synthesis of single-walled carbon nanotubes by a fluidized-bed method, *Chem. Phys. Lett.*, **384**, pp. 98–102.
8. Monthioux, M., Burteaux, B., Claye, A., Fischer, J. E., Luzzi, D. E., and Smith, B. W. (2001). Sensitivity of single-wall carbon nanotubes to chemical processing: an electron microscopy investigation, *Carbon*, **39**, pp. 1251–1272.

9. Salvetat, J.-P., Boguin, F., Bonard, J. M., Bonnamy, S., Briggs, G. A. D., Burnham, N. A. Forró, L., Kulik, A. J., Motonier, K., and Stöckli, T. (1999). Elastic modulus of ordered and disordered multiwalled carbon nanotubes, *Adv. Mater.*, **11**(2), pp. 161–165.
10. Bianco, A., and Prato, M., Tagmatarchis, N., and Tasis, D. (2006). Chemistry of carbon nanotubes, *Chem. Rev.*, **106**(3), pp. 1105–1136.
11. Kim, D. S., Geckeler, K. E., and Nepal, D. (2005). Individualization of single-walled carbon nanotubes: is the solvent important, *Small*, **1**(11), pp. 1117–1124.
12. Nepal, D., and Geckeler, K. E. (2006). pH-Sensitive dispersion and debundling of single-walled carbon nanotubes: lysozyme as a tool, *Small*, **2**(3), pp. 406–412.
13. Shin, J.-Y., Geckeler, K. E., and Premkumar, T. (2008). Dispersion of single-walled carbon nanotubes by using surfactants: are the type and concentration important, *Chem. Eur. J.*, **14**, pp. 6044–6048.
14. Hu, H., Haddon, R. C., Hamon, M. A., Itkis, M. E., Kamaras, K., and Zhao, B. (2003). Sidewall functionalization of single-walled carbon nanotubes by addition of dichlorocarbene, *J. Am. Chem. Soc.*, **125**(48), pp. 14893–14900.
15. Holzinger, M., Abraha, J., Hennrich, F., Hirsch, A., Graupner, R., Kappes, M., Ley, L., and Whelan, P. (2003). Functionalization of single-walled carbon nanotubes with (R-) oxycarbonylnitrenes, *J. Am. Chem. Soc.*, **125**(28), pp. 8566–8580.
16. Brunetti, F. G., Diaz-Ortiz, A., Filippone, S., Giordani, S., Herrero, M. A., Meneghetti, M., Munoz, J. D. M., Prato, M., Ruaro, G., and Vazquez, E. (2007). Reversible microwave assisted cycloaddition of aziridines to carbon nanotubes, *J. Am. Chem. Soc.*, **129**(47), pp. 14580–14581.
17. Tunckol, M., Durand, J., Fantini, S., Malbosc, F., and Serp, P. (2013). Effect of the synthetic strategy on the non-covalent functionalization of multi-walled carbon nanotubes with polymerized ionic liquids, *Carbon*, **57**, pp. 209–216.
18. Gawande, M. B., Shelke, S. N., Varma, R. S., and Zboril, R. (2014). Microwave-assisted chemistry: synthetic applications for rapid assembly of nanomaterials and organics, *Acc. Chem. Res.*, **47**, pp. 1338–1348.
19. Zhu, Y.-J., and Chen, F. (2014). Microwave-assisted preparation of inorganic nanostructures in liquid phase, *Chem. Rev.*, **114**(12), pp. 6462–6555.

20. Vazquez, E., and Prato, M. (2009). Carbon nanotubes and microwaves: interactions, responses, and applications, *ACS Nano*, **3**(12), pp. 3819–3824.
21. Reeja-Jayan, B., Harrison, K. L., Manthiram, A., Wang, C.-L., Yang, K., and Yilmaz, A. E. (2012). Microwave-assisted low-temperature growth of thin films in solution, *Nat. Sci. Rep.*, **2**(1003), pp. 1–8.
22. Tompsett, G. A., Conner, W. C., and Yngvesson, K. S. (2006). Microwave synthesis of nanoporous materials, *Chem. Phys. Chem.*, **7**, pp. 296–319.
23. Liu, J., Dossot, M., Ehrhardt, J. J., Fort, Y., Mcrae, E., Vigolo, B., and Zubiri, M. R. I. (2007). Efficient microwave-assisted radical functionalization of single-wall carbon nanotubes, *Carbon*, **45**(4), pp. 885–891.
24. Chajara, K., Andersson, C. H., Grennberg, H., Lu, J., and Widenkvist, E. (2010). The reagent free, microwave-assisted purification of carbon nanotubes, *New J. Chem.*, **34**(10), pp. 2275–2280.
25. Adam, D. (2003). Microwave chemistry: out of the kitchen, *Nature*, **421**(6923), pp. 571–572.
26. Liu, S. W., and Wightman, J. P. (1971). Decomposition of simple alcohols, ethers and ketones in a microwave discharge, *J. Appl. Chem. Biotechnol*, **21**(6), pp. 168–172.
27. Kappe, C. O. (2004). Controlled microwave heating in modern organic synthesis, *Angew. Chem., Int. Ed.*, **43**, pp. 6250–6284.
28. Kitchen, H. J., Carassiti, L., Drysdale, T. D., Gregory, D. H., Harrison, A., Kennedy, J. L., Kingman, S. W., Tapia-Ruiz, N., Vallance, S. R., and Whittaker, A. G. (2014). Modern microwave methods in solid-state inorganic materials chemistry: from fundamentals to manufacturing, *Chem. Rev.*, **114**, pp. 1170–1206.
29. Sutton, W. H. (1989). Microwave processing of ceramic materials, *Am. Ceram. Soc. Bull.*, **68**(2), pp. 376–386.
30. Metaxas, A. C., and Meredith, R. J. (1983). *Industrial Microwave Heating* (Institution of Engineering and Technology, London, UK).
31. Clark, D. E., and Sutton, W. H. (1996). Microwave processing of materials, *Annu. Rev. Mater. Sci.*, **26**, pp. 299–331.
32. Bykov, Y. V., Rybakov, K. I., and Semenov, V. E. (2001). High-temperature microwave processing of materials, *J. Phys. D: Appl. Phys.*, **34**(13), pp. R55–R75.
33. Lin, Y., Baggett, D. W., Connell, J. W., Kim, J. W., and Siochi, E. J. (2011). Instantaneous formation of metal and metal oxide nanoparticles on carbon nanotubes and graphene via solvent-free microwave heating, *ACS Appl. Mater. Interfaces*, **3**, pp. 1652–1664.

34. Murali, S., Piner, R. D., Ruoff, R. S., Stoller, M. D., Velamakanni, A., and Zhu, Y. (2010). Microwave assisted exfoliation and reduction of graphite oxide for ultracapacitors, *Carbon*, **48**, pp. 2106–2122.
35. Hetzel, R., Lovingood, D., Manning, T., Phillips, D., and Strouse, G. (2012). Production of fullerenes by microwave synthesis, *Fuller. Nanotub. Carbon Nanostr.*, **20**, pp. 99–108.
36. Liu, C., Li, W., Li, F., Liu, W., Tian, F., and Zhang, P. (2011). One-step synthesis of surface passivated carbon nanodots by microwave assisted pyrolysis for enhanced multicolor photoluminescence and bioimaging, *J. Mater. Chem.*, **21**, pp. 13163–13167.
37. Nadagouda, M. N., Polshettiwar, V., and Varma, R. S. (2009). Self-assembly of palladium nanoparticles: synthesis of nanobelts, nanoplates and nanotrees using vitamin B-1, and their application in carbon-carbon coupling reactions, *J. Mater. Chem.*, **19**, pp. 2026–2031.
38. Harris, P. J. F. (1999). *Carbon Nanotubes and Related Structures* (Cambridge University Press, Cambridge, UK).
39. Liu, W. W., Chai, S. P., Hashim, U., and Mohamed, A. R. (2014). Synthesis and characterization of graphene and carbon nanotubes: a review on the past and recent developments, *J. Ind. Eng. Chem.*, **20**, pp. 1171–1185.
40. Saito, R., Dresselhaus, G., and Dresselhaus, M. S. (1198). *Physical Properties of Carbon Nanotubes* (Imperial College Press, UK).
41. Ishigami, M., Chen, S., Cummings, J., and Zettl, A. (2000). A simple method for the continuous production of carbon nanotubes, *Chem. Phys. Lett.*, **319**, pp. 457–459.
42. Sano, N., Wang, H., Alexandru, I., Amaratunga, G. A., and Chhowalla, M. (2001). Synthesis of carbon onions in water, *Nature*, **414**, pp. 506–507.
43. Lange, H., Huczko, A., Kroto, H. W., Sioda, M., Walton, D. R. M., and Zhu, Y. Q. (2003). Nanocarbon production by arc discharge in water, *Carbon*, **41**, pp. 1617–1623.
44. Shi, Z. J., Gu, Z. N., Iijima, S., Lian, Y. F., Yue, K. T., Zhang, S. L., Zhang, Y., Zhou, X. H., and Zhou, L. X. (1999). Mass-production of single-wall carbon nanotubes by arc-discharge method, *Carbon*, **37**, pp. 1449–1453.
45. Shi, Z. J., Gu, Z. N., Iijima, S., Li, H. D., Lian, Y. F., Liao, F. H., Yue, K. T., Zhang, Y., Zhang, S. L., and Zhou, X. H. (2000). Large scale synthesis of single-wall carbon nanotubes by arc-discharge method, *J. Phys. Chem. Solids*, **61**, pp. 1031–1036.

46. Saito, Y., Kawabata, K., and Okuda, M. (1995). Single-layered carbon nanotubes synthesized by catalytic assistance of rare-earths in a carbon arc, *J. Phys. Chem.*, **99**, pp. 16076–16079.
47. Thess, A., Colbert, D. T., Dai, H., Fischer, J. E., Kim, S. G., Lee, R., Lee, Y. H., Nikolaev, P., Petit, P., Rinzler, A. G., Robert, J., Smalley, R. E., Scuseria, G. E., Tomanek, D., and Xu, C. (1996). Crystalline ropes of metallic carbon nanotubes, *Science*, **273**, pp. 483–487.
48. Aïssa, B., El Khakani, M. A., and Therriault, D. (2011). On-substrate growth of single-walled carbon nanotube networks by an all-laser processing route, *Carbon*, **49**, pp. 2795–2808.
49. Little, R. B. (2003). Mechanistic aspects of carbon nanotube nucleation and growth, *J. Cluster Sci.*, **14**, pp. 135–185.
50. Moisala, A., Kauppinen, E. I., and Nasibulin, A. G. (2003). The role of metal nanoparticles in the catalytic production of carbon nanotubes, *J. Phys. Condens. Matter*, **15**, pp. S3011–S3035.
51. Dupuis, A. C. (2005). The catalyst in the CCVD of carbon nanotubes: a review, *Prog. Mater. Sci.*, **50**, pp. 929–961.
52. Li, Q., Liu, Z., Yan, H., and Zhang, J. (2004). Effect of hydrocarbons precursors on the formation of carbon nanotubes in chemical vapor deposition, *Carbon*, **42**, pp. 829–835.
53. Takagi, D., Hibino, H., Homma, Y., Kobayashi, Y., and Suzuki, S. (2007). Carbon nanotube growth from semiconductor nanoparticles, *Nano Lett.*, **7**, pp. 2272–2275.
54. Liu, H., Chiashi, S., Chokan, T., Homma, Y., Ohno, H., and Takagi, D. (2008). Growth of single-walled carbon nanotubes from ceramic particles by alcohol chemical vapor deposition, *Appl. Phys. Express*, **1**, pp. 014001–014003.
55. Liu, H., Chiashi, S., Chokan, T., Homma, Y., and Takagi, D. (2010). Investigation of catalytic properties of Al_2O_3 particles in the growth of single-walled carbon nanotubes, *J. Nanosci. Nanotechnol.*, **10**, pp. 4068–4073.
56. Liu, H., Chiashi, S., Homma, Y., and Takagi, D. (2010). The growth of single-walled carbon nanotubes on a silica substrate without using a metal catalyst, *Carbon*, **48**(1), pp. 114–122.
57. Huang, S., Cai, Q., Chen, J., Qian, Y., and Zhang, L. (2009). Metal-catalyst-free growth of single-walled carbon nanotubes on substrates, *J. Am. Chem. Soc.*, **131**, pp. 2094–2095.
58. Liu, B., Cheng, H. M., Gao, L., Jiang, C., Li, S., Liu, C., Pei, S., and Ren, W. (2009). Metal-catalyst-free growth of single-walled carbon nanotubes, *J. Am. Chem. Soc.*, **131**, pp. 2082–2083.

59. Liu, B., Cheng, H. M., Jiang, C., Li, F., Liu, C., Ren, W., Sun, C., Tang, D. M., Yin, L. C., Yu, W. J., and Zhang, L. (2011). Importance of oxygen in the metal-free catalytic growth of single-walled carbon nanotubes from SiOx by a vapor–solid–solid mechanism, *J. Am. Chem. Soc.*, **133**, pp. 197–199.
60. Steiner III, S. A., Baumann, T. F., Bayer, B. C., Blume, R., Hart, A. J., Hofmann, S., Moberlychan, W. J., Schlögl, R., Shaw, E. L., Wardle, B. L., and Worsley, M. A. (2009). Nanoscale zirconia as a nonmetallic catalyst for graphitization of carbon and growth of single- and multiwall carbon nanotubes, *J. Am. Chem. Soc.*, **131**, pp. 12144–12154.
61. Lin, J. H., Bachmatiuk, A., Büchner, B., Chen, C. S., Chen, H. W., Ma, H. L., Rümmeli, M. H., and Zeng, Z. Y. (2011). Growth of carbon nanotubes catalyzed by defect-rich graphite surfaces, *Chem. Mater.*, **23**(7), pp. 1637–1639.
62. Kreupl, F., Duesberg, G. S., Graham, A. P., Liebau, M., Seidel, R., and Unger, E. (2004). Carbon nanotubes for interconnect applications, *IEEE International Electron Devices Meeting, Technical Digest*, pp. 683–686.
63. Li, J., Cassell, A., Han, J., Meyyappan, M., Ng, H. T., Stevens, R., and Ye, Q. (2003). Bottom-up approach for carbon nanotube interconnects, *Appl. Phys. Lett.*, **82**(15), pp. 2491–2493.
64. Biercuk, M. J., Fischer, J. E., Hyun, J. K., Johnson, A. T., Llaguno, M. C., and Radosavljevic, M. (2002). Carbon nanotube composites for thermal management, *Appl. Phys. Lett.*, **80**(15), pp. 2767–2769.
65. Liu, C. H., Fan, S. S., Huang, H., and Wu, Y. (2004). Thermal conductivity improvement of silicone elastomer with carbon nanotube loading, *Appl. Phys. Lett.*, **84**(21), pp. 4248–4250.
66. Koziol, K., Cunniff, P., Moisala, A., Motta, M., Sennett, M., Vilatela, J., and Windle, A. (2007). High-performance carbon nanotube fiber, *Science*, **318**(5858), pp. 1892–1895.
67. Vigolo, B., Bernier, P., Coulon, C., Journet, C., Pailler, R., Penicaud, A., Poulin, P., and Sauder, C. (2000). Macroscopic fibers and ribbons of oriented carbon nanotubes, *Science*, **290**(5495), pp. 1331–1334.
68. Lee, P. L., Ling. Y. C., and Sun, Y. C. (2009). Magnetic nano-adsorbent integrated with lab-on valve system for trace analysis of multiple heavy metals, *J. Anal. At. Spectrom.*, **24**(3), pp. 320–327.
69. Lee, P. L., Chiu, Y. K., Ling. Y. C., and Sun, Y. C. (2010). Synthesis of a hybrid material consisting of magnetic iron-oxide nanoparticles and carbon nanotubes as a gas adsorbent, *Carbon*, **48**(5), pp. 1397–1404.

70. Zanello, L.P., Haddon, R. C Hu, H., and. Zhao, B. (2006). Bone cell proliferation on carbon nanotubes, *Nano Lett.*, **6**(3), pp. 562–567.

71. Deokar, A. R., Chang, C. C., Lin, L. Y., and Ling, Y. C. (2013). Single-walled carbon nanotube coated antibacterial paper: preparation and mechanistic study, *J. Mater. Chem. B*, **1**, pp. 2639–2646.

72. Ghule, A. V., Chang, J. Y., Kathir, K. M., Kumar, T. K. S., Tzing, S. H., Ling, Y. C., and Yu, C. (2007). Carbon nanotubes prevent 2,2,2 trifluoroethanol induced aggregation of protein, *Carbon*, **45**(7), pp. 1586–1589.

73. Lin, W., Chen, M. X., Ding, Y., Moon, K. S., Shang, J. T., Wong, C. P., and Zhang, S. J. (2011). Microwave makes carbon nanotubes less defective, *ACS Nano*, **4**(3), pp. 1716–1722.

74. Paton, K. R., and Windle, A. H. (2008). Efficient microwave energy absorption by carbon nanotubes, *Carbon*, **46**, pp. 1935–1941.

75. Imholt, T. J., Dyke, C. A., Hasslacher, B., Perez, J. M., Price, D. W., Roberts, J. A., Scott, J. B., Tour, J. M., Wadhawan, A., and Ye, Z. (2003). Nanotubes in microwave fields: light emission, intense heat, outgassing, and reconstruction, *Chem. Mater.*, **15**, pp. 3969–3970.

76. Su, J. W., Gwo S., and Lin, K. J. (2009). Well-aligned multi-walled carbon nanotubes emitting natural white-light under microwave irradiation, *Chem. Commun.*, pp. 6777–6779.

77. MacKenzie, K., Dunens, O., and Harris, A. T. (2009). A review of carbon nanotube purification by microwave assisted acid digestion, *Sep. Purif. Technol.*, **66**(2), pp. 209–222.

78. Gude, V. G., Deng, S., Guerra1, E. M., Nirmalakhandan, N., and Patil, P. (2013). Microwave energy potential for biodiesel production, *Sustain. Chem. Processes*, **1**(5), pp. 1–31.

79. Wadhawan, A., Garrett, D., and Perez, J. M. (2003). Nanoparticle-assisted microwave absorption by single-wall carbon nanotubes, *Appl. Phys. Lett.*, **83**(13), pp. 2683–2685.

80. Ye, Z., Deering, W. D., Krokhin, A., and Roberts, J. A. (2006). Microwave absorption by an array of carbon nanotubes: a phenomenological model, *Phys. Rev. B*, **74**, p.075425.

81. Chou, Y. H., Brydson, R., Douthwaite, R. E., Hondow, N., Mitchell, R., and Thomas, C. I. (2012). Microwave plasma synthesis of lanthanide zirconates from microwave transparent oxides, *Dalton Trans.*, **41**, pp. 2472–2476.

82. Brooks, D. J., and Richard Douthwaite, R. E. (2004). Microwave-induced plasma reactor based on a domestic microwave oven for bulk solid state chemistry, *Rev. Sci. Instrum.*, **75**(12), pp. 5277–5279.

83. Baughman, R. H., Heer, W. A., and Zakhidov, A. A. (2002). Carbon nanotubes-the route toward applications, *Science*, **297**, pp. 787–792.

84. Bandow, S., Eklund, P. C., Rao, A. M., Smalley, R., Thess, A., and Williams, K. A. (1997). Purification of single-wall carbon nanotubes by microfiltration, *J. Phys. Chem. B*, **101**(44), pp. 8839–8842.

85. Hou, P. X., Cheng, H. M., and Liu, C. (2008). Purification of carbon nanotubes, *Carbon*, **46**(15), pp. 2003–2025.

86. Ando, Y., Iijima, S., Inoue, S., and Zhao, X. (2002). Mass production of multiwalled carbon nanotubes by hydrogen arc discharge, *J. Cryst. Growth*, **237–239**, pp. 1926–1930.

87. Shi, Z., Gu, Z., Iijima, S., Lian, Y., Liao, F., Zhang, Y., and Zhou, X. (1999). Purification of single-wall carbon nanotubes, *Solid State Commun.*, **112**, pp. 35–37.

88. Ling, Y. C., and Deokar, A. (2011). Microwave-assisted preparation of carbon nanotubes with versatile functionality. In *Carbon Nanotubes Applications on Electron Devices*, Marulanda, J. M., ed. (InTech, Europe and China), pp. 127–142.

89. Lobach, A. S., Obraztsova. E. D., Spitsina, N. G., and Terekhov, S. V. (2002). Comparative analysis of various methods of purification of single-walled carbon nanotubes, *Phys. Solid State*, **44**(3), pp. 475–477.

90. Liu, Y. Q., Gao, L., Jiang, L. Q., Kajiura, H., Li, Y. M., Noda. K., Sun, J., Wang, Y., and Zheng, S. (2007). A multi-step strategy for cutting and purification of single-walled carbon nanotubes, *Carbon*, **45**(10), pp. 1972–1978.

91. Tsai, P. A., Chiu, W. M., Kuo, H. Y., and Wu, J. H. (2013). Purification and functionalization of single-walled carbon nanotubes through different treatment procedures, *J. Nanomater.*, **2013**, pp. 1–9.

92. Vazquez, E., Georgakilas, V., and Prato, M. (2002). Microwave-assisted purification of HIPCO carbon nanotubes, *Chem. Commun.*, pp. 2308–2309.

93. Qiu, H., Akasaka, T., Maeda, Y., and Yang, J. (2012). Diameter-selective purification of carbon nanotubes by microwave-assisted acid processing, *Sep. Purif. Technol.*, **96**, pp. 182–186.

94. Ko, C. J., Chen, H. L., Chu, T. C., Ko, F. H., and Lee, C. Y. (2004). Highly efficient microwave-assisted purification of multiwalled carbon nanotubes, *Microelectron. Eng.*, **73–74**, pp. 570–577.

95. Sun, H., Ajayane, P. M., Jia, Y., Ning, C., Sun, J., Vajtai, R., Wang, K., Wei, J., Wu, D., Yang, S., Zhao, Y., Zhu, H., and Zhu, J. (2014). Effect of microwave

irradiation on carbon nanotube fibers: exfoliation, structural change and strong light emission, *RSC Adv.*, **4**, pp. 15502–15506.

96. Wu, Z., Hamilton Jr., R. H., Holian, A., Mitra, S., and Wang, Z. (2014). Oxidation debris in microwave functionalized carbon nanotubes: chemical and biological effects, *Carbon*, **68**, pp. 678–686.
97. Meng, L., Fu, C., and Lu, Q. (2009). Advanced technology for functionalization of carbon nanotubes, *Prog. Nat. Sci.*, **19**(7), pp. 801–810.
98. Brunetti, F. G., Alfonsi, J., Diaz-Ortiz, A., Herrero, M. A., Meneghetti, M., Munoz, J. M., Prato, M., and Vazquez, E. (2008). Microwave-induced multiple functionalization of carbon nanotubes, *J. Am. Chem. Soc.*, **130**(25), pp. 8094–8100.
99. Economopoulos, S. P., Iijima, S., Pagona, G., Tagmatarchis, N., and Yudasaka, M. (2009). Solvent-free microwave-assisted Bingel reaction in carbon nanohorns, *J. Mater. Chem.*, **19**(39), pp. 7326–7331.
100. Li, J., and Grennberg, H. (2006). Microwave-assisted covalent sidewall functionalization of multiwalled carbon nanotubes, *Chem. Eur. J.*, **12**, pp. 3869–3875.
101. Raghuveer, M. S., Agrawal, S., Bishop, N., and Ramanath, G. (2006). Microwave-assisted single-step functionalization and in situ derivatization of carbon nanotubes with gold nanoparticles, *Chem. Mater.*, **18**(6), pp. 1390–1393.
102. Druzhinina, T. S., Hoeppener, S., and Schubert, U. S. (2010). Microwave-assisted fabrication of carbon nanotube AFM tips, *Nano Lett.*, **10**, pp. 4009–4012.
103. Bhandavat, R., Kuhn, W., Lehman, J., Mansfield, E., and Singh, G. (2012). Synthesis of polymer-derived ceramic Si(B)CN-carbon nanotube composite by microwave-induced interfacial polarization, *ACS Appl. Mater. Interfaces*, **4**, pp. 11–16.
104. Mazloumi, M., Nazar, L. F., Rangom, Y., Shadmehr, S., and Tang, X. (2013). Fabrication of three-dimensional carbon nanotube and metal oxide hybrid mesoporous architectures, *ACS Nano*, **7**(5), pp. 4281–4288.
105. Chen, Y. H., Krishnan, S. S., Li, C., Mitra, S., Muthukumar, V. S. Sai, S. S. S., Venkataramaniah, K., and Wang, Y. B. (2009). Microwave-assisted solid-state grafting of multi-walled carbon nanotubes on polyurethane for the synthesis of a composite with optical limiting properties, *J. Mater. Chem.*, **19**(36), pp. 6568–6572.

106. Zhang, W. M., Chen, J., Ma, Z. F., Swiegers, G. F., and Wallace, G. G. (2010). Microwave assisted synthesis of Pt/CNT nanocomposite electrocatalysts for PEM fuel cells, *Nanoscale*, **2**(2), pp. 282–286.

107. Rahsepara, M., Kim, H., and Pakshira, M. (2013). Synthesis of multiwall carbon nanotubes with a high loading of Pt by a microwave-assisted impregnation method for use in the oxygen reduction reaction, *Electrochim. Acta*, **108**, pp. 769–775.

Chapter 8

Microwave Synthesis Porous Zeolitic Metal–Organic Framework Materials

Hossein Kazemian[a,b] and Sohrab Rohani[a]

[a] *Department of Chemical and Biochemical Engineering, Western University, London, ON N6A5B9, Canada*

[b] *University of Northern British Columbia, Prince George, BC, Canada*

srohani@uwo.ca, hossein.kazemian@uwo.ca

8.1 Introduction

Porous zeolitic materials are one of the most premising categories of materials particularly for application in separation, purification, and catalytic processes. Nowadays, crystalline microporous zeolitic materials contribute to a very large segment of the global economy (>$350 billion). They are utilized in a wide range of applications in different industries such as petrochemical, detergent, water and wastewater treatment purification, and gas separation and storage. Recently, extensive research has been focused on using the molecular building block approach to synthesize 3D metal-organic frameworks (MOFs). Metal–ligand coordination strategies are used to link the molecular components for construction of these porous compounds. A number of such frameworks have been

Microwave Engineering of Materials: From Mesoscale to Nanoscale
Edited by Erwann Guénin

ISBN 978-981-4669-42-9 (Hardcover), 978-981-4669-43-6 (eBook)
www.panstanford.com

found to exhibit desirable zeolitic properties such as microporosity of the framework, extraordinary adsorption, and selective catalytic activity. These materials generally consist of metal–oxygen or metal–nitrogen polyhedra containing divalent (e.g., Zn^{2+}, Cu^{2+}) or trivalent (e.g., Al^{3+}, Cr^{3+}) cations. Metal cations interconnected with a variety of organic linker molecules result in tailored nanoporous materials. With appropriate choice of linkers, it is possible to fine-tune the size, shape, and chemical functionality of materials to improve the properties. This unique structural feature offers revolutionary opportunities in H_2 storage, CH_4 and CO_2 separation and storage, chemical sensors, chiral pharmaceutical synthesis and separation, and catalysis. The zeolitic-like framework organometallic materials are similar to zeolites in many aspects such as high crystallinity, tunable hydrophil(phob)icity, and acid(basic)ity. They are, however, different in many other aspects such as higher surface area/pore volume (up to 6000–9000 m^2/g in comparison to 200–900 m^2/g for zeolites) and much more diverse chemistry (e.g., many metals/metal clusters and organic linkers can contain functionality). Despite the abovementioned extraordinary capabilities of MOFs, particularly for H_2 and CO_2 storage, one of the main obstacles toward commercial application of the materials is relatively high expense of the final product because of their costly manufacturing process. Conventional approaches for synthesis of MOFs are very expensive because of high-energy consumption (e.g., typical synthesis might takes several hours or days at high temperature of 100°C–250°C) and expensive organic linkers and solvents. Therefore development of a novel synthesis process by means of a new energy source, such as microwaves (MWs) and ultrasound, to decrease the crystallization time (seconds or minutes) and to remarkably reduce the energy consumption for the highly efficient manufacturing of MOFs in a techno-economically viable manner is very desirable.

8.2 Metal–Organic Frameworks

Porous materials are an important category of chemical compounds renowned for their excellent characteristics for different applications such as gas storage and separation, shape-/size-selective

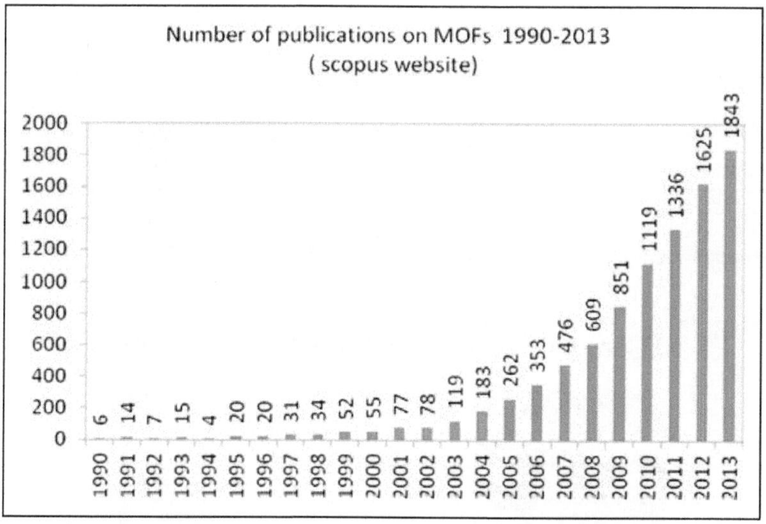

Figure 8.1 Rapid increase in publications on MOFs (scopus.com).

solid catalysis, and drug delivery [1–6]. Porous zeolitic compounds have made significant contribution to the global economy. They are being used in diverse industries and processes such as oil refining, the petrochemical industry, the detergent industry, water and wastewater treatment and purification, environmental remediation processes, gas separation, and storage processes [7].

Since this class of porous solids emerged in the early 1990s, remarkable advancements have been made in their chemistry. These materials have attracted a great deal of attention in the past two decades, and the exponential increase in the number of publications on MOFs during the recent years is remarkable (Fig. 8.1).

Before becoming known as MOFs they were more generally referred to as "coordination polymers." This term correctly identifies that these materials contain small clusters of inorganic polyhedra that are connected together by ligand monomers through coordination bonds; however, it is a somewhat simplistic definition that does not consider the final structure, or morphology, of the materials. The discovery that these hybrid solids contain a large dimensionality, forming chains (1D), layers (2D), or frameworks (3D), led to their classification as MOFs [8]. This classification

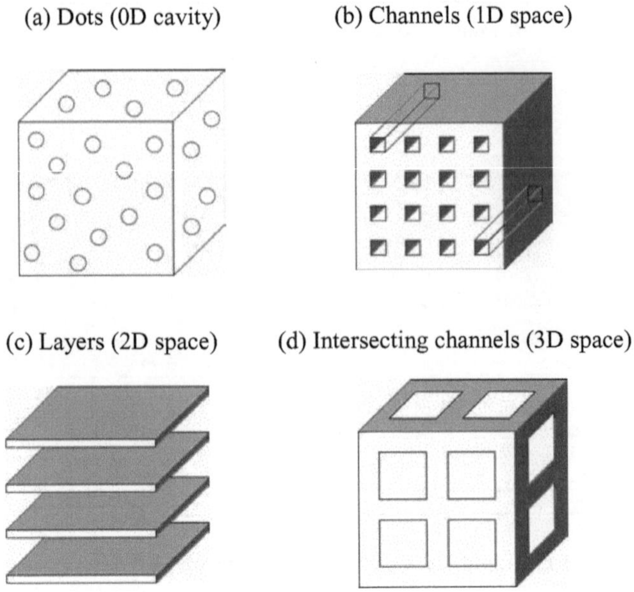

Figure 8.2 Classes of porous structures on the basis of spatial dimensions. Reprinted from Ref. [10]. Copyright (c) 2004, with permission from John Wiley and Sons.

defines them as materials with strong bonding and a high degree of robustness, linking units that are available for modification, and geometrically well-defined structures [9].

In general, the porosity of every type of porous structure belongs to one of the following four categories: 0D cavities, 1D channels, 2D layers, or 3D intersecting channels (Fig. 8.2). Zero-dimensional cavities (dots) are completely enclosed by surrounding wall molecules, preventing entrapped guest molecules from escaping. MOFs can be designed to have channels, layers, or intersecting channels, each of which can accommodate and exchange guest molecules [10].

This chemical versatility that MOFs provide is the sole reason why they continue to receive so much interest from various domains in the scientific community, including physics, chemistry, and materials science. Before they were discovered, there existed two types of porous materials, organic and inorganic materials. MOFs combine both organic and inorganic groups within their structures; thus they represent a new type of porous materials referred to as hybrid solids. The two largest subclasses of inorganic solids are the aluminophosphates and aluminosilicates. Zeolites are members of the aluminosilicate subclass, whose structures are composed of interconnected tunnels or cages within which water molecules or metal ions are inserted. Their porosity is provided by the elimination of these species from the pores. The aluminophosphates are composed of tetrahedral Al^{3+} and P^{5+} ions linked by corner-sharing oxygen atoms [10]. Structurally, MOFs are very similar to inorganic porous solids in that both of their 3D skeletons are composed of secondary building units (SBUs). In place of the inorganic tetrahedral species found in the structure of inorganic porous solids, though, MOFs contain organic linker molecules. The properties of MOFs also resemble those of zeolites, such as the adsorption of gases and shape-selective catalysis of both materials. However, MOFs offer a much wider range of structures and properties such as high porosity, pore dimensionality, and functionality of metal ions and organic linkers [11].

The MOF structure contains two central components, inorganic metal centers and organic linkers. The two characteristics of the metal centers and linkers that are important for defining the final structure of the framework are the number and orientation of their binding sites (coordination numbers and coordination geometries). MOFs can be constructed from a great variety of transition metal ions, with coordination numbers ranging from two to seven. Many different coordination geometries may be formed as a result of this large selection of metals, including linear, T- or Y-shaped, tetrahedral, square planar, square pyramidal, trigonal bipyramidal, octahedral, trigonal prismatic, and pentagonal bipyramidal [10]. Furthermore, considering the large number of organic linkers that can be associated with the metals and the different possible linkages, binding strength, and directionality, the number of possible combinations is seemingly infinite. Consequently, many different MOF framework topologies exist, forming either microporous (pore diameter $\leq$ 2 nm) or mesoporous (pore diameter between 2 and 50 nm) channels. The existence of organic and inorganic species in the MOF structure also presents an opportunity to incorporate both hydrophilic and hydrophobic parts within the pores, thus allowing for control over the adsorption properties [8].

Owing to the infinite number of possible combinations between inorganic and organic parts, soon after the discovery of MOFs a rationalization became necessary for organizing the structures into classes. The inorganic clusters that are linked to the organic component were observed to resemble the shapes of simple geometric figures (diamonds, sodalites, etc.). These became known as SBUs, which, by sharing of vertices through the organic linkers, build up the final solid. The majority of MOFs, including the Materials of Institute Lavoisier (MIL) family, incorporate a carboxylate functionality to chelate the metal ions and lock them into rigid metal–oxygen–carbon clusters. The first carboxylate MOFs designed to resemble this structure were MOF-2 and MOF-5. The MOF-5 structure, as an example, is composed of octahedral SBUs, each of which is formed by four ZnO_4 tetrahedra. Benzene dicarboxylate linkers to give an extended 3D cubic framework (Fig. 8.3) with an exceptionally high porosity and stability join these SBUs together [12].

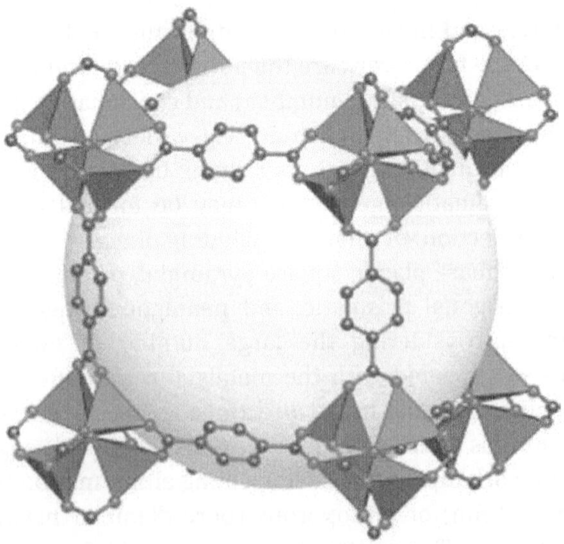

Figure 8.3 The MOF-5 structure and its topology. Reprinted by permission from Macmillan Publishers Ltd: *Nature* (Ref. [12]), copyright (2003).

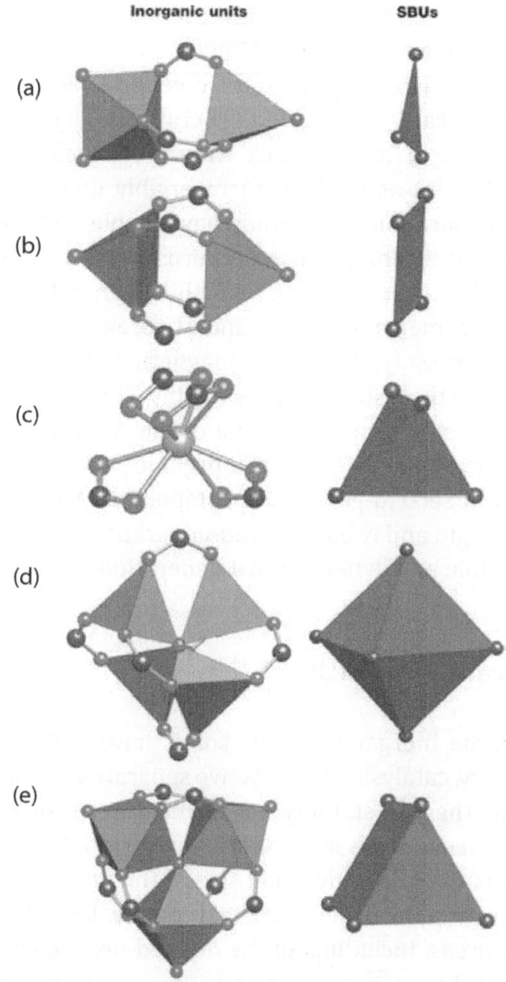

Figure 8.4 Examples of SBUs from carboxylate MOFs. Reprinted by permission from Macmillan Publishers Ltd: *Nature* (Ref. [12]), copyright (2003).

Figure 8.4 shows some of the SBUs that are formed in carboxylate MOFs. The inorganic units are defined by metal–oxygen polyhedra (such as the ZnO_4 tetrahedra in MOF-5). The carboxylate atoms that are bound to these polyhedra define the vertices of the SBUs, which are held together by the organic linkers (such as the benzene dicarboxylate linkers in MOF-5) [12].

Porous coordination compounds in general can be further classified into three categories, called generations. The first-generation compounds have a microporous framework held together only by guest molecules and collapse irreversibly upon guest removal. The second-generation compounds have stable and robust porous frameworks that exhibit permanent porosity when guest molecules are evacuated from the pores. The third-generation compounds have flexible frameworks that respond to external stimuli (light, temperature, pressure, electric or magnetic field, guest molecules, etc.) and change their structures reversibly. The stimulus can induce either an expansion or a contraction of the cell volume and induce movements larger than 10 Å. Most inorganic porous solids fall under the category of second-generation compounds. MOFs, on the other hand, can be rigid and robust second-generation compounds as well as highly flexible and dynamic third-generation compounds [10].

8.3 Applications of MOFs

For a long time inorganic porous solids have had applications in petrochemistry, catalysis, and selective separation as a result of their porosity, high thermal stability, and surface area. However, until the discovery of mesoporous materials, an important limitation of these materials was the relatively small size of the pores. MOFs provided a breakthrough, since they have potentially unlimited porosities and surface areas, including all the desired properties of inorganic solids. Currently, we are living in a time when energy problems must be addressed by the development of sustainable and greener alternatives to existing technologies. Owing to the tunability of their structures and the increasingly efficient and greener synthesis techniques, MOFs can have many applications in these areas. MOFs have electronic, optical, and magnetic applications, as well as applications in the areas of catalysis, gas adsorption/separation/storage, and drug adsorption/delivery [8].

In the field of catalysis, the advantage of MOF catalysts over other nanoporous materials such as zeolites and activated carbon stems from their inorganic–organic hybrid nature and the tunability of their structures. Not only may the organic linker provide catalytic sites in addition to the metal catalytic sites, but it may also be used as a scaffold to bind catalytic complexes. The synthetic flexibility of MOFs allows for a greater control of the pore size and environment, allowing for higher selectivity [8]. MOF catalysis mainly includes cyanosilylation, the Diels–Alder reaction, hydrogenation, esterification, and CO oxidation; however, many others exist [8].

MOFs have also been used as templates for nanoparticles in the area of guest-based catalysis. MOF-5 encapsulated with Pd, Cu, Ru, Cu, ZnO, and Au nanoparticles by various methods has been used to catalyze such reactions as cyclo-octene hydrogenation, methanol production from synthesis gas, and the oxidation of 1-phenylethanol [13]. On the basis of previous results from inorganic mesoporous materials, the surface functionalization of MOF pores should also be a promising research area for catalytic applications; however, this area has yet to be explored.

The domain of gas adsorption/separation/storage has become increasingly important as worldwide fossil fuel reserves continue to decrease. MOFs may provide a solution to this problem for their capacity to adsorb high amounts of important gases like CO_2, H_2, O_2, CH_4, CO, and NO_x in their pores. The ease of evacuating solvents from the pores of MOFs, coupled with their high surface areas and tunable pore sizes, makes them particularly attractive in this field. So far, the use of MOFs in the separation of mixtures of gases has been confined primarily to capturing H_2, CH_4, and CO_2 [8]. However, progress has also been made in the removal of other toxic and environmentally hazardous gases as well. Weireld and coworkers [14] investigated the hydrogen sulphide (H_2S) adsorption properties of several MOFs. H_2S is known to be a very challenging adsorbate due to the adsorbent decomposition and difficult regeneration that usually occur upon uptake. However, both MIL-47(V) and MIL-53(Al, Cr) display little degradation and reversible uptake, and MIL-53 materials display step-wise adsorption, making them attractive materials for the purification of natural gases. Yaghi and coworkers [15] compared the adsorption of various harmful gases by several MOF materials to an activated carbon material, Calgon BPL carbon.

For each gas, at least one of the MOFs studied exceeded the adsorption potential of BPL carbon. For ammonium, the MOFs displayed a 6- to 105-fold adsorption enhancement. Another group showed that the zinc clusters in MOF-5 actually coordinate to ammonia, thus not only increasing the adsorption capacity but also serving as a degradation pathway [15].

Collectively, these results are an indication that MOFs are promising adsorbents for harmful gases and with further research may have applications in large-scale industrial processes.

In the field of drug adsorption/delivery the size of the pores becomes very important. The main drawback of organic porous solids is that their pore sizes are too small to allow for successful incorporation of most organic molecules. With their tunable porosities, MOFs are able to provide a solution to this problem. Horcajada et al. [16] showed that the rigid mesoporous MOFs, MIL-100 and MIL-101, were able to achieve both high drug loading and controlled release. MIL-100 and MIL-101 both exhibited high uptake (0.35 and 1.4 g, respectively, of ibuprofen per gram of MOF) and 3–6 days' release times under simulated physiological conditions. Even at these high loading values there was no observed loss of crystallinity or decomposition of the frameworks.

Despite the fact that microporous solids are limited to the incorporation of small organic molecules, Horcajada et al. [17] also demonstrated the adsorption and delivery capabilities of a flexible microporous MOF, MIL-53(Fe). This material adsorbed 0.21 g ibuprofen/g MOF and completely released the drug after three weeks. The sustained release profile was attributed to the flexibility of the framework, which is able to reversibly open and close its pores upon adsorption of guest species. This property allows MIL-53(Fe) to maximize host–guest interactions. More recently, this group investigated the delivery of nine drugs from five MIL materials. MIL-100 was able to achieve 25% loading by weight of the anticancer drug busulfan, approximately five times higher than the best system of polymer nanoparticles (5–6 wt%). Studies on human leukemia and human multiple myeloma cells in culture showed that busulfan-loaded MIL-100 exhibited very similar cytotoxic activity to busulfan in its free form. Empty MIL-100 nanoparticles exhibited a complete absence of cytotoxicity in the same cell lines. In vitro HIV virus replication was significantly inhibited by MIL-100 loaded with the anti-HIV drug AZT-TP [18]. The MIL family is a unique candidate in the field of drug adsorption/delivery and is ready for more extensive in vivo testing.

8.4 Microwaves for Chemical Synthesis

MWs lie between infrared waves and radio waves on electromagnetic spectrum. The use of 2450 MHz MWs is more prevalent due to the fact that energy absorption is maximum for water at this frequency and also magnetrons at this frequency are more widely available in commercial MW and chemistry equipment [19]. MWs, like all electromagnetic radiation, have an electrical component as well as a magnetic component. The MW portion of the electromagnetic spectrum is characterized by wavelengths between 1 mm and 1 m, corresponding to frequencies between 100 and 5000 MHz. As many noticeable studies are reported in scientific literature, the MW-heating technique has gained high expectation in the synthesis of chemicals, including material synthesis, organic synthesis, and conversion of energy resources with high reaction and energy efficiency. Of the four frequencies available for industrial, scientific, or medical applications (i.e., 915, 2450, 5800, and 22,125 MHz), 2450 MHz is the most commonly used because it has the right penetration depth (~1.75 inches) to be effective with laboratory-scale applications and synthesis. At this frequency, the energy supplied to the system is equal to 0.037 kcal/mol. The typical bond strength of most organic compounds ranges from 80 to 120 kcal/mol and even the weakest intramolecular bonds (van der Waals forces) are 48 kcal/mol. The MW only serves to rotate the molecules and, therefore, will neither break nor form any additional bonds. One of the advantages of using MW energy is that the rate enhancement achieved does not depend on thermal conduction. With conventional heating, as a reactor is slowly heated, the thermal energy is slowly transferred from the reactor wall to the bulk of the reacting mixture. That heat must then be transferred again through each layer of liquid until the temperature is uniform. The mechanism of heat transfer in MW energy transfer is different. Reactor materials for MW synthesis must be transparent to MW energy (e.g., glass), enabling the reactant molecules to absorb the energy directly, resulting in more efficient energy transfers. Heat is not used to drive the reaction, but MW energy is directly applied to the reacting species. Thermal energy is produced only as a by-product. Extensive research has been carried out on chemical processing (mostly organic synthesis) at 2.45 GHz, but we can find only few studies reported at 915 MHz. It is proved that 2.45 GHz is more selective than 915 MHz and that even higher MW frequencies would be desirable. However, in terms of physical and power penetration factors, 915 MHz MW is a more desirable frequency, which is also a little less expensive and slightly more energy efficient for large systems than is 2.45 GHz [20].

It is useful to consider the quantum energy of MWs in relation to other forms of electromagnetic energy to recognize that the energy delivered by MWs is insufficient for breaking covalent chemical bonds. This information can help to narrow speculation on the mechanisms for enhancement in specific reactions. There are two specific mechanisms of interaction between materials and MWs, (1) dipole interactions and (2) ionic conduction. Both mechanisms require effective coupling between components of the target material and the rapidly oscillating electrical field of the MWs. Dipole interactions occur with polar molecules, in which the polar ends of a molecule tend to align themselves and oscillate in step with the oscillating electrical field of the MWs. Collisions and friction between the moving molecules result in heating. Broadly, the more polar a molecule, the more effectively it will couple with (and be influenced by) the MW field.

Ionic conduction is different from dipole interactions because ions in solution do not have a dipole moment. They are charged species that are distributed and can couple with the oscillating electrical field of the MWs. The rate of MW heating of an ionic solution is relevant to the concentration of ions in solution.

It has long been known that molecules undergo excitation with electromagnetic radiation. The interaction of the electromagnetic field with materials is utilized in household MW ovens to warm up and cook food. Chemists (particularly organic chemists and recently material chemists) are using MWs as a novel reaction methodology [21, 22]. Some of the first examples gave amazing results, which led to extensive interest in this novel technique. The water molecule is the target for MW ovens in the kitchen, in which, like any other molecule with a dipole momentum, it absorbs MW radiation. Absorbed MW radiation is converted into heat with high efficiency, in which superheating becomes possible even at ambient pressure. Particularly, enormous accelerations in reaction time can be achieved if superheating happens in closed vessels under high pressure; a reaction that takes several hours under conventional conditions can be completed over the course of minutes and even seconds with extraordinary higher yields of production. As mentioned before, excitation with MW radiation results in the molecules aligning their dipoles within the external field. Strong agitation, provided by the reorientation of molecules, in phase with the electrical field excitation, causes intense internal heating. The question of whether a nonthermal process is operating can be answered simply by comparing the reaction rates when the reaction is carried out under irradiation versus under conventional heating. In fact, no nonthermal effect has been found in the majority of reactions, and the acceleration is attributed to superheating alone. However, it has been found that nonthermal effects play a role in some reactions [23].

Conventional heating usually involves the use of a thermal energy, which heats the walls of the reactor by convection or conduction. The core of the sample takes much longer to achieve the target temperature (particularly for heating large reactors). MW heating is able to heat the target compounds without heating the entire reactors' constructing materials, which saves time and energy. Sufficient heating of thin objects throughout their volume (instead of through their outer surface) is possible by MWs, which produces more uniform heating. Different compounds convert MW radiation to heat in various extents. This selectivity allows some parts of the object to be heated more quickly or more slowly than others (particularly the reactor itself).

MW enhancement can take several forms, for example, reaction rates can be accelerated, yields can be improved, energy usage can be reduced, milder reaction condition can be achieved, and reaction pathways can be selectively activated or suppressed. Some typical

Table 8.1 Comparison of conventional synthesis of materials vs. MW heating

Parameters	Conventional solvothermal	Microwave
Synthesis time	Hours to several days	Seconds to hours
Scale	mg–kg	mg–kg
Reaction control	Minimal (autoclave settings)	High (real time)
Online monitoring	Difficult	Temperature, pressure, and visual
Particle size and distribution	Wide, difficult to control	Narrow, easy to control
Product yield	Very low to low	Very high

factors of synthesis of zeolitic materials by means of conventional and MW heating sources are compared in Table 8.1.

However, the design parameters of an MW reactor must be considered very precisely. Usually due to uneven absorption of MWs by different materials, the MW field can cause nonuniform and localized superheating, which makes the design of special MW reactors, particularly at high pressures, a very critical issue.

MW application for the synthesis of nonmaterials is an extensive research area, with dynamically increasing number of published papers, especially in recent years. There is a range of approaches where MWs can help to produce better powders; however, this requires advanced control and instrumentation of processing conditions, as well as careful characterization of nanoparticles using widely accepted procedures. The MW-driven hydrothermal (HT) and solvothermal reactions provide fast heating of the reactants, which may accelerate the reaction and lead to products with higher purity. MW heating is particularly useful for reactions carried out under elevated pressures; because of contactless delivery of energy, high energy density is achieved. Moreover, short heating times lead to the formation of weakly agglomerated or nonagglomerated nanoparticles with high crystallinity and narrow grain size distribution. This progress is possible because of application of specially designed reactors suitable for pressure- and temperature-controlled conditions [24]. The energy transfer in MWs is due to the interaction of dielectric molecules. Dielectric heating occurs when dipoles of polar molecules follow the inversion of the alternating electric field. The dipole molecules in the irradiated field try to orient them with the magnetic field; as the magnetic field oscillates the molecules also follow the oscillations. Their motion is resisted by interparticle interactions and electric resistance. The induced energy dissipates in the form of heat through molecular friction, resulting in uniform distribution of heat compared to conventional heating. Furthermore, beyond thermal effects, nonthermal effects have also been observed. It has been shown that MW irradiation elevates the boiling point of the solvent. This is called the superheating effect caused due to retardation of nucleation in MW radiation. However, this nucleation retardation can be overcome by stirring the reaction mixture [25]. The activation energy is also observed to decrease during MW irradiation; this decrease in activation energy, compared to conventional heating, might be due to higher atomic mobility in the crystal lattice and can be given by Eqs. 8.1 and 8.2 [26]

$$K = A \exp\left(\frac{Q}{RT}\right) \tag{8.1}$$

$$A = \gamma\lambda^2\Gamma \tag{8.2}$$

where γ is the number of neighbor jump sites in the crystal lattice, λ is the jump distance, and Γ is the jump frequency. The number of jump sites and the jump distance depend on the crystal structure and not on the heating source. It is concluded that the pre-Arrhenius factor is different for MW-irradiated reactions due to a higher number of jump frequencies [26]. MW irradiation can also be used for nonpolar molecules by addition of susceptors. These materials efficiently absorb MWs and then transfer the absorbed energy as thermal energy to the molecules in the vicinity that are weak at absorbing MWs. Silicon carbide and ionic liquids have been shown to increase the overall MW absorption and significantly reduce the required MW energy [27, 28].

Arafat et al. [29] produced zeolite Y and ZSM-5 using pure chemicals and well-established oxide molar ratios of the precursor solution. It was reported that the reaction time of zeolite Y production from pure chemicals was reduced from 10 h to 50 h to 10 min using MWs instead of conventional heating, whereas ZSM-5 conversion was achieved in 30 min compared to several days. Fast dissolution of the gel in the MWs is associated with relatively fast crystallization of zeolite from the precursor gel [29]. MW heating produced purer zeolite crystals due to nucleation occurring at 120°C, while crystallization occurred at 100°C; therefore nucleation and crystal growth of other types of zeolite did not happen. The rapid synthesis of zeolite may indicate the absence of an induction period; however, this cannot be concluded from the given data [29].

Experiments conducted by Inada et al. [30] showed that MW-induced heating played a positive role in the production of zeolite from coal fly ash. They noted that irradiation schedule should be controlled over the course of the reaction to achieve the best results. It was further hypothesized that MWs assist in the dissolution of alumina and silica in the reaction mixture, thus assisting the reaction [30]. It was reported that the activation time of the production of zeolite from coal fly ash using MWs was reduced from 24 h to 48 h to 30 min [31]. Different researchers use different approaches and the choice of a chemical system to study. Many reports on the successful application of MW energy for manufacturing of different zeolitic materials prove that MW energy should be considered as an effective way to prepare uniform materials in a controllable manner at larger scale [32, 33]. Physicochemical properties of titania, zirconia, and hafnia nanopowders synthesized by means of HT, HT-MW, and HT-ultrasonic (HT-UTS) methods at different temperatures and times have been compared [34]. The results revealed that titania, zirconia, and hafnia crystallization is substantially enhanced and that the percentage of the thermally stable zirconia phase (m-ZrO_2) in the HT-MW and HT-UTS processes increases compared to conventional HT synthesis. They concluded that a common factor in both cases is likely the uniformity of heating of treated suspensions. Local overheating in the reaction mixture, which appears during both ultrasonication and MW treatment, can also play a role in accelerating the HT processes.

8.5 MOF Synthetic Approaches

To date, the majority of studies on these materials have been devoted to discovering new types of MOFs and demonstrating their potential applications in various fields. However, detailed investigations about the role of reaction conditions during synthesis and scale-up have been ignored. Although successful synthesis of many different MOFs on the lab scale (i.e., micrograms to grams) is reported, just a few are produced in larger amounts (i.e., kilograms). The average time for synthesis of most of these porous compounds is several days at relatively high crystallization temperatures of up to 300°C. The traditional methods of MOF synthesis are both slow and on a small scale. Currently, very little understanding of the crystal growth process and the effects of the synthesis condition is known. Researchers rely on a trial-and-error method for MOF synthesis.

In spite of the well-known extraordinary properties of MOF materials, one of the main obstacles of industrialization of these relatively new porous materials is the energy and time of the production process. Therefore, for the large scale, production of MOF materials is highly desirable.

Thus far, the majority of MOFs have been synthesized by hydro- or solvothermal methods using conventional electric (CE) heating [8] requiring reaction times as long as several days. However, because of their huge industrial potential [1, 35], a fundamental understanding of the synthesis of these materials is vital if they are to offer viable applications in industry. The synthesis times must be reduced, and the overall energy efficiency must be improved. To this end, alternative techniques such as solvent-free methods [36], electrochemical methods [37], US [38, 39], and MW irradiation [40–42] have been reported. Numerous zeolitic-like materials such as MOF compounds have been produced using MW-assisted synthesis.

Reaction conditions in MW-assisted synthesis may vary considerably with respect to the type of MOF under study. The MW apparatus and synthesis parameters such as irradiation power and time, temperature, and, consequently, pressure and the reacting mixture composition can be changed; however, others are specific to the system under investigation. Typically, MW synthesis of MOFs is conducted around 100°C and irradiation time is systematically varied in order to optimize the best conditions for the synthesis of the desired MOF product. The power of MW irradiation and how it is generated, delivered, and controlled throughout the reaction is the most important parameter in MW heating. This depends mainly on the device used in the specific experiment.

MW irradiation is a particularly promising alternative technique due to the minimization of energy and optimization of reaction conditions. MW irradiation is characterized by accelerated reactions, as a consequence of the intense localized heating, reducing reaction times from days and hours by classical heating to minutes and seconds. Since the magnitude of heating depends on the dielectric properties of the molecules, generated energy is supplied to the material directly and uniformly. This allows the whole material to be heated rapidly and simultaneously, resulting in homogeneous nucleation, fast crystallization, vast reductions in particle size (due to the increased nucleation rates), and higher yields.

Recently, a novel hybrid synthesis technique involving initial UTS irradiation followed by MW irradiation for rapid synthesis of different MOFs has been reported [43, 44]. The remarkable effects of UTS energy (20 kHz) are attributed to acoustic cavitation, the phenomenon involving the formation, growth, and implosive collapse of cavitation bubbles generated in a liquid by the ultrasound wave [23]. Accelerated reactions result from the violent collapse of thousands of these tiny micrometer-sized hotspots, generating extremely high temperatures (>5000 K), pressure (>2000 atm), and heating and cooling rates (>10^{11} K/s) [23]. In addition, UTS irradiation can lead to homogeneous nucleation. Despite these impressive capabilities, very little effort has been invested to understanding the beneficial effects of MW and UTS irradiation for MOF synthesis. Compared with IRMOF-1 synthesized by means of CE heating, crystallites synthesized using the combined UTS and MW approach were more cubic shaped and much smaller (size reduction by an order of magnitude of approximately 10), had a narrower particle size distribution (5–15 μm), and had similar high surface areas (~2470 m^2/g) [38]. Additionally, it was demonstrated that the accelerated synthesis of MIL-53(Fe) by UTS and MW irradiation at relatively low temperatures decreased the crystallization rate in the order UTS >MW >>CE [37]. These results suggest that synthesis performed under UTS and MW conditions will be very promising for producing MOFs. Successful synthesis of a MOF requires the reaction of a bridging ligand with a metal ion having more than one vacant or labile site. Syntheses are typically carried out under solvothermal conditions (>100°C) from a mixture containing the ligand and metal salt for 12–48 h [42]. Most commonly used solvents include water, alcohols, dialkyl formamides, and pyridine. Some of the more labile metal ions, which are most often used include Cu^{+}, Cu^{2+}, Ag^{+}, Cd^{2+}, Zn^{2+}, Co^{2+}, and Ni^{2+}. However, one of the difficulties in using labile metal ions is that they do not have a strong preference for a given geometry, leading to a lack of control and predictability over the structure of the network obtained. Flexibility in a bridging ligand having a number of possible conformations can also contribute to this problem. Another difficulty is that a single metal coordination geometry can propagate into more than one type of framework topology [45].

For example, more than 100 different topologies are possible for linking tetrahedral building blocks together [12]. Of course, during synthesis there are other important structure-determining factors affecting product crystallinity and yield, including the pH, concentration, solvent polarity, and temperature. It has been shown that only subtle changes in these parameters can lead to significant changes in product crystallinity and yields or the formation of entirely new phases. Therefore, a careful selection of reaction conditions is important.

Temperature is an especially important factor. Under solvothermal conditions the dielectric properties of the solution change, leading to weakened intermolecular interactions between the solvent molecules. For example, at 180°C the pH of water becomes 5.5; thus, an extrapolation of the conditions applied at room temperature is not valid since the chemistry of the solution is not the same at elevated temperatures. An excellent example of the temperature influence was demonstrated by the synthesis of cobalt (II) succinates between room temperature and 250°C. Over this temperature range, seven different materials were synthesized. As the temperature increased, the number of coordinated water molecules per cobalt atom decreased (when $T \leq 100$°C), the edge-sharing connectivity for the Co^{2+} octahedral increased, the coordination numbers for carboxylate groups increased, and above 100°C hydroxide groups were incorporated into the phases. The dimensionality of the structures was also changed. One-dimensional structures were isolated at 60°C (with isolated Co^{2+} octahedral structure for the inorganic part) and at 100°C (with isolated inorganic trimmers). The framework became 2D at 150°C in which 14-membered rings of the corner-sharing Co^{2+} octahedral structure were obtained. The frameworks were 3D above 190°C and 250°C, and the inorganic subnetwork was always 2D, but the connectivity of the M–O–M bonds was higher at higher temperature [8]. The nature of the metallic salt also greatly influences the final products. Owing to their own redox characteristics, counterions of metal salts can influence the reaction conditions. The Zn^{2+}/1,4-benzenedicarboxylic acid (BDC)/solvent reaction system resulted in the synthesis of MOF-5 with NO_3^-,$CH3CO_2^-$, and O_2^- as the counterion of the salt. However, when SO_4^- was used an unknown phase was synthesized [46]. Recently, significant steps have been taken in developing more efficient methods of MOF synthesis. The conventional solvothermal method can yield high-quality crystals, but it suffers from long reaction times and is difficult to scale above ~1 g. Therefore, researchers have been investigating new and efficient alternative synthesis techniques capable of reducing production costs by decreasing reaction times and improving energy efficiency. MW heating is a method that has already been applied to inorganic porous compounds as a means of accelerating the reaction rates. Crystallization of these materials using conventional solvothermal synthesis methods usually requires several days because of slow heat transfer. MW heating has proven to be a much more efficient synthesis technique, resulting in rapid crystallization, homogeneous nucleation, vast reductions in particle size/size distribution, higher yields, and facile morphology control and efficient evaluation of process variables [42]. The rate enhancements under MW irradiation are thought to be a consequence of intense localized heating that can be rapidly attained in the MW field. This is referred to as the thermal/kinetic effect, but specific MW effects are also responsible for the accelerations, which cannot be achieved by conventional heating. There have also been suggestions of the existence of nonthermal MW effects that result from a direct interaction of the electric field with dipolar molecules in the reaction medium [21]. The few studies that have investigated MW heating for MOF synthesis suggest promising results. Conventional synthesis methods of MIL-100 and MIL-101 require heating for 96 h at 220°C. However, Chang and coworkers [47] have shown that under MW irradiation these MOFs can be formed in less than 1 h. Another study demonstrated the accelerated reactions of a cubic nickel glutarate, MIL-77, which previously required several hours or days, depending on the reaction temperature. MW irradiation produced the more stable tetragonal nickel glutarate within a few minutes [48]. Morris and coworkers [41] have reported the synthesis of an anionic MOF by conventional and MW conditions using an ionic liquid 1-ethyl-3-methylimidazolium bromide as the solvent and template. Their results indicate that MW irradiation produces MOFs that are purer in phase and have higher crystallinity.

8.6 MOF Synthesis Using the Microwave and Ultrasonic Hybrid Technique

Ultrasound energy in the 20–100 kHz range is used in chemical synthesis. Ultrasound does not interact directly with molecules to induce the chemical change, as its typical wavelength (in the millimeter range) is too long compared to the molecules. However, it causes cavitation, which produces extremes of local temperatures and pressures in the liquid where the reaction happens. Furthermore, it breaks up solids and removes passivating layers of inert material to give a larger surface area for the reaction to occur over it. These phenomena make the reaction faster. The chemical enhancement of reactions by ultrasound has been explored and has beneficial applications in mixed-phase synthesis, materials chemistry, and biomedical uses. Because cavitation can only occur in liquids, chemical reactions are not used in the UTS irradiation of solids or gas–solid systems.

For example, in chemical kinetics, it has been observed that ultrasound can greatly enhance chemical reactivity in a number of systems by as much as a million-fold acting as a catalyst by exciting the atomic and molecular modes of the system (such as the vibrational, rotational, and translational modes). In addition, in reactions that use solids, ultrasound breaks up the solid pieces from the bubbles using the energy released by cavitation due to collapsing bubbles. This gives the solid reactant a larger surface area for the reaction to proceed, increasing the observed rate of reaction.

Sonochemistry can be performed by using a bath (usually used for UTS cleaning) or with a high-power probe.

Another method that accelerates reactions and produces small and homogeneous crystals is sonochemical synthesis. There have only been a few investigations using this method, but the results are promising. The accelerated reactions under ultrasound conditions are a result of a phenomenon termed "acoustic cavitation." This process involves the formation, growth, and implosive collapse of cavitation bubbles generated in the liquid by the ultrasound wave. Extremely high local temperatures (>5000 K), pressure (>2000 atm), and heating and cooling rates ($>10^{11}$ K/s) are generated [42].

MOF-177 and a Zn-2,2′-bipyridine-5,5′-dicarboxylate MOF synthesized by UTS irradiation produced crystals very similar to those obtained from conventional heating and MW irradiation. In the case of MOF-177, crystals had superior properties to those synthesized by MW irradiation [42]. Furthermore, synthesis of $Zn_3(BTC)_2 \cdot 12H_2O$ by this method allowed the control of the crystal size distribution over 50–900 nm, depending on the reaction time [38]. Despite these results the beneficial effects of MOF sonochemical synthesis have yet to be fully explored.

Recently, a novel hybrid synthesis technique involving initial UTS irradiation followed by MW irradiation for rapid synthesis of IRMOF-1 was reported by our group [44]. Compared with IRMOF-1 synthesized by means of CE heating, crystallites synthesized using this approach were more cubic shaped and much smaller (size reduction by a magnitude of approximately 10), had a narrower particle size distribution (5–15 μm), and had comparable high surface areas ($\sim$2470 m^2/g) [44].

To develop rapid and energy-efficient synthesis techniques utilizing UTS and MW irradiation, MW-assisted crystallization of MIL-53(Fe), a structurally flexible and nontoxic iron(III) benzenedicarboxylate MOF was studied by our group [43]. Syntheses were carried out under CE, MW, and UTS conditions to gain an understanding of the effects of multiple synthesis procedures on the product yield and crystallinity. Within each reaction procedure, two-level multifactorial designs were used to study the dependence of the two responses on each influential factor. Multifactorial designs are more efficient than studying one factor at a time, because they allow the effect of several factors and even interactions between them to be determined within a set of experiments. The structure of MIL-53(Fe) is composed of parallel *trans* corner-sharing iron(III) octahedral chains, each of which are crosslinked by BDC anions to form a 1D lozenge-shaped pore channel system [17]. MIL-53(Fe) only opens its pores in the presence of guest molecules; therefore, unlike other MOFs it does not have a high surface area. MIL-53 has the formula $M^{III}(OH)\cdot(O_2C-C_6H_4-CO_2)\cdot H_2O$ ($M = Al^{3+}$, Cr^{3+}, or Fe^{3+}) [17]. MIL-53(Fe) crystals synthesized by means of MW and UTS irradiation were compared with those synthesized by CE heating. The XRPD patterns (Fig. 8.5) of the most highly crystallized as-synthesized samples confirm that the crystal phase of the products is MIL-53(Fe), as evidenced by their monoclinic symmetry (*C2/c*, no. 15). Figure 8.6 illustrates typical scanning electron microscopy (SEM) images of crystallized MIL-53(Fe) obtained from each of the synthesis methods. Under CE conditions a bimodal distribution of particle sizes was obtained [43]. Two different crystal morphologies can be seen coexisting: large elongated triangular prism-shaped crystals ranging in length from 25 to 250 μm, in addition to much smaller hexagonal bipyramidal particles, approximately 2.5 μm in diameter (Fig. 8.6a). Despite the wide size distribution, these phases should be MIL-53(Fe), depending on the similarity of their X-ray powder diffraction (XRPD) patterns with the other synthesized samples (Fig. 8.5). This nonuniform size distribution can likely be explained by simultaneous nucleation and crystal growth as a result of slow heat transfer and poor temperature distribution under CE heating.

MIL-53(Fe) crystals synthesized under MW and UTS conditions produced small and homogeneous crystals, which is a clear indication of both the crystal-phase purity and the efficiency of these two synthesis methods. Despite the different scales used to create the SEM images, the particles synthesized from both methods can be seen to have approximate dimensions of 0.5–1.5 μm and hexagonal bipyramidal morphologies (Fig. 8.6b,c). This size reduction is typical of crystals synthesized under MW or UTS conditions, a phenomenon that can be attributed to uniform and fast nucleation. Furthermore, small MIL-53(Fe) crystals are especially effective in the fields of diffusion, catalysis, and drug adsorption/delivery.

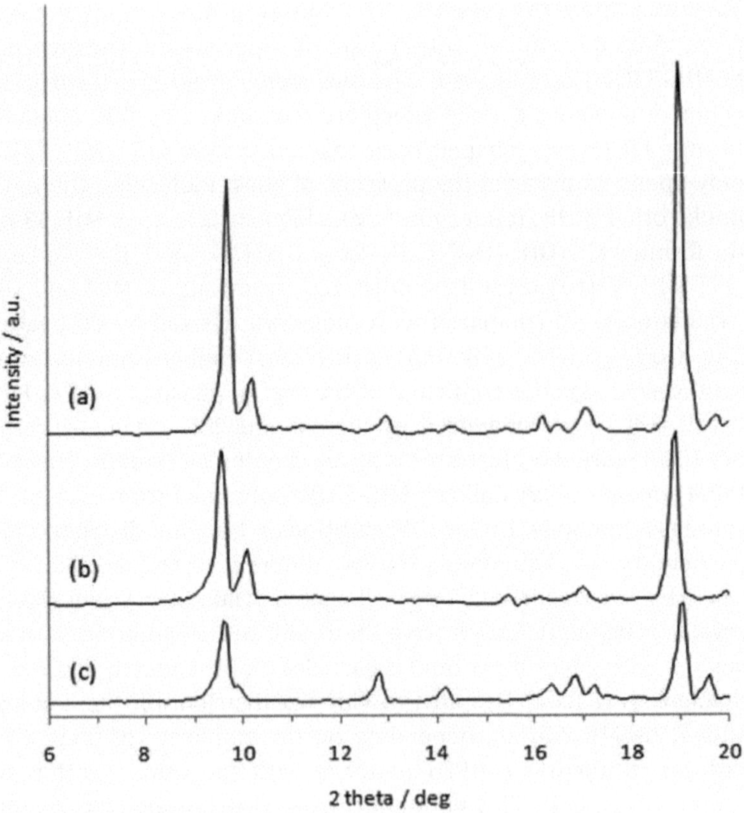

Figure 8.5 Typical XRPD patterns of fully crystallized MIL-53(Fe) synthesized by (a) CE for 18 h at 150°C, (b) MW for 30 min at 150°C, and (c) UTS for 15 min at 60% power.

The analysis of variance (ANOVA) approach was used to determine which synthesis variables significantly influenced MIL-53(Fe) yield and crystallinity utilizing models created by Design-Expert 7.1.5. All of the models used for the MW and CE synthesis methods are statistically significant (p-value <0.05). It was found that the time of exposure to MW energy (F-value $= 107.48$) and MW temperature (F-value $= 324.03$) strongly influence product crystallinity. As irradiation time and temperature increase, product crystallinity increases. MW power (F-value $= 13.99$) and the type of

Figure 8.6 Typical SEM images of fully crystallized MIL-53(Fe) synthesized by (a) CE for 18 h at 150°C (25 μm scale), (b) MW for 30 min at 150°C (5 μm scale), and (c) UTS for 15 min at 60% power (10 μm scale).

reaction vessel (F-value = 42.54) both have a much less significant influence on product crystallinity. Likewise, for the CE synthesis method both the time of conventional heating (F-value = 54.50) and the CE temperature (F-value = 109.65) strongly influence product crystallinity. The final model equations for each of the methods are shown in Table 8.2.

MW temperature (F-value = 29.92) and the type of reaction vessel (F-value = 26.46) were the factors most strongly influencing product yield. Much higher yields were obtained when the temperature was 150°C and the pressurized glass vessel was used. Comparing the results from each of the three synthesis methods revealed that UTS irradiation resulted in much lower yields than CE heating and MW irradiation. This is a surprising result, especially considering that higher yields are common when using MW and UTS energies instead of conventional solvothermal methods. However, this result can likely be explained by the collapse of cavitation

Table 8.2 The final model equations for product crystallinity using CE heating and MW and UTS irradiation

Model Equations
$\log10(RC_1) = 1.37 + 0.25A_1 + 0.36B_1 + 0.16B_1C_1 - 0.13A_1B_1C_1$
$\log10(RC_2) = 1.32 + 0.16A_2 + 0.28B_2 - 0.057C_2 - 0.10D_2 + 0.054A_2B_2 - 0.034A_2C_2$ $+0.050A_2D_2 + 0.081B_2C_2 + 0.043B_2D_2 - 0.088A_2B_2C_2 + 0.11A_2B_2D_2$
$RC_3 = 87.58 + 2.64A_3 - 1.24B_3 - 0.81A_3B_3$

RC_1, CE relative crystallinity; A_1, CE time; B_1, CE temperature; C_1, CE reaction vessel; RC_2, MW relative crystallinity; A_2, MW time; B_2, MW temperature; C_2, MW power; D_2, MW reaction vessel; RC_3, UTS relative crystallinity; A_3, UTS power; and B_3, UTS time.

Table 8.3 Final model equations for product yield from CE heating and MW and UTS irradiation

Model Equations
$Y_1 = 47.61 + 10.64A_1 + 7.01B_1 - 8.56A_1B_1$
$Y_2 = 30.46 + 12.58B_2 + 11.82D_2$
$Y_3 = 19.84 + 2.04A_3 + 5.48B_3 - 0.39A_3B_3$

Y_1, CE yield; A_1, CE time; B_1, CE temperature; Y_2, MW yield; B_2, MW temperature; D_2, MW reaction vessel; Y_3, UTS yield; A_3, UTS power; and B_3, UTS time.

bubbles close to or on the walls of the vial, which causes a microjet of liquid to be forced to the surface at speeds upwards of 200 m/s. Since UTS experiments were performed in an open vial much of the reaction mixture had actually escaped as mist, most likely as a result of this phenomenon. The final model equations for each of the synthesis methods are shown in Table 8.3.

Utilizing the MW synthesis method, it is found that nucleation had occurred in as little as 30 s. The nucleation time was determined by visual observation (when the reaction mixture became cloudy). Increasing the reaction time to 10 min yielded highly crystalline materials; the most crystalline materials were obtained after 30 min. Nucleation was first observed after 5 min under UTS conditions. Since the effects of UTS time and UTS power on product crystallinity were shown to be insignificant we can conclude that after only 7 min the products have reached their maximum crystallinity for this

Figure 8.7 SEM image of MIL-53(Fe) synthesized by UTS for 7 min at 40% power.

method. The SEM image of this product (Fig. 8.7) shows crystals similar to those in Fig. 8.6c, confirming the ANOVA results that time of exposure and power level are insignificant. Also, the XRPD spectrum is identical to that obtained after 10 min of MW irradiation at 150°C (Fig. 8.8). To the best of our knowledge, these reaction times are the shortest reported for MIL-53(Fe) synthesis. The onset of nucleation under CE heating occurred after 20 min, which was 40 times longer than that observed for MW irradiation. The most crystallized products were obtained after 15 h (30× and 128× longer than for MW and UTS irradiation, respectively), indicating that UTS and MW irradiations are quicker and more efficient alternatives to CE heating. It has generally been accepted that in the majority of cases, rate enhancements under MW conditions can be explained by high reaction temperatures that can rapidly be attained in an MW field [15]. This is referred to as the thermal/kinetic effect. The specific MW effects caused by dielectric heating must also be considered, and these include (1) the superheating effect, (2) the formation of microscopic hotspots, (3) more uniform heating, and (4) selective heating. These accelerations (which are still currently the subject of considerable debate) cannot be achieved by conventional heating. There have also been suggestions of the

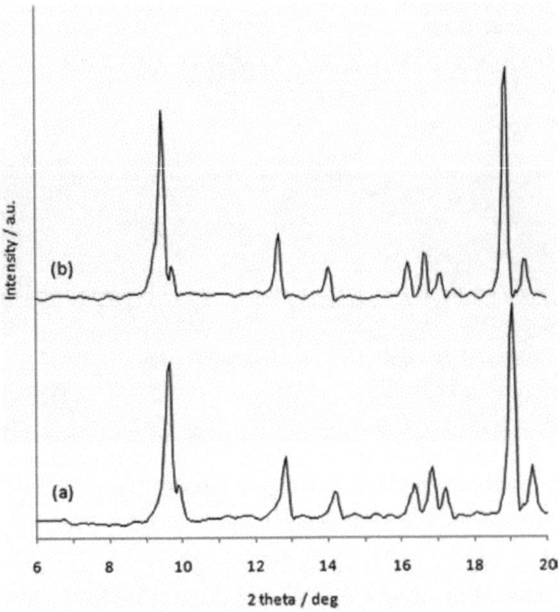

Figure 8.8 XRPD patterns of MIL-53(Fe) synthesized by (a) UTS for 7 min at 40% power and (b) MW for 10 min at 150°C.

existence of nonthermal MW effects that result from a direct interaction of the electric field with dipolar molecules in the reaction medium. Accelerated reactions observed under UTS irradiation result from acoustic cavitation and the implosive collapse of formed cavitation bubbles.

To illustrate the efficiency of MW and UTS syntheses we compared the total energy consumption of these two methods to that of CE heating. For this, we determined the energy required to (1) induce nucleation and (2) produce the most crystalline materials by each method. The MW operated on a 120 V power source at 15 A. This was used to calculate the instantaneous power usage P (watts), as shown:

$$P = VI \tag{8.3}$$

where V is the voltage (V) and I is the current (A). The energy consumption, E, in joules (J) was then estimated using Eq. 8.4:

$$E = Pt \tag{8.4}$$

where P is the constant power (watts) and t is the time (seconds). The MW required approximately 54 and 3240 kJ of energy for nucleation and maximum crystallization, respectively. As determined from ANOVA, the UTS time and the UTS power were not significant factors affecting product crystallinity; therefore, the method requiring the least amount of energy will be considered. The ultrasound also operated on a 120 V power source at 15 A, which corresponded to 540 and 756 kJ of consumed energy for nucleation and maximum crystallization, respectively. Compared to CE heating, these energy requirements are very low. The oven operated on a 115 V power source at 12 A and required approximately 2500 kJ of energy to reach the temperature set point (150°C). With a power rating of 1.2 kW, an energy consumption of 4320 kJ was required to run the oven for 1 h at the set point. This corresponded to an energy consumption of 3900 and 67,000 kJ for nucleation and maximum crystallization, respectively. Considering these energy requirements, as well as the fact that both MW and UTS irradiation produced small, homogeneous crystals compared to the two populations of crystals obtained from CE heating, it is evident that these two technologies are quicker, more efficient, and greener alternatives to conventional synthesis methods.

8.7 Conclusions

MW energy heats material in different mechanism compared to the conventional heating. MWs can enhance chemical reactions in several forms, including accelerating the reaction rate, increasing the product yield, decreasing the energy consumption, improving the results at milder reaction condition, and activating the selective reaction pathways. Zeolitic and zeolitic-like compounds such as MOF materials can be synthesized by MW irradiation at higher yield and enhanced quality. Zeolites and MOF crystals can be synthesized quickly in a couple of minutes by means of the MW-assisted synthesis technique. MOF crystals produced by MW and UTS conditions are smaller and more homogeneous, whereas those produced by CE heating may contain different morphologies and crystallite sizes. The observed size reduction of crystals synthesized under MW and UTS conditions can be attributed to the fast and uniform nucleation. The nonuniform size distribution of particles produced from CE heating is likely a result of the method's inherently slow heat transfer and poor temperature distribution within the reaction mixture. For both MW irradiation and CE heating methods, the exposure time and temperature were shown to have a significant influence on product crystallinity. However, for the UTS method neither the exposure time nor the power had a significant influence on crystallinity, suggesting that the products reached their maximum crystallinity in this method very shortly after nucleation.

Therefore, at present, MW heating seems to become the most promising way to synthesize MOFs and other zeolitic materials at shorter crystallization times, yielding more homogenous and finer crystallite sizes, properties that are crucial for applications and commercialization of these materials. We invite every material chemist to explore MW heating in the hope of discovering new methodologies for facile production of these materials and new products.

References

1. Lim, K. L., Kazemian, H., Yaakob, Z., and Daud, W. R. W. (2010). Solid state materials and methods for hydrogen storage: a critical review, *Chem. Eng. Technol.*, **33**(2), pp. 213–226.
2. Wang, Z., Chen, G., and Ding, K. (2009). Self-supported catalysts, *Chem. Rev.*, **109**(2), pp. 322–359.
3. Li, J.-R., Kuppler, R. J., and Zhou, H. C. (2009). Selective gas adsorption and separation in metal-organic frameworks, *Chem. Soc. Rev.*, **38**, pp. 1477–1504.
4. Morris, R. E., and Wheatley, P. S. (2008). Gas storage in nanoporous materials, *Angew. Chem., Int. Ed. Engl.*, **47**, pp. 4966–4981.
5. Borns, D. J. (2006). Oil and gas industry applications of gas flow in porous media. In *Theory and Applications of Transport in Porous Media*, Jacob Bear, Senior ed. (Springer, Dordrecht, the Netherlands), **20**, pp. 407–418.
6. McEnaney, B., Alain, E., Yin, Y. F., and Mays, T. J. (2001). porous carbons for gas storage and separation, *NATO Sci. Ser., Ser. E*, **374**, pp. 295–318.

7. Sabouni, R., Kazemian, H., and Rohani, S. (2013). Mathematical modeling and experimental breakthrough curves of carbon dioxide adsorption on metal organic framework CPM-5, *Environ. Sci. Pollut. Res.,* **47**(16), pp. 9372–9380.
8. Ferey, G. (2008). Hybrid porous solids: past, present, future, *Chem. Soc. Rev.,* **37**, pp. 191–214.
9. Rowsell, J. L. C., and Yaghi, O. M. (2004). Metal-organic frameworks: a new class of porous materials, *Microporous Mesoporous Mater.,* **73**, pp. 3–14.
10. Kitagawa, S., Kitaura, R., and Noro, S. (2004). Functional porous coordination polymers, *Angew. Chem., Int. Ed.,* **43**, pp. 2334–2375.
11. Hong, D., Hwang, Y. K., Serre, C., Ferey, G., and Chang, J. (2009). Porous chromium terephthalate MIL-101 with coordinatively unsaturated sites: surface functionalization, encapsulation, sorption and catalysis, *Adv. Funct. Mater.,* **19**, pp. 1537–1552.
12. Yaghi, O. M., O'Keeffe, M., Ockwig, N. W., Chae, H. K., Eddaoudi, M., and Kim J. (2003). Reticular synthesis and the design of new materials, *Nature,* **423**, pp. 705–714.
13. Meek, S. T., Greathouse, J. A., and Allendorf, M. D. (2011). Metal-organic frameworks: a rapidly growing class of versatile nanoporous materials, *Adv. Mater.,* **23**, 249–267.
14. Hamon, L., Serre, C., Devic, T., Loiseau, T., Millange, F., Ferey, G., and Weireld, G. D. (2009). Comparative study of hydrogen sulfide adsorption in the MIL-53(Al, Cr, Fe), MIL-47(V), MIL-100(Cr), and MIL-101(Cr) metal–organic frameworks at room temperature, *J. Am. Chem. Soc.,* **131**, pp. 8775–8777.
15. Britt, D., Tranchemontagne D., and Yaghi O. M. (2008). Metal-organic frameworks with high capacity and selectivity for harmful gases, *Proc. Natl. Acad. Sci. U S A,* **105**, pp. 11623–11627.
16. Horcajada, P., Serre, C., Vallet-Regí, M., Sebban, M., Taulelle, F., and Férey, G. (2006). Metal–organic frameworks as efficient materials for drug delivery,*Angew. Chem., Int. Ed.,* **45**, pp. 5974–5978.
17. Horcajada, P., Serre, C., Maurin, G., Ramsahye, N. A., Balas, F., Vallet-Regí, M., Sebban, M., Taulelle, F., and Férey, G (2008). Flexible porous metal-organic frameworks for a controlled drug delivery, *J. Am. Chem. Soc.,* **130**, pp. 6774–6780.
18. Horcajada, P., Chalati, T., Serre, C., Gillet, B., Sebrie, C., Baati, T., Eubank, J. F., Heurtaux, D., Clayette, P., Kreuz, C., Chang, J. S., Hwang, Y. K., Marsaud, V., Bories, P. N., Cynober, L., Gil, S., Férey, G., Couvreur, P., and

Gref, R (2010). Porous metal–organic-framework nanoscale carriers as a potential platform for drug delivery and imaging, *Nat. Mater.*, **9**, pp. 172–178.

19. Li, Y., and Yang, W. J. (2008). Microwave synthesis of zeolite membranes: a review, *Membr. Sci.*, **316**, pp. 3–17.
20. Industrial Mineral Background Paper #14. (1990). *Microwaves and Minerals; I. Technology Review, II. Tests of Ontario's Industrial Minerals* (Atomic Energy of Canada, Research Company, Chalk River, Ontario, and Voss Associates Engineering, Victoria, British Columbia).
21. Kappe, C. O. (2004). Controlled microwave heating in modern organic synthesis, *Angew. Chem., Int. Ed.*, **43**, pp. 6250–6284.
22. Martín-Gil, J., Martín-Gil, F. J., José-Yacamán, M., Carapia-Morales, L., and Falcón-Bárcenas, T. (2005). Microwave-assisted synthesis of hydrated sodium uranyl oxonium silicate, *Polish J. Chem.*, **79**, pp. 1399–1403.
23. de la Hoz, A., Diaz-Ortiz, A. L., and Moreno, A. (2005). Microwaves in organic synthesis. Thermal and non-thermal microwave effects, *Chem. Soc. Rev.*, **34**, pp. 164–178.
24. Park, J., Chan Kim, B., Soo Park, S., and Chan Park, H. (2001). Conventional versus ultrasonic synthesis of zeolite 4A from kaolin, *J. Mater. Sci. Lett.*, **20**(6), pp. 531–533.
25. Perreux, L., and Loupy, A. (2001). A tentative rationalization of microwave effects in organic synthesis according to the reaction medium, and mechanistic considerations, *Tetrahedron*, **57**, pp. 9199–9223.
26. Binner, J. G. P., Hassine, N. A., and Cross T. E. (1995). The possible role of the pre-exponential factor in explaining the increased reaction rates observed during the microwave synthesis of titanium carbide, *J. Mater. Sci.*, **30**, pp. 5389–5393.
27. Glasnov, T. N., and Kappe, C. O. (2007). Microwave-assisted synthesis under continuous-flow conditions, *Macromol. Rapid Commun.*, **28**, pp. 395–410.
28. Leadbeater, N. E., and Torenius, H. M. (2002). A study of the ionic liquid mediated microwave heating of organic solvents, *J. Org. Chem.*, **67**, pp. 3145–3148.
29. Arafat, A., Jansen, J. C., Ebaid, A. R., and Van Bekkum, H. (1993). Microwave preparation of zeolite Y and ZSM-5, *Zeolites*, **13**, pp. 162–165.
30. Inada, M., Tsujimoto, H., Eguchi, Y., Enomoto, N., and Hojo, J. (2005). Microwave-assisted zeolite synthesis from coal fly ash in hydrothermal process, *Fuel*, **84**, pp. 1482–1486.

31. Querol, X., Alastuey, A., López-Soler, A., Plana, F., Andrés, J. M., and Juan, R. (1997). A fast method for recycling fly ash: microwave-assisted zeolite synthesis, *Environ. Sci. Technol.*, **31**, pp. 2527–2533.
32. Hwang, Y. K., Lee, U. H., Chang, J. S., Kwon, Y. U., and Park, S. E. (2005). Microwave-induced fabrication of MFI zeolite crystal films onto various metal oxide substrates, *Chem. Lett.*, **34**, pp. 1596–1597.
33. Yoo, Y., and Jeong, H. K. (2008). Rapid fabrication of metal organic framework thin films using microwave-induced thermal deposition, *Chem. Commun.*, **21**, pp. 2441–2443.
34. Suslick, K. S. (1990). Sonochemistry, *Science*, **247**, pp. 1439–1445.
35. Mueller, U., Schubert, M., Teich, F., Puetter, H., Schierle-Arndt, J., and Pastré K., (2006). Metal–organic frameworks: prospective industrial applications, *J. Mater. Chem.*, **16**, pp. 626–636.
36. Rybak, J., Schellenberg, I., Pöttgen R., and Müller-Buschbaum, K. Z. (2010). MOFs by transformation of 1D-coordination polymers II: the homoleptic divalent rare earth 3D benzotriazolate $^{3}_{\infty}$[Eu(Btz)$_2$] initiating from $^{1}_{\infty}$[Eu(Btz)$_2$(BtzH)$_2$], *Anorg. Allg. Chem.*, **636**, pp. 1720–1725.
37. Haque, E., Khan, N., Park, J. J., and Hung, J. (2010). Synthesis of a metal–organic framework material, iron terephthalate, by ultrasound, microwave, and conventional electric heating: a kinetic study, *Eur. Chem. J.*, **16**, pp. 1046–1052.
38. Qiu, L., Li, Z., Wu, Y., Wang, W., Xu, T., and Jiang, X. (2008). Facile synthesis of nanocrystals of a microporous metal–organic framework by an ultrasonic method and selective sensing of organoamines, *Chem. Commun.*, **31**, pp. 3642–3644.
39. Sabouni, R., Kazemian, H., and Rohani, S. (2010). A novel combined manufacturing technique for rapid production of IRMOF-1 using ultrasound and microwave energies, *Chem. Eng. J.*, **165**, pp. 966–973.
40. Klinowski, J., Almeida Paz, F. A., Silva, P., and Rocha, J. (2011). Microwave-assisted synthesis of metal–organic frameworks, *Dalton Trans.*, **40**, pp. 321–330.
41. Lin, Z., Wragg, D. S., and Morris, R. E. (2006). Microwave-assisted synthesis of anionic metal–organic frameworks under ionothermal conditions, *Chem. Commun.*, **19**, pp. 2021–2023.
42. Leonelli, C., and Mason, T. J. (2010). Microwave and ultrasonic processing: now a realistic option for industry, *Chem. Eng. Process.*, **49**, pp. 885–900.

43. Gordon, J., Kazemian, H., and Rohani, S. (2012). Rapid and efficient crystallization of MIL-53(Fe) by ultrasound and microwave irradiation, *Microporous Mesoporous Mater.,* **162**, pp. 36–43.

44. Sabouni, R., Kazemian, H., and Rohani, S. (2012). Microwave synthesis of CPM-5 metal organic framework, *Chem. Eng. Technol.,* **35**(6), pp. 1085–1092.

45. James, S. L. (2003). Metal-organic frameworks, *Chem. Soc. Rev.,* **32**, pp. 276–288.

46. Bauer, S., Bein, T., and Stock, N. (2005). High-throughput investigation and characterization of cobalt carboxy phosphonates, *Inorg. Chem.,* **44**, pp. 5882–5889.

47. Jhung, S. H., Lee, J. H., and Chang, J. S. (2005). Microwave synthesis of a nanoporous hybrid material, chromium trimesate, *Bull. Korean Chem. Soc.,* **26**, pp. 880–881.

48. Jhung, S. H., Lee, J., Forster, P. M., Ferey, G., Cheetham, A. K., and Chang, J. (2006). Microwave synthesis of hybrid inorganic–organic porous materials: phase-selective and rapid crystallization, *Chem. Eur. J.,* **12**, pp. 7899–7905.

Chapter 9

Microwave-Assisted Synthesis of Metallic Nanoparticles

Stephany García, Graham W. Piburn, and Simon M. Humphrey
Department of Chemistry, The University of Texas at Austin, 2.204 Welch Hall, 105 E. 24th St. A5300, Austin, TX 78712-1224, USA
smh@cm.utexas.edu

9.1 Introduction

The fundamental properties and practical applications of metallic nanoparticles (MNPs) continue to generate extensive interest in both academic and industrial settings. The magnetic [1–3], optical [3–5], and catalytic [6, 7] properties of MNPs with dimensions below 10 nm display behaviors that can differ greatly from the corresponding bulk metals, because quantum size confinement effects become pronounced in such materials [8]. Small MNPs also exhibit inherently high surface-area-to-volume ratios [9] and are therefore atom economical for surface-specific applications [10, 11]. The spectroscopic, electronic, and chemical properties of MNPs are also directly affected by particle morphology, which correlates with specific surface structure. From an applications standpoint, it has

Microwave Engineering of Materials: From Mesoscale to Nanoscale
Edited by Erwann Guénin

ISBN 978-981-4669-42-9 (Hardcover), 978-981-4669-43-6 (eBook)
www.panstanford.com

therefore become important to achieve fine synthetic control over both the average size and the shape of MNPs.

Classical top-down approaches to preparing MNPs such as mechanical grinding of bulk metal are ideal for large-scale production, but they are severely limited by the failure to produce very small particles and by the size and morphological nonuniformity of the resulting MNPs [12–14]. Conversely, bottom-up approaches—which rely on the controlled reduction of molecular metal precursors to construct larger clusters and particles—usually suffer from scalability issues but, importantly, offer vastly improved control over MNP size and shape [14–16]. Indeed, it has been shown that extremely small (>1.5 nm) MNPs can be obtained, in which small clusters of atoms (e.g., Ag_4, Rh_{55}, Pt_{147}) display atomic packing arrangements that differ from the bulk metal [17–21]. Arguably the most common type of bottom-up synthetic approach involves the solvent-based in situ chemical reduction of readily available metal precursors (e.g., metal halides, nitrates, acetates, simple organometallic species, etc.) in the presence of suitable reducing agents [14] or alternatively by using the solvent itself as a sacrificial reductant (e.g., the polyol method) [22, 23]. It is also common for surface passivation ligands, or capping agents, to be employed. These are commonly organic molecules, polymers, or bulky ions that bind weakly to the surfaces of the growing MNPs, thus inhibiting particle-on-particle aggregation [24, 25]. In some cases, capping agents have been shown to preferentially bind to certain faces of growing MNPs, which pacifies growth along a certain atomic directions and can result in enhanced morphological control of the resulting structures [26–28]. Physical parameters such as reaction time and temperature have also been shown to play vital roles in the synthesis of defined MNPs. Some less obvious reaction parameters such as precursor addition rate or reaction stirring rate are also important in ensuring reproducibility between individual reactions [29, 30]. Furthermore, the kinetics that govern the initial phases of MNP nucleation are quite different from the kinetics that control subsequent growth at existing nucleates to yield larger MNPs [16, 31]. These physical effects have been thoroughly examined in a number of seminal reports, the details of which are beyond the scope of our discussion here. However, it is important to acknowledge that the coupling of these factors with the vast library of solvents, metal precursors, and additives available provides a wealth of synthetic possibilities. A multitude of methods for the preparation of a vast array of MNPs have been reported in a relatively short period of time. Unfortunately, many existing methods are not generally applicable to the synthesis of a broad range of MNPs, and many are fraught with difficulties concerning reproducibility.

The identification of truly systematic, scalable, and widely applicable methods for the reproducible bottom-up synthesis of MNPs has become an important consideration. Some notable recent advances have been achieved in this arena by replacement of conventional (convective) heating as the source of external energy input in MNP synthesis with faster, more economical, and more versatile irradiation methods. Sonochemical synthesis and gamma ray irradiation are just two such examples [32, 33]. Over the past decade, microwave-assisted heating has continued to grow in popularity, as demonstrated by the number of new reports of microwave-assisted MNP synthesis (Fig. 9.1).

Microwave-assisted heating in wet MNP synthesis provides several discernible advantages over convective heating. First, dielectric heating affords more uniform heating of the reaction medium, as well as ensuring more efficient heat transfer to reactants within the reaction vessel, thus reducing crystallization (particle

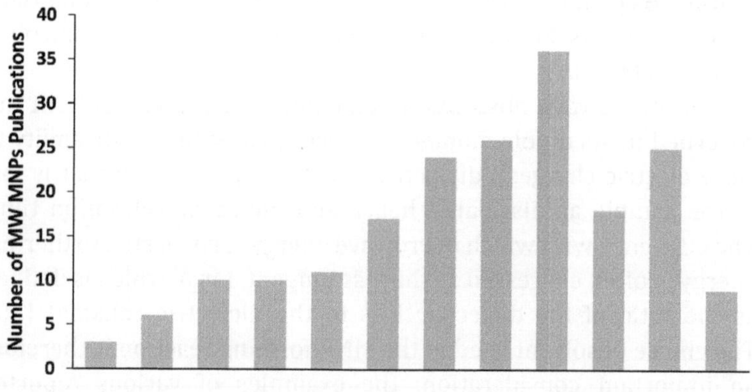

Figure 9.1 Number of publications related to the synthesis of MNPs using microwave irradiation per year from January 2003 to May 2014.

nucleation) times [34, 35]. This is particularly true for the use of polar solvents as reaction media, since they strongly rotationally couple to microwaves, which results in rapid energy transfer [36]. Many common inorganic precursors (such as the metal precursors themselves) are also known to strongly couple with microwaves via dipole polarization or ionic conduction [35]. As a result, thermal induction periods can also be significantly shortened using microwave heating. Second, microwave irradiation is predicted to generate localized high-temperature regions in solution (a.k.a., hot spots), which are thought be at least several tens of degrees hotter than the average bulk solvent temperature [1, 38]; reduction of metal ions is likely to proceed at much faster rates within such zones [34]. As discussed above, earlier studies of nanocrystal formation have identified a clear distinction between the nucleation and growth stages in a reaction mechanism. Synthetic control over both phases is required in order to access uniform nanostructures with narrow size distributions [39]. To prepare highly uniform nanoparticles (NPs), it is equally necessary to prevent further nucleation during subsequent overgrowth processes, which would otherwise lead to the formation of MNPs with bimodal size distributions [40]. The rapid heating rates provided by microwave heating should therefore result in the reduction of bulk thermal gradients in solution, while concomitantly providing much higher (superheated) localized reaction temperatures than cannot be achieved by conventional heating [34]. Thus, microwave-assisted heating presents potentially ideal conditions for the homogeneous nucleation and controlled growth of MNPs.

The microwave absorption capabilities of a given solvent are governed by its dipole moment, dielectric constant (or its ability to store electric charges), dielectric loss (or input energy that is lost to the sample as dissipated heat), and dielectric relaxation time. The efficiency with which microwave energy is converted to thermal energy is often expressed as the loss tangent, tan δ, which is defined as the ratio of the dielectric loss to the dielectric constant [36]. The choice of solvent used in the MNP-forming reaction is therefore an important consideration; the examples of various reported syntheses that are described in this chapter provide a perspective on the role of solvents. Since MNP formation in solution is a heterogeneous process, the microwave absorption properties of the solid–liquid interface (or more specifically, the MNP–solvent interface) is particularly important [41]. It is believed that the accessibility of higher temperatures can lower thermodynamic and kinetic barriers at this interface, thus facilitating growth and formation of particular MNP morphologies [42]. Selective heating and energy variation of intermediates also lead to surface structure control, in a similar fashion to how certain selective capping agents have been utilized under conventional heating. Even small variations in microwave absorption characteristics at different crystallographic faces of a growing MNP can induce preferential growth of certain facets to yield a wide range of kinetically favored structures. Enhanced single-particle crystallinity is also often observed in inorganic nanostructures that are synthesized using microwave dielectric heating, whereas conventionally prepared MNPs are more susceptible to polycrystallinity and twinning [29, 43–45]. This effect has been partly attributed to the selective decrease in energy of intermediate species in solution, caused by the higher effective temperatures experienced by both the metal precursors and the reaction medium [35, 40].

In this chapter, pertinent literature examples of microwave-assisted MNP syntheses will highlight how microwave-assisted heating can facilitate the synthesis of stable and well-defined MNPs with superior resulting properties. Particular experimental methodologies and apparatus that involve microwave-assisted heating to obtain specific MNPs in a highly reproducible fashion are also discussed.

9.2 Equipment

Microwaves can be adequately provided for most small-scale MNP syntheses using a simple magnetron device akin to what is installed in a regular domestic microwave oven. Such magnetrons usually generate photons with a frequency of 2.45 GHz (λ = 12.2 cm), which roughly translates to an energy value of 1 J/mol. This energy is much too low to cleave chemical bonds, but it can excite molecular rotations through dipole polarization or, in the case of ionic species, can cause translation known as ionic conduction. The use of actual domestic microwave ovens for research purposes is strongly discouraged for safety reasons. However, a number of compact microwave reactors specifically designed for laboratory use are now available, which feature modifications that are designed to withstand harsher chemical environments and also protect the user from harm in the case of potential reaction breaching and/or venting. Commercially available microwave reactor systems also feature fine temperature and pressure control, which is intrinsically important for reproducibility in MNP synthesis.

A number of secondary and perhaps less obvious factors should also be considered when deciding which type of microwave reactor to use in MNP synthesis: compact bench-top devices that are designed with a small cavity that fits round-bottomed flasks and glass vials are perhaps most popular for individual reactions, while larger box-type reactors that more closely resemble household microwave ovens are usually preferred for scale-up applications. The Discover and MARS systems manufactured by CEM Corporation are two such examples. However, the homogeneity of microwave irradiation in each cavity is quite different and can impose different outcomes for a given reaction. The microwave density in a small cavity reactor is directional, while microwaves inside a large-cavity reactor are not as homogeneous throughout the cavity. Some commercially available reactors employ so-called multimode heating, where the microwave cavity is larger than the wavelength of the microwaves. The interference of the waves leads to the simultaneous creation of high-energy and low-energy fields and nodes within the microwave cavity. The specified reaction temperature can be maintained by toggling between on and off states of the magnetron using a fixed power output. Other reactors employ an alternative continuous mode of heating, in which the reaction temperature is maintained by constant irradiation with modulated power. In some cases, materials that are poor absorbers of microwave energy (or are only available in small quantities) are heated using a single-mode resonant cavity arrangement, in which the microwave power and frequency are both tuned to ensure optimal heating. Each of these methods of maintaining a constant reaction temperature typically uses only a fraction of the reactor's maximum available power, which is often in the vicinity of 1600 W; this much power is generally only utilized during initial and/or rapid heating steps. This is especially helpful if the materials need to be heated to high temperatures in a controlled fashion [46]. Single-mode ovens with options for batch or continuous processing have only been designed for research purposes, given that this experimental setup is more expensive [47]. It is clear that MNP syntheses do not always proceed in an identical fashion using different heating modes [48].

Modern commercially available microwave systems are equipped with built-in magnetic stirrers and advanced pressure and temperature sensors that allow for accurate control of reaction parameters throughout the course of a reaction, as shown in Fig. 9.2 [35]. A more thorough review of instrumentation currently available for microwave ovens can be found in the book recently published by Kappe, Dallinger, and Murphree [49].

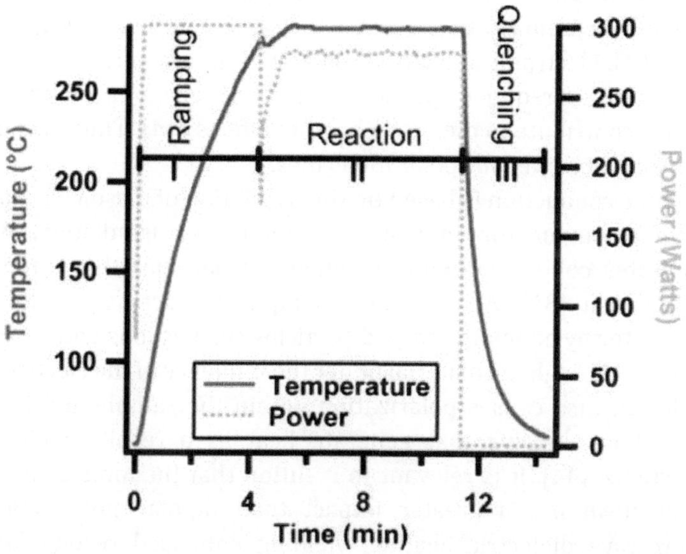

Figure 9.2 Temperature (°C) and power (W) profiles of a typical nanoparticle synthesis reaction (InGaP synthesis). During the reaction period (II), power output is regulated by the microwave reactor in order to maintain a constant temperature throughout the process. Reprinted with permission from Ref. [42], Copyright 2005 American Chemical Society.

9.3 Common Reagents and Their Compatibility with Microwave Heating

As previously discussed, the two most important mechanisms of microwave heating in MNP synthesis are dipolar polarization and ionic conduction [50]. Electromagnetic radiation produces an oscillating field that causes the dipoles and ions to continually attempt to realign in the electric field. Depending on the frequency of the irradiation and the time it takes for the dipoles and ions to align and misalign with the field, the molecules might have the ability to release heat via molecular friction or dielectric loss [44, 51]. To produce resistive heating in the reaction medium, the dipoles should lag behind the electric field oscillation but the frequency of the field should not be that much greater than the response time of the dipoles [35]. In other words, the dipoles should still have time to realign themselves with the field. This type of heating is commonly referred to as dielectric heating, and it refers to the input of microwave energy that is lost to the sample by being dissipated as heat [44]. Electronic and atomic polarizations are much faster (10^{-9} Hz) than the frequency of the microwaves, which means that they do not contribute to dielectric heating effects [44]. Thus, this effect is mostly observed for polar molecules.

Ionic conduction is based on the oscillation of dissolved charged particles under the influence of microwave irradiation, which generates collisions among neighboring particles that ultimately create heat [35]. An irradiated sample can act as an electrical conductor by enabling charged particles such as ions and electrons to move through the material under the influence of the electric field, which in turn causes polarization within the sample and induces consistent currents that generate heat as a result of electrical resistance [44]. It is relevant to mention that the ionic conduction mechanism has a greater impact than dipolar polarization in microwave dielectric heating, yielding enhanced results for the synthesis of NPs in ionic liquids (ILs).

A material with a high dielectric loss factor will efficiently convert electromagnetic radiation into heat and will in turn rapidly heat. This means that the loss tangents of solvents depend on the relaxation times of the molecules, which are clearly related to the nature of the functional groups within and the volume of the molecules. Having these criteria in mind, it is possible to categorize solvents on the basis of their microwave absorbing properties. Solvents with a loss tangent greater than 0.5 are considered to be high absorbing, while solvents with loss tangents between 0.1 and 0.5 are categorized as medium absorbing. Solvents with a loss tangent of less than 0.1 are low absorbing and may sometimes be referred to as microwave transparent (nonabsorbing).

It is helpful to have components in a chemical reaction that strongly couple with microwave energy in order to potentially achieve higher heating rates than those attained using convective heating. Organic and inorganic solvents with low molecular weights and high dipole moments effectively couple with microwaves with a frequency of 2.45 GHz [49]. Furthermore, it is beneficial to carry out reactions in solvents with dielectric constants that vary from that of the reactants in order to induce selective dielectric heating. Selective heating enhances energy transfer directly to the reactants, thus triggering an instantaneous internal temperature increase [52, 53]. The associated rapid temperature change lowers the activation energy of a given reaction compared to conventional heating methods and enhances the reaction rate accordingly [54]. This direct heating also minimizes thermal gradient effects [53, 55] that lead to poor MNP nucleation processes and broadened size distributions [40]. Figure 9.3 depicts this homogeneous heating effect [56]. Similarly, if metal precursors with large microwave absorption cross sections (relative to the solvent) are used, high effective reaction temperatures can be achieved, leading to the rapid decomposition of the precursors used and to the creation of highly supersaturated solutions. These conditions are ideal to promote nucleation and growth of MNPs in solution and favor the formation of smaller nanocrystals [57–59]. Hence, microwave-assisted synthetic methods can yield MNPs with narrow size distributions by reducing the formation of agglomerates; the starting materials are rapidly consumed, and the chemical reaction is governed by fast kinetics [44].

The use of highly absorbing chemical reagents and solvents are fortunately commonplace in the fabrication of inorganic NPs using convective heating, especially in polyol reactions [50, 59, 60].

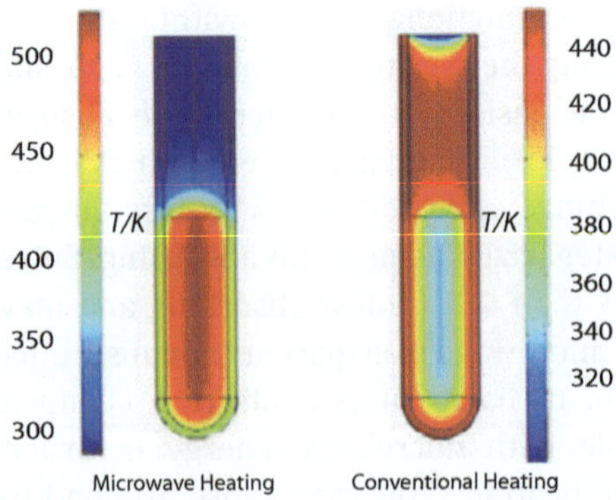

Figure 9.3 Temperature profile of reaction vessel after 60 s of (left) microwave irradiation exposure and (right) convective heating using an oil bath. Microwave heating simultaneously increases the temperature of the entire reaction medium; while using conventional heating, the reaction medium that is in contact with the vessel wall achieves higher temperatures first compared to the rest of the reaction (temperature gradient). From Ref. [56]. Copyright © 2004. Reproduced with permission of John Wiley & Sons, Inc.

For example, precursor metal salts (halides, nitrates, acetates, etc.) couple strongly with microwaves via ionic conduction [47, 51]. Polyalcohols such as ethylene glycol, propanediols, and butanediols have permanent dipoles that grant them high loss tangents, which makes them ideal candidates for microwave synthesis [35, 61–65]. In the polyol process, the solvents decompose homogeneously at elevated temperatures to yield reducing species for the reduction of the metal ions [66]. This mechanism provides a certain level of control over the size distribution and shape of the NPs being synthesized. ILs have also been utilized widely as solvents in MNP syntheses; these undergo ionic conduction and are therefore heated effectively by microwaves [67]. Exposure of metal powders to microwave fields results in very high temperatures that can generate hot spots, which dramatically affect the reaction heating rates of the metals and other organic and inorganic substrates [45, 49, 68]. Hot spots are considered as an undesired effect in some reactions [69, 70]. However, they are very advantageous in the synthesis of MNPs as they are believed to provide enhanced nucleation zones. In reactions where all components are effectively heated by microwave irradiation, high reaction temperatures and improved reaction kinetics can be achieved with relatively low power input.

9.4 Direct Microwave-Assisted Synthesis of Unsupported MNPs

The direct synthesis of unsupported MNPs—whose surfaces are pacified with capping agents—is a popular strategy because such MNPs can be easily isolated by precipitation and subsequently redispersed in a number of alternative solvent systems. The chemical identity of the particular surface passivation agent employed in a given synthesis commonly determines which solvents can be used to form stable MNP dispersions. Hence, this synthetic strategy offers attractive flexibility for a range of subsequent applications, such as controlled deposition onto secondary support materials in the construction of heterogeneous catalyst composites [71, 72]. Alternatively, MNPs synthesized in strongly reducing solvents can be isolated and transferred into biologically compatible solutions for use in biomedical applications such as DNA tagging [73–75].

The controlled synthesis of unsupported MNPs has attracted a great deal of attention in the past two decades due to their tunable chemical properties that greatly differ from bulk metal surfaces and for their high surface-area-to-volume ratios that provide enhanced catalytic activities [76, 77]. The vast majority of previously reported methods utilize conventional heating to facilitate reduction of metal precursors by using one-pot protocols that involve the chemical reduction of metal salts in aqueous solution or by using reducing organic solvents (e.g., alcohols, *N*, *N*-dimethylformamide [DMF]). Similar methods have recently been studied using microwave irradiation as the mode of heating, with the intention of determining how microwaves affect MNP formation in solution.

9.4.1 *Monometallic MNPs*

Perhaps the best starting point for this discussion is the synthesis of Au NPs, which represents one of the most well-studied systems to date. The microwave-assisted preparation of Au NPs of various sizes and shapes was reported by Kundu, Peng, and Liang, who used 2,7-dihydroxynaphthalene (2,7-DHN) to reduce Au^{3+} ions in aqueous solution [78]. It was shown that variation of the ratio of metal precursor to capping agent (cetryltrimethylammonium bromide [CTAB]) as well as the concentration of 2,7-DHN afforded selectively a range of particle shapes such as trigonal prisms, rods, or nanospheres (cuboctahedra). Under microwave irradiation, it was possible to obtain large (20+ nm) but well-defined Au NPs in only 90 s. El-Shall and coworkers proposed that the size and morphology of Au NPs obtained under microwave irradiation can be tailored by adjustment of the concentration of Au^{3+} ions in solution, the microwave exposure time, and the ratio of capping agents in a mixture of oleic acid and oleylamine, as depicted in Fig. 9.4 [58]. It was suggested that the presence of polar carboxylate functional groups in oleic acid increased the growth rate and also enhanced the formation of more anisotropic shapes through the formation of stronger oleic acid–Au NP capping interactions at particular faces of the growing crystals. In the absence of oleic acid, only spherical Au NPs were obtained. A previously reported (and somewhat counterintuitive) observation was also confirmed in this work, whereby increasing the heating duration, the ramping rate, and the reaction temperature would afford smaller Au NPs with narrower size distributions [79]. In comparison to conventional synthesis methods, the physical effects of microwave heating have been found to strongly influence the size and morphology of Au NPs in a number of ways.

The synthesis of Ag NPs has also been widely studied, due in large part to their potential antibacterial properties. Microwave-based procedures to obtain Ag NPs have focused on green reactions. For example, uniform Ag NPs have been prepared using basic amino acids as reducing agents and soluble starch as a surface stabilizer and by irradiation for only 10 s [80]. Near-monodisperse Ag NPs have also been prepared using carboxymethyl cellulose sodium (CMS) as both reducing and capping agent in a green microwave-assisted synthesis [81]. A detailed review of these types of methodologies has been reported by Sharma, Yngard, and Lin [82]. NPs of Ag and Au have also been prepared with the use of organic solvents such as DMF and organic polymer

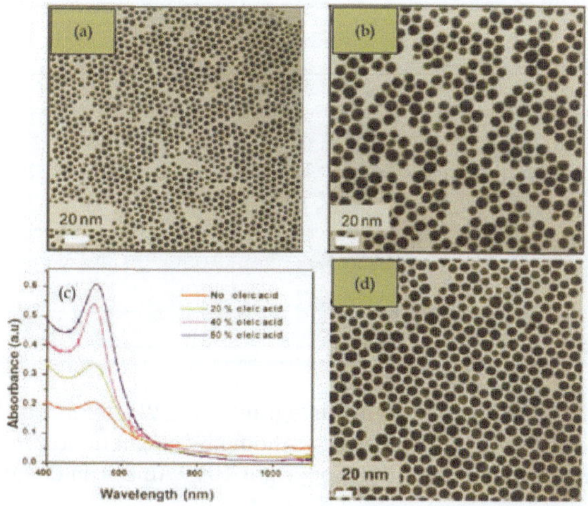

Figure 9.4 TEM images of Au NPs prepared by microwave irradiation methods after 60 s exposure using a fixed concentration of $HAuCl_4$ precursor, (a) pure oleylamine, and (b) a 1:1 ratio of oleic acid:oleylamine as capping agents. (c) The UV-visible spectrum displays a clear change in plasmon resonance of Au NPs as a function of varying ratios of oleic acid to oleylamine. Shifts in the plasmon resonance of these nanostructures are strongly related to changes in average Au NP size and distribution. (d) TEM image of Au NPs prepared using 9 mM of Au^{3+} ions in the presence of 3:1 oleic acid/oleylamine. Reprinted with permission from Ref. [58], Copyright 2010 American Chemical Society.

stabilizing agents, such as the ubiquitous poly(*N*-vinylpyrrolidone) (PVP). It has been shown that small (ca. 3.5 nm) and near-monodisperse Ag and Au NPs are preferentially formed in the presence of PVP and DMF under microwave irradiation, compared to conventional heating, which yields much larger structures. It was suggested by Pastoriza-Santos and Liz-Marzán that complexation of metal ions by PVP and their subsequent reduction occur under favorable conditions due to microwave heating because both PVP and DMF contain polar functional groups that couple strongly with microwaves [83].

It has similarly been shown that small (3.6 nm) and near-monodisperse Pt NPs can be formed rapidly using dielectric heating because the metal precursors employed are also strongly coupled

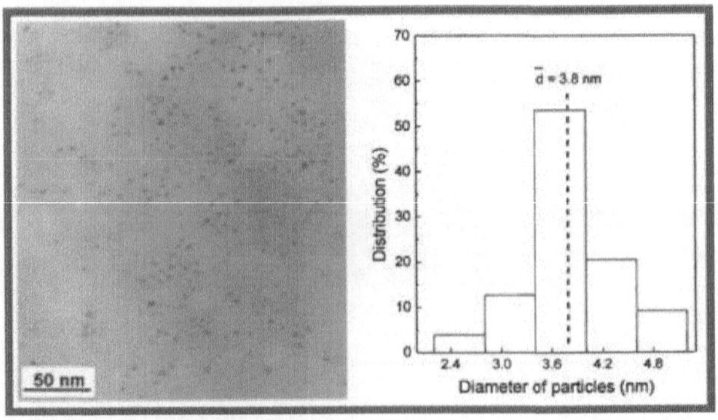

Figure 9.5 TEM image and histogram of PVP-stabilized Pt colloids synthesized by microwave heating without any magnetic stirring. Reprinted with permission from Ref. [84], Copyright 1999 American Chemical Society.

with microwaves [84, 85]. Such is the case for the reduction of H_2PtCl_6; this convenient and widely available Pt precursor induces such a high degree of heating homogeneity under microwave irradiation that even stirring of the reaction is not always necessary to obtain defined products (see Fig. 9.5) [82]. Size-controlled synthesis of Pt NPs has also been achieved by microwave heating of aqueous solutions of the same Pt^{4+} precursor in the presence of 3-thiophenemalonic acid [86].

Microwave irradiation has also been exploited in the rapid formation of monometallic Ru, Rh, Pd, Cu, and Ni NPs by exposure of cationic molecular precursors to repeated cycles of microwave heating that aid in metal ion reduction by DMF [53]. In most cases, the particles are stabilized with monomeric capping agents such as oleic acid/oleylamine mixtures. In the same study, variation of microwave irradiation time allowed the resulting mean MNP size to be adjusted.

ILs are excellent absorbers of microwaves since they are highly polarizable and exhibit large ionic conductivities. For this reason, they have been used in microwave-assisted preparation of MNPs. A one-step procedure has been reported by Ren et al., in which Au NPs are afforded via microwave heating in a variety of ILs (e.g., imidazolium- and pyridium-based ILs with different anions such as PF_6^-, $[Tos]^-$ (*p*-toluene sulfonate), or Br^-) and notably in the absence of any other solvents or reducing agents [67]. It was shown that the anionic components of the ILs employed were most influential in the resulting topologies of the MNPs; it is expected that negatively charged fragments interact most closely with the MNP surfaces in the formation of charge bilayers [87]. It was also established that the reaction temperature and the heating regime were key parameters in the control of the morphology and size of these Au nanoclusters.

Organic solvents such as polyols that have the ability to serve as both the reducing agent and the reaction medium are broadly used in microwave-assisted chemical synthesis of MNPs due to their high microwave absorption capabilities. 3*d* MNPs comprising Cu, Ni, and Co have been fabricated using a microwave-assisted hydrothermal process developed by Komarneni et al. [63]. They observed that the kinetics associated with the formation of these MNPs increased by at least an order of magnitude compared to conventional, convective heating techniques. Microwave-assisted reactions also promoted reduction of the 3*d* metal precursors at lower temperatures than were required to reduce the same metals under conventional heating. Suryawanshi et al. employed a similar methodology for the preparation of Ru NPs (Fig. 9.6a) [64]. Parameters such as reaction temperature, time, and irradiating power were systematically varied in order to determine their individual effects on the products. Lower reaction temperatures led to the formation of larger Ru NPs. High irradiation power (400 and 500 W) led to rapid heating rates and favored the formation of very small Ru nuclei. The interaction between stabilizer and Ru NP surface was, in turn, weakened as the effective concentration of capping agent was reduced with the increase in number of Ru seeds; this ultimately led to agglomeration. Conversely, low irradiation power (100 and 200 W) increased the induction period required for the onset of Ru NP nucleation, which, in turn, shortened the actual reaction time and resulted in the formation of fewer, larger Ru NPs. Control over the size of Ni NPs can also be achieved through the passivation of NP surfaces using a combination of capping ligands: dodecylamine and PVP capping agents have been jointly utilized for the stabilization of uniform Ni NPs with an average diameter of 63 nm, generated using a polyol approach in a microwave reactor (Fig. 9.6b) [65].

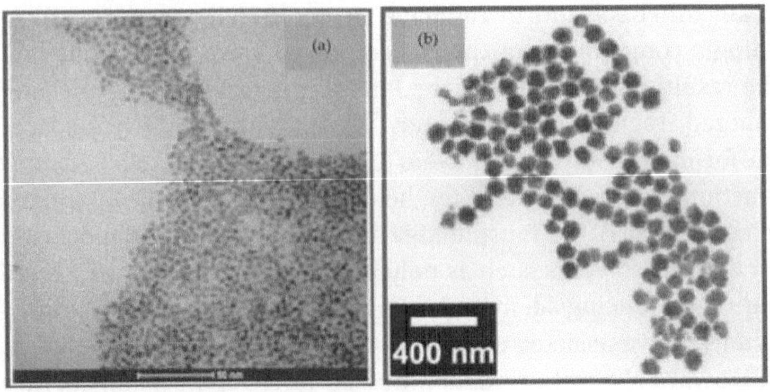

Figure 9.6 TEM images of PVP-capped microwave-synthesized (a) Ru NPs and (b) Ni NPs with narrow size distributions. From Ref. [64]. Copyright © 2013. Reproduced with permission of John Wiley & Sons, Inc., and From Ref. [65]. Copyright © 2006. Reproduced with permission of John Wiley & Sons, Inc.

9.4.2 *Heterobimetallic MNPs*

Bimetallic NPs have attracted significant attention in the past few years due to the enhanced optical and catalytic properties that are obtained due to *d*-band intermixing and quantum size confinement effects [88]. Bimetallic MNPs can be broadly divided into two categories, alloyed MNPs in which two metals are mixed throughout, and core–shell MNPs in which the two metals are segregated at a common interface. Both types of MNPs usually display surface plasmon band energies that differ relative to their monometallic counterparts. Bimetallic MNPs can also exhibit enhanced surface catalytic properties because the original reactivity of metal A can be tuned by interaction with metal B to achieve optimal resulting reactivity for a given reaction. This effect can even be observed for combinations of metals in which one of the metals is not catalytically active in isolation. Thus, bimetallic MNPs can provide advantages for the preparation of cost-effective nanomaterials, in which expensive and rare metals can be diluted with more earth-abundant metals.

A number of important physical parameters facilitate or restrict the generation of heterobimetallic nanospecies. Crystal packing parameters are intrinsically important for the preparation of both core–shell and alloy MNPs. Metals with common crystal packing arrangements are more likely to form ordered interfaces or to form mixed alloy phases. Even so, combinations of metals that share the same crystal structure but have significantly different lattice parameters (corresponding to their respective atomic radii) might not form stable alloys or core–shell structures. Some combinations of metals will lead to metastable arrangements, while other combinations of metals may be entirely thermodynamically unstable (otherwise termed immiscible metals). So-called strain effects may be induced when a few monolayers of metal A are deposited over metal B, in which there is a lattice mismatch between A and B. Such considerations are identical to those encountered in bulk alloy chemistry [89], but it should be noted that combinations of metals that are thermodynamically unstable in the bulk can often be rendered stable (or metastable) on the nanoscale [90, 91]. This section provides a summary of known examples of microwave-assisted methodologies that have been employed to prepare heterobimetallic MNPs for specific applications.

9.4.2.1 Core–shell MNPs

Microwave-assisted synthesis of core–shell NPs has been performed in a two-step process that involves the formation of monometallic core MNPs followed by overgrowth of a shell layer using a secondary metal precursor (Fig. 9.7) [53]. Since no two metals have exactly the same lattice parameter, strain properties are usually imposed in core–shell MNPs for which the shell thickness is less than five or six monolayers. MNPs with thicker shells show greatly reduced strain effects because atoms on the surface are sufficiently distant from lattice mismatching effects at the bimetallic interface. In instances where the shell comprises only one to two monolayers, electronic interactions can also be observed, even though the metals are not actually alloyed [92, 93].

The benefits of microwave heating have been established in the synthesis of Cu–Ag core–shell MNPs [94]. These heterostructures have been shown to display excellent temporal and thermal stability against oxidation because the Ag shell provides an effective physical barrier to oxygen, while there is a beneficial electronic interaction between Cu and Ag at the interface. Another example is the

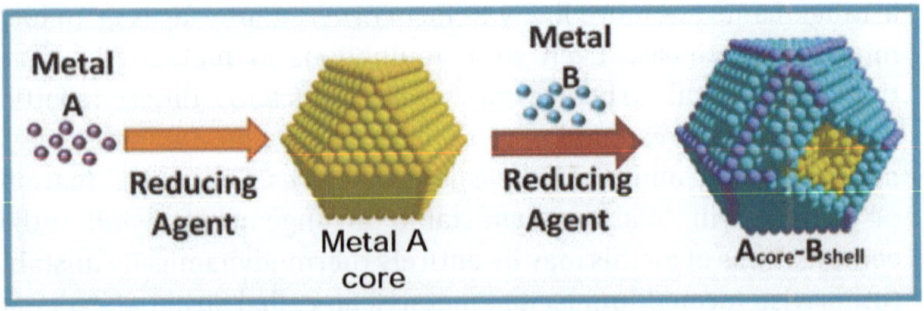

Figure 9.7 Schematic representation of the proposed two-pot growth mechanism of core–shell MNPs in which the metal core is first synthesized and isolated. Metal ions of the target shell metal are then reduced and deposited on the existing nuclei.

microwave-assisted preparation of PVP-capped Ag–Au core–shell structures by reaction in ethylene glycol or glycerol [95]. Although Ag and Au are miscible metals across all composition ranges in the bulk, microwave irradiation allows for the reduction rate of these metal ions to be managed, which has been shown to facilitate better structural control in the formation of MNPs with segregated structures. Under the reported experimental conditions, a proportion of the metal ions were not reduced prior to nucleation. This was thought to lower the probability of atoms colliding in solution to form alloyed AgAu nuclei, thus favoring the generation of true core–shell structures.

Energy barriers can also be lowered using microwave irradiation to provide access to more unusual heterostructures. For instance, Au–Rh core–shell MNPs are considered to be thermodynamically unstable structures since Au has a lower surface energy than Rh ($1.5\ J/m^2$ vs. $2.8\ J/m^2$) [96] under oxidizing conditions (such as air), so Au prefers to be on the surface. Dielectric heating has allowed for the less preferred Au–Rh configuration to be achieved in a two-step fashion through the synthesis and isolation of Au seeds, which were subsequently re-dispersed in a second reaction and used as seed points for the isotropic overgrowth of Rh shells of varying thicknesses [97]. It was also shown that even after the particles were exposed to a reducing environment (flowing H_2 gas), there was no inversion of the shell and core, presumably because Rh and Au are

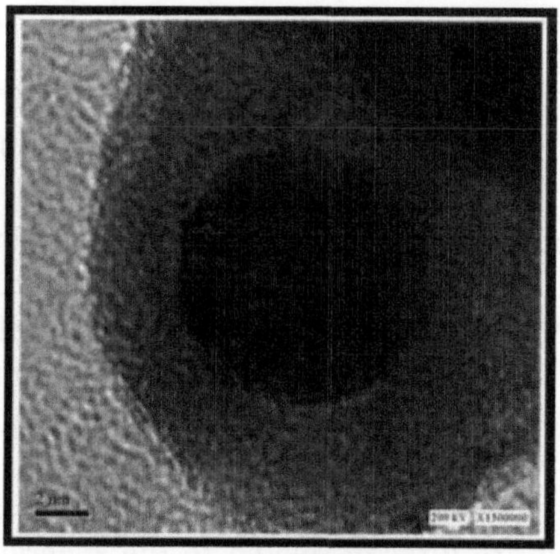

Figure 9.8 HRTEM image of a single AuPd core–shell MNP synthesized using microwave irradiation. Reprinted with permission from Ref. [98], Copyright 2004 American Chemical Society.

immiscible, and therefore Au cannot easily migrate through the Rh layer, which itself acts as a barrier.

It is possible to generate core–shell structures in a direct one-pot reaction that involve co-addition of both metal precursors, if the reduction potentials of the metal ions are significantly different: large differences in the corresponding rates of metal ion reduction result in control over the nucleation and growth events, which result in the spontaneous assembly of ordered core–shell MNPs. Such is the case of Au–Pd core–shell MNPs as synthesized by Harpeness and Gedanken [98]. The reduction potential of Au^{3+} is 0.015 V more positive than Pd^{2+} (relative to standard hydrogen electrode [SHE]). Au ions undergo reduction first to form cores that are then encapsulated with Pd metal, without the competing formation of any monometallic Pd NPs (Fig. 9.8).

9.4.2.2 Alloy MNPs

Converse to the previous example, MNPs with randomly alloyed structures are usually favored from mixtures of metal ions in solution that undergo reduction at similar rates, leading to binary nucleation [53]. Alloy MNPs can also be achieved through successive reduction of two metal ions in the presence of a surface passivation agent. Generally, the atoms of the less noble metal transfer electrons to the more noble metal atoms, which in turn aids in the complete reduction of the latter into metallic clusters [99]. Subsequent interdiffusion of the metals in solution then leads to alloy formation. A unique property of bimetallic alloy MNPs relates to the fact that different metals display different binding strengths toward particular capping agents. Hence, the size and shape of alloyed MNPs can be controlled due to selective adsorption of the surface stabilizing agents in ways that differ from the preparation of monometallic MNPs [53].

Bimetallic alloy MNPs display enhanced optical and catalytic properties that are directly related to the synergistic effects that arise due to electronic interactions of neighboring heteroatoms in the alloy. Electronic mixing leads to charge transfer between metals as their *d*-band centers are shifted, which ultimately alters substrate adsorption kinetics and surface binding energies [100]. It is believed that microwave-assisted synthesis facilitates the formation of bimetallic MNPs with enhanced properties by lowering the alloying temperature [101] and by providing good mixing of the two components [102]. Furthermore, as observed in other methodologies for both monometallic and core–shell structures, reactions rates are dramatically increased with dielectric heating.

Bimetallic Pt_xRu_{100-x} NPs ($x = 25$–75) have been synthesized using a simple polyol method coupled with microwave heating. These MNPs were shown to exhibit superior catalytic properties in the methanol oxidation reaction and when utilized in direct methanol fuel cells, compared to their monometallic counterparts [99, 101]. The microwave-assisted methodologies employed displayed excellent control and tunability of the alloy's physicochemical properties such as the number of *d*-band vacancies, which in turn resulted in MNPs with highly tunable catalytic properties. Pt and Ru form metastable alloys in the bulk, and they do not share the same crystal structures (face-centered cubic and hexagonal close packed, respectively). It is believed that the mechanism for the formation of the alloy Pt_xRu_{100-x} NPs is on the basis of the formation of initial Pt-rich nuclei that undergo atom substitution through Ru diffusion, which results in alloyed seed particles that subsequently engage in microwave-assisted growth (Fig. 9.9).

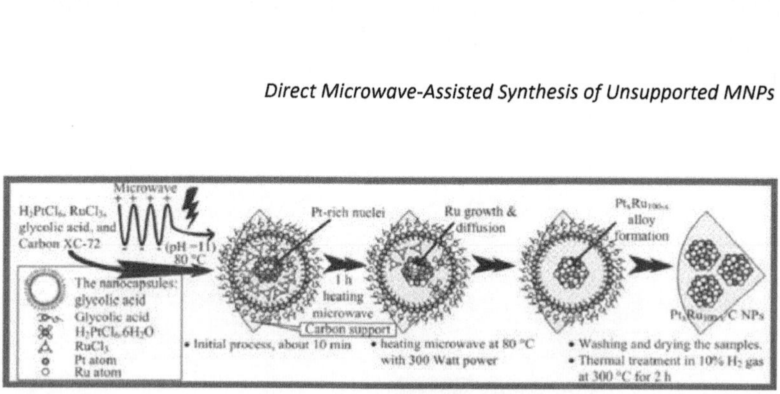

Figure 9.9 Schematic representation of the formation of PtRu alloy NPs via a diffusion process aided by dielectric heating. From Ref. [102]. Copyright © 2012. Reproduced with permission of John Wiley & Sons, Inc.

PdAu alloys are popular in catalysis for a number of reactions involving CO oxidation [53]. The microwave-assisted preparation of PdAu alloy NPs has been used to prepare catalysts for the oxidation of methanol and other alcohols and carboxylic acids (e.g., ethanol and formic acid) for potential use in fuel cell technologies. Mehta and Gupta have synthesized isotropic PdAu NPs with average diameters of 5 nm, in reaction times as short as 20 s [103]. Microwave-synthesized PdAu NPs have also been supported on ceria and used as catalysts for similar oxidation reactions; other bimetallic alloy MNPs such as PtAu, PtRh, RhCu, and PdCu have also all been prepared using the same microwave-assisted protocol (Fig. 9.10) [53]. This example shows that microwave-assisted methodologies are both simple and versatile, enabling the synthesis of a wide range of supported and unsupported bimetallic nanoalloys for a variety of applications.

Microwave-assisted polyol methods have been applied to prepare FeRu NPs of different sizes and compositions, which have shown extremely high catalytic activities in the water–gas shift reaction due synergistic effects between the metals [104]. The relative metallic compositions were observed to be an influential factor in the catalytic performance of these systems.

Although metals that are close in proximity in the periodic table have similar properties, the observed synergistic effects are usually greatest for bimetallic NPs of this type. Such is the case for RhPd NPs that have been synthesized through a modified polyacrylamide sol–gel method under microwave irradiation [105]. Impurity-free RhPd NPs were generated with ethylenediaminetetraacetic acid (EDTA) as

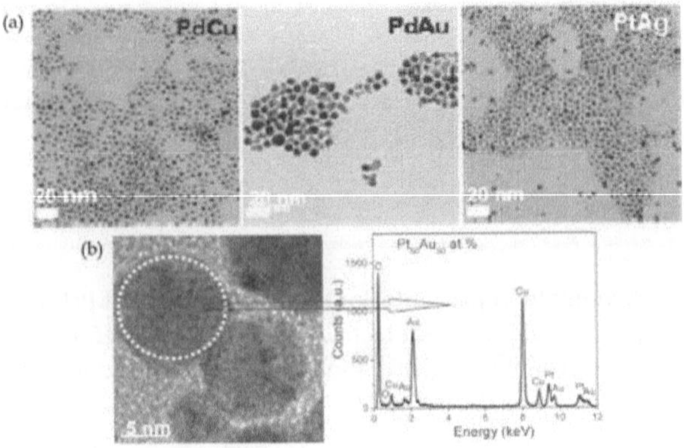

Figure 9.10 TEM images of (a) PdCu, PdAu, and PtAg NPs prepared utilizing a microwave irradiation method. (b) HRTEM image and EDX spectrum that shows atomic composition of a single PtAu NP, confirming that an alloyed structure is indeed formed using this methodology. Reprinted with permission from Ref. [53], Copyright 2009 American Chemical Society.

the initial stabilizing agent, which was removed postsynthetically via thermal treatment. A sol–gel method was employed in this case because it provided additional control over the particle size through control over the reaction pH, coupled with heat treatment. Microwave heating has also been coupled with similar polyacrylamide methods for the synthesis of cuboctahedral-alloyed AgAu NPs with different Ag:Au compositions [106].

Pd- and Pt-containing composites have also attracted significant attention due their enhanced catalytic activity and selectivity in the hydrogen evolution reaction using visible light and in hydrogenation reactions. Patel et al. have synthesized both PVP-stabilized PtAg NPs and PdAg NPs through a microwave-assisted polyol method [95, 98]. They conducted several experiments in an attempt to understand the mechanism of formation of these alloyed NPs; when the viscous solvent glycerol was employed as the reaction medium, it was possible to monitor the diffusion process of the metal atoms spectrophotometrically by decreasing the rate of diffusion of the metals. Very small PdPt NPs with an average size of 1.05 nm have also been synthesized for hydrogenation purposes, which have been successfully employed in the hydrogenation of palm olein at ambient temperature and atmospheric pressure [107]. Compared to syntheses performed under conventional heating or sonication, the microwave-prepared PdPt NPs were formed three times faster and were also catalytically more highly active.

9.4.3 *Ex situ Supported MNP Catalysts*

Enhancement of catalytic activity of MNPs has been observed in several instances as a direct consequence of microwave synthesis. A number of microwave-synthesized monometallic and core–shell and alloy MNPs have been deposited onto various support media in an ex situ manner to generate stable and easily recyclable composite catalyst materials, which have been studied in a range of catalytic applications, in both liquid- and vapor-phase settings. For example, microwave-synthesized Rh, Pd, and Pt NPs [29] and Au–Rh core–shell NPs [97] have been supported on amorphous silica by simple incipient wetness impregnation methods, which do not induce any measurable morphological changes to the MNPs. The resulting composite catalysts were shown to act as superior hydrogenation catalysts in the room temperature, vapor-phase hydrogenation of cyclohexene using H_2 gas. The supported catalysts were highly resistant to sintering or agglomeration under these catalytic conditions after several repeated cycles of hydrogenation (Fig. 9.11). Similarly, microwave-synthesized Cu and Ag NPs have been supported on silica-coated hollow polystyrene beads [108], silica beads [109], and activated carbon granules [110] and tested for a number of catalytic applications. One specific example is the immobilization of AuCu NPs on TiO_2 to serve as catalysts for selective oxidation of 5-hydroxymethyl-2-furfural [111]. These alloy NPs were presynthesized by microwave-assisted reduction of Au^{3+} and Cu^{2+} precursors in the presence of β-D-glucose in an aqueous medium. Catalysts for the water–gas shift reaction have also been generated by supporting preformed FeRu NPs on $MgAl_2O_4$ at room temperature using potassium acetate as a promoter [104]. These are only a few examples of a larger number of methodologies that have been employed in the ex situ deposition of preformed, microwave-synthesized MNPs onto support media to create recyclable catalyst materials.

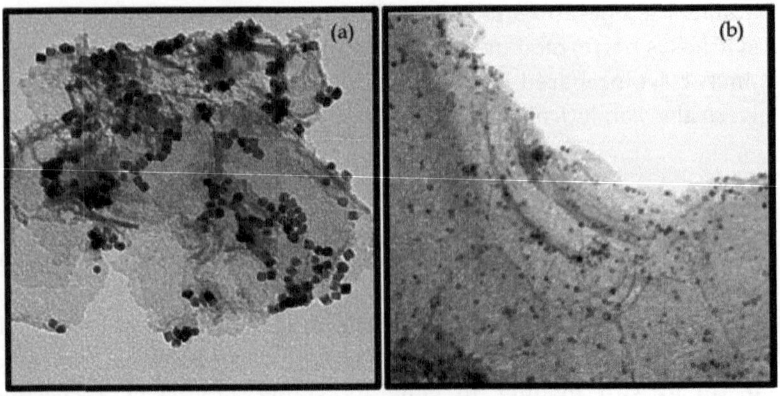

Figure 9.11 TEM images of (a) cubic Rh NPs and (b) AuRh core–shell NPs synthesized using microwave heating and supported on amorphous silica using a simple ex situ method. Reprinted with permission from Ref. [29], Copyright 2012 American Chemical Society, and Reproduced in part from Ref. [97], Copyright (2013), with permission of the Royal Society of Chemistry.

In alternative works, others have demonstrated how conventionally synthesized MNPs can be supported on a variety of substrates utilizing microwave irradiation to aid in the deposition process. For example, a hybrid catalyst consisting of multiwalled carbon nanotubes (MWCNTs) decorated with conventionally prepared Rh NPs was prepared by direct mixing of MWCNTs with tridecylamine-capped Rh NPs followed by microwave exposure for 1 min at 350 W (Fig. 9.12) [112]. Similar methodologies performed using convective heating involve tedious synthetic procedures and require harsher experimental conditions. Another pertinent example involves the incorporation of preformed Ag NPs into a mesoporous ZnO material with the aid of microwave heating under an Ar atmosphere, again with retention of the original structures of the Ag NPs [113].

9.4.4 *In situ Microwave-Assisted Synthesis of MNPs Inside Support Media*

The key advantages provided by microwave-assisted chemistry in the synthesis of MNPs that have been explored above (e.g.,

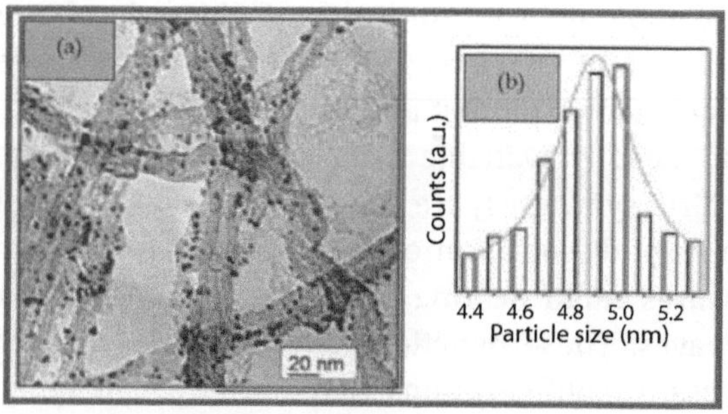

Figure 9.12 (a) TEM image and (b) histogram of Rh NPs with narrow size distribution supported on multiwalled nanotubes synthesized using dielectric heating. Reprinted with permission from Ref. [112], Copyright 2008 American Chemical Society.

short reaction times and control over MNP size and morphology) have recently been extended to the direct, in situ preparation of mono- and bimetallic MNPs at or inside support media. This one-pot strategy negates the need to isolate unsupported MNPs and to deposit them onto support materials in separate steps, which, in some instances, can cause problems such as agglomeration or uneven spatial deposition. The majority of examples of in situ microwave-assisted synthesis of supported MNPs involve the use of chemically inert supports, which prevent agglomeration and improve recyclability/recoverability of the MNPs. In some cases, chemically reactive supports have been used to enhance the overall catalytic properties of the MNP–support composites via synergistic effects (e.g., oxidation resistance and spillover) [114–116]. Microwave-assisted syntheses of such composite materials are applicable to many different polar solvents and metal precursors, and as such, there exist a large number of examples of syntheses of supported MNPs using microwave-assisted methods, some of which have been previously reviewed by Campelo et al. [117]. In this final section, some of the most recently identified in situ, one-pot synthetic methods for the generation of even dispersions of defined MNPs on support materials will be explored.

The majority of such examples in the literature utilize the polyol process to assist in the in situ reduction of molecular metal precursors that are predispersed alongside the support matrices.

The most commonly used solvent is ethylene glycol due to its high dielectric constant (41.4 at 298 K) [35] and effective reduction ability, which allows for fast and uniform heating using microwave irradiation. Readily available metal precursors (akin to those used in ex situ MNP syntheses) are commonly dissolved in polar solvents that encourage dissociation of metal cations from their co-anions, which ensures rapid heating by ionic conduction in the presence of microwaves. These beneficial effects have also been shown to allow access to configurations that can be challenging to achieve using conventional heating methods. For example, the decoration of single-walled carbon nanotubes (SWCNTs) and MWCNTs with noble metal ions, clusters, and NPs at room temperature has been thoroughly studied in recent times. Nadagouda et al. recently reported the facile synthesis of Ag and Pd NPs on CNTs wrapped with carboxymethyl cellulose (CMC) using a microwave method, which could not be achieved under conventional heating [46].

Supported MNPs are ordinarily synthesized by the initial reduction of metal salts, followed by nucleation and growth of the initial nucleates upon the support itself. In most cases, the support material is simply suspended in the reaction medium and does not undergo any chemical transformations, other than providing preferential nucleation and growth sites for the formation of discrete MNPs. Pt NPs have been directly synthesized at SWCNTs and Vulcan carbon [118] to yield ideal composites (Fig. 9.13), whereas analogous materials prepared using convective heating yielded unacceptably wide Pt NP size distributions. These Pt NP–containing composites have been applied in methanol electrooxidation [119], selective hydrogenation [120], and fuel cell applications [121]. Au, Ag, and Pd NPs have also been prepared inside the cylindrical mesopores of ordered mesoporous silica SBA-12 particles in very short reaction times (<2 min) using microwave-based methods; their catalytic properties have been assessed in reactions involving the liquid phase oxidation of styrene and benzyl alcohol with hydrogen peroxide [122]. The in situ preparation of Fe NPs on MCM-41, starch, cellulose, and carbonaceous supports has also been demonstrated in ethanol, yielding catalysts that were active in the liquid-phase oxidation of alcohols [123]. Fe^{2+} is typically difficult to reduce toward the generation of Fe NPs because it has a significantly

Figure 9.13 TEM images of Pt NPs supported on SWCNTs synthesized in situ using microwave heating and ethylene glycol for the reduction of the Pt^{2+} precursor. The SWCNTs group into micron-sized clumps (a and b). Although most particles are approximately 1–2 nm (c), some are as large as 10 nm (d) or smaller than 1 nm (e). Reprinted with permission from Ref. [120], Copyright 2001 American Chemical Society.

lower reduction potential (−0.41 V) compared to other cations that are commonly used to prepare MNPs (e.g., Ag^{+} = 0.79 V; Pd^{2+} = 0.98 V; Ru^{2+} = 0.45 V) [123]. However, the formation of supported Fe NPs mediated by ethanol as the sacrificial reductant has been successfully achieved under microwave irradiation [123]. El-Shall and coworkers have also extensively investigated the preparation of a range of supported monometallic NPs, including Au and Pd NPs on oxides such as CeO_2, MgO, and SiO_2 [124, 125]. Microwave irradiation has also afforded a number of heterometallic-supported MNPs: binary PtRu NPs and ternary PtRuIr NPs have been prepared as nanocomposites supported on carbon (Fig. 9.14) and utilized in

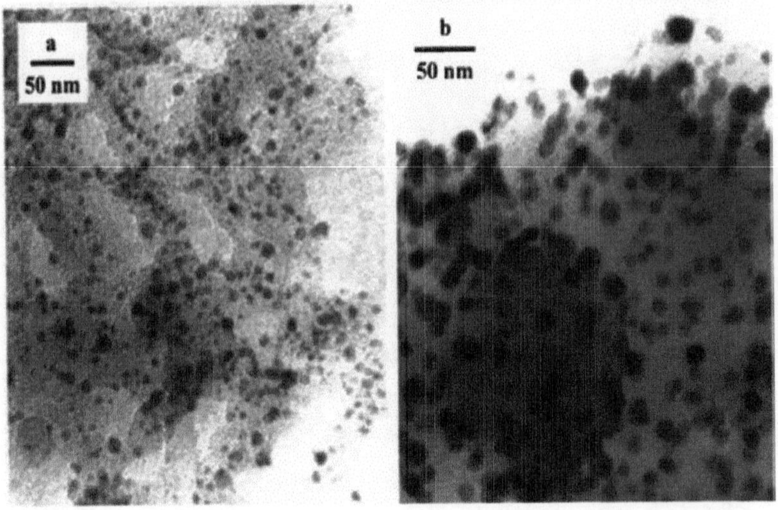

Figure 9.14 Bright field TEM images of PtRu NPs supported on carbon, synthesized in situ using dielectric heating and one round (a) or four rounds (b) of precursor addition. Reprinted with permission from Ref. [127], Copyright 2001 American Chemical Society.

proton exchange membrane fuel cells [126] and in direct methanol fuel cells [127].

It is even possible to generate supported MNP composites in a simultaneous fashion. This is commonly achieved via multiple concomitant chemical reactions that involve metal ion reduction and monomer polymerization to form the support or via chemical modifications to a preformed support material [53, 128]. An example of the latter is provided by the preparation of graphene-supported Pd NPs, which can be accessed by the simultaneous microwave-assisted reduction of Pd^{2+} and deoxygenation of graphite oxide using hydrazine hydrate [128]. In this particular setting, microwave heating has been shown to provide the required control over the extent of graphite oxide reduction, since the irradiating power and reaction time can be easily adjusted to optimize the parallel reduction processes. Another good example is the one-pot synthesis of mono- and bimetallic MNPs of Cu, Pd, CuPd, RuCu, AuPd, AuRu, PtRh, PdRh, and AuPt supported on CeO_2, which is generated in situ (Fig. 9.15) [53]. It is also possible to achieve well-dispersed MNPs

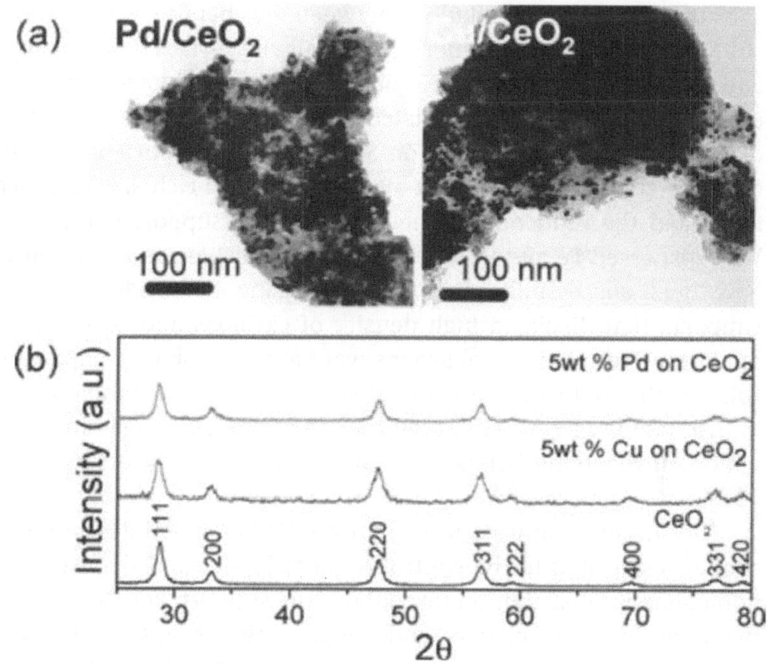

Figure 9.15 TEM images (a) and XRD patterns (b) of Pd and Cu NPs synthesized in situ with ceria supports using microwave irradiation. All three XRD patterns match well with ceria standards, indicating fine dispersal of the metals on the CeO_2 particles. Reprinted with permission from Ref. [53], Copyright 2009 American Chemical Society.

within polymer matrices and thin films via concomitant metal ion reduction and monomer polymerization. Examples of this type of process include the formation of homogeneously dispersed Pt, Ag, and Cu nanocomposites within polyacrylamide resins. These were obtained after the polymerization of an acrylamide monomer and simultaneous reduction of the corresponding metal salt in ethylene glycol under microwave heating [129]. Wada et al. have reported the formation of well-dispersed Ag NPs in a polymer matrix via direct synthesis in the presence of reactive methacrylate monomers under microwave heating, followed by microwave-induced polymerization of the methacrylate monomers [130].

Some important correlations between the chemical nature of the support material used (or more specifically, its surface functionality) and the ability to effectively reduce metal cations have also been elucidated via microwave-assisted chemistry. In the case of MNPs of Au, Ag, and Pd supported on SBA-12, it was reported that not only the ethanolic solvent but also the hydroxyl-rich silica surface facilitated the reduction of the MNPs on the support [115, 131]. Well-dispersed Fe and Pt NPs anchored on carbonaceous supports have been successfully prepared by utilizing chemically modified supports that display a high density of carboxyl and/or carbonyl moieties. These functional groups were found to play an important role in the reduction process through coordinative binding of individual metal ions from solution and by ion exchange reactions, which consequently facilitated favorable nucleation events [60, 123, 132].

The in situ synthesis of supported MNPs sometimes imposes challenges related to the control of particle size and the extent of homogeneous distribution throughout the support. For this reason, the addition of stabilizers and capping agents has been found to enhance the outcome of several of the previously described synthetic schemes. For instance, the use of PVP and polyalcohols with established microwave-assisted polyol methods has resulted

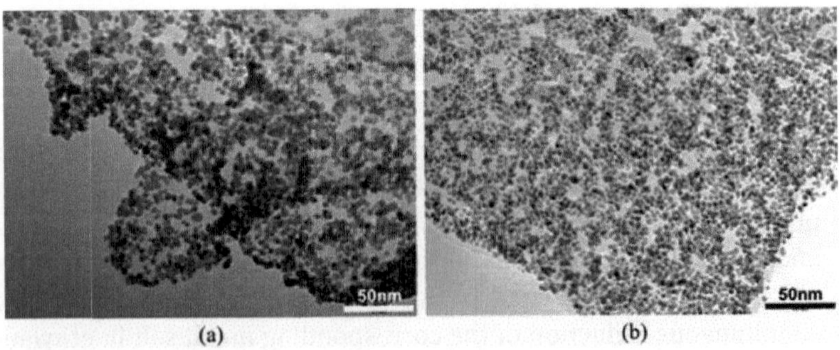

Figure 9.16 TEM images of Pt NPs supported on exfoliated graphite nanoplatelets by in situ reduction of the Pt precursor by microwave irradiation in ethylene glycol (a) and ethylene glycol with added room-temperature ionic liquids (b). Reproduced with permission from Ref. [135], American Chemical Society.

in the formation of composites with more uniform MNPs that have narrower size distributions [133]. Similarly, standard synthetic methods used to prepare supported Au and Pd NPs mentioned above have been modified to include surface passivation agents such as polyethylene glycol and PVP to further stabilize the NPs and to prevent agglomeration [124, 125]. It is believed that the rapid and uniform heating achieved using microwave irradiation allows for even faster reduction of the metal on the support. Near-monodisperse Pt NPs supported on carbon have been obtained using acetate ions that promote favorable electrostatic interactions [118]. Excellent morphological control has also been demonstrated using room-temperature ionic liquids like 1-butyl-3-methylimidazolium hexafluorophosphate to form smaller Pt NPs on exfoliated graphite platelets in situ than is possible in ethylene glycol alone (Fig. 9.16) [134].

References

1. Magnani, M., Galluzzi, L., and Bruce, I. J. (2006). The use of magnetic nanoparticles in the development of new molecular detection systems, *J. Nanosci. Nanotechnol.,* **6**, pp. 2302–2311.
2. Morneta, S., Vasseura, S., Grassteb, F., Veverkac, P., Goglioa, G., Demourguesa, A., Portiera, J., Pollertc, E., and Duguet, E. (2006). Magnetic nanoparticle design for medical applications, *Prog. Solid State Chem.,* **34**, pp. 237–247.
3. Kim, T., Kim, D. S., Lee, B. Y., Kim, Z. H., and Hong, S. (2009). A "nanoprism" probe for nano-optical applications, *Adv. Mater.,* **21**, pp. 1238–1242.
4. Zhang, L. Z., and Tang, G. Q. (2004). Preparation, characterization and optical properties of nanostructured ZnO thin films, *Opt. Mater.,* **27**, pp. 217–220.
5. Sun, Y., and Wiederrecht, G. P. (2007). Surfactantless synthesis of silver nanoplates and their application in SERS, *Small,* **3**, pp. 1964–1975.
6. Daniel, M. C., and Astruc, D. (2004). Gold nanoparticles: assembly, supramolecular chemistry, quantum-size-related properties, and applications toward biology, catalysis, and nanotechnology, *Chem. Rev.,* **104**, pp. 293–346.

7. Roucoux, A., Schultz, J., and Patin, H. (2002). Reduced transition metal colloids: a novel family of reusable catalysts?, *Chem. Rev.,* **102**, pp. 3757–3778.
8. Volokitin, Y., Sinzig, J., de Jong, L. J., Schmid, G., Vargaftik, M. N., and Moiseev, I. I. (1996). Quantum-size effects in the thermodynamic properties of metallic nanoparticles, *Nature,* **384**, pp. 621–623.
9. Gilbert, B., Huang, F., Zhang, H., Waychunas, G. A., and Banfield, J. F. (2004). Nanoparticles: strained and stiff, *Science,* **305**, pp. 651–654.
10. White, R. J., Luque, R., Budarin, V. L., Clark, J. H., and Macquarrie, D. J. (2009). Supported metal nanoparticles on porous materials, *Chem. Soc. Rev.*, **38**, pp. 481–494.
11. Porter, L. A., Choi, H. C., Ribbe, A. E., and Buriak, J. M. (2002). Controlled electroless deposition of noble metal nanoparticle films on germanium surfaces, *Nano Lett.*, **2**, pp. 1067–1071.
12. Shen, L., Bao, N., Yanagisawa, K., Domen, K., Gupta, A., and Grimes, C. A. (2006). Direct synthesis of ZnO nanoparticles by a solution-free mechanochemical reaction, *Nanotechnology,* **17**, pp. 5117–5123.
13. Wang, J. S., Yin, S., Komatsu, M., Zhang, Q. W., Saito, F., and Sato, T. (2004). Photo-oxidation properties of nitrogen doped $SrTiO_3$ made by mechanical activation, *Appl. Catal. B: Environ.*, **52**, pp. 11–21.
14. Eutis, S., and El-Sayed, M. A. (2006). Why gold nanoparticles are more precious than pretty gold: noble metal surface plasmon resonance and its enhancement of the radiative and nonradiative properties of nanocrytals of different shapes, *Chem. Soc. Rev.*, **35**, pp. 209–217.
15. Murray, C. B., Norris, D. J., Bawendi, M. G. (1993). Synthesis and characterization of nearly monodisperse CdE ($E = S$, Se, Te) semiconductor nanocrystallites, *J. Am. Chem. Soc.*, **115**, pp. 8706–8715.
16. Burda, C., Chen, X., Narayanan, R., and El-Sayed, M. A. (2005). Chemistry and properties of nanocrystals of different shapes, *Chem. Rev.*, **105**, pp. 1025–1102.
17. Schmid, G., Klein, N., Korste, L., Kreibig, U., and Schönauer, D. (1984). Large transition metal clusters, 2. Synthesis and properties of $Rh_{55}[P(tert\text{-}Bu)_3]_{12}Cl_{20}$, *Inorg. Chim. Acta*, **85**, pp. 97–102.
18. Schmid, G., and Huster, W. (1986). Large transition metal clusters, IV. Ru_{55}-, Rh_{55}-, and Pt_{55}-Clusters, *Z. Naturforsch. Teil B*, **41**, 1028–1033.
19. Kawazoe, Y., Kondow, T., and Ohno, K. (2002). *Clusters and Nanomaterials: Theory and Experiment* (Springer, Berlin).
20. Templeton, A. C., Wuelfing, W. P., and Murray, R. W. (2000). Monolayer-protected cluster molecules, *Acc. Chem. Res.*, **33**, pp. 27–36.

21. Kim, Y.-G., Oh, S.-K., and Crooks, R. M. (2004). Preparation and characterization of 1–2 nm dendrimer-encapsulated gold nanoparticles having very narrow size distributions, *Chem. Mater.*, **16**, pp. 167–172.
22. Fievet, F., Lagier, J. P., Blin, B., Beaudoin, B., and Figlarz, M. (1989). Homogeneous and heterogeneous nucleations in the polyol process for the preparation of micron and submicron size metal particles, *Solid State Ionics*, **32–33**, pp. 198–205.
23. Fievet, F., Lagier, J. P., and Figlarz, M. (1989). Preparing monodisperse metal powders in micrometer and submicrometer sizes by the polyol process, *MRS Bull.*, **14**, pp. 29–34.
24. Ahmadi, T. S., Wang, Z. L., Green, T. C., Henglein, A., and El-Sayed, M. A. (1996). Shape-controlled synthesis of colloidal platinum nanoparticles, *Science*, **272**, 1924–1925.
25. Li, M., Schnablegger, H., and Mann, S. (1999). Coupled synthesis and self-assembly of nanoparticles to give structures with controlled organization, *Nature*, **402**, pp. 393–395.
26. Sau, T. K., and Murphy, C. J. (2004). Room temperature, high-yield synthesis of multiple shapes of gold nanoparticles in aqueous solution, *J. Am. Chem. Soc.*, **126**, pp. 8648–8649.
27. Bakshi, M. S., Thkur, P., Sachar, S., Kaur, G., Banipal, T. S., Possmayer, F., and Petersen, N. O. (2007). Aqueous phase surfactant selective shape controlled synthesis of lead sulfide, *J. Phys. Chem. C*, **111**, pp. 18087–18098.
28. Kuhn, J. N., Tsung, C.-K., Huang, W., and Somorjai, G. A. (2009). Effect of organic capping layers over monodisperse platinum nanoparticles upon activity for ethylene hydrogenation and carbon monoxide oxidation, *J. Catal.*, **265**, pp. 209–215.
29. Dahal, N., Garcia, S., Zhou, J., and Humphrey, S. M. (2012). Beneficial effects of microwave-assisted heating in noble metal nanoparticle synthesis, *ACS Nano*, **6**, pp. 9433–9446.
30. Oliveira, M. M., Ugarte, D., Zanchet, D., and Zarbin, A. J. G. (2005). Influence of synthetic parameters on the size, structure and stability of dodecanethiol-stabilized silver nanoparticles, *J. Colloid Interface Sci.*, **292**, pp. 429–435.
31. Pamplin, B. R. (1975). *Crystal Growth* (Pergamon Press, New York).
32. Fujimoto, T., Terauchi, S.-Y., Umehara, H., Kojima, I., and Henderson, W. (2001). Sonochemical preparation of single-dispersion metal nanoparticles from metal salts, *Chem. Mater.*, **13**, pp. 1057–1060.

33. Zhang, Z., Nenoff, T. M., Leung, K., Ferreira, S. R., Huang, J. Y., Berry, D. T., Provencio, P. P., and Stumpf, R. (2010). Room-temperature synthesis of Ag-Ni and Pd-Ni alloy nanoparticles, *J. Phys. Chem. C*, **114**, pp. 14309–14318.
34. Tsuji, M., Hashimoto, M., Nishizawa, Y., Kubokawa, M., and Tsuji, T. (2005). Microwave-assisted synthesis of metallic nanostructures in solution, *Chem. Eur. J.*, **11**, pp. 440–452.
35. Bilecka, I., and Niederberger, M. (2010). Microwave chemistry for inorganic nanomaterials synthesis, *Nanoscale*, **2**, pp. 1358–1374.
36. Hayes, B. L. (2002). *Microwave Synthesis: Chemistry at the Speed of Light* (CEM, Matthews).
37. Zhang, X., Hayward, D. O., and Mingos, D. M. P. (1999). Apparent equilibrium shifts and hot-spot formation for catalytic reactions induced by microwave dielectric heating, *Chem. Commun.*, pp. 975–976.
38. Zhao, X., Huang, K., Yan, L., and Yao, Y. (2009). A preliminary study on numerical simulation of microwave heating process for chemical reaction and discussion of hotspot and thermal runaway phenomenon, *Sci. China Phys. Mech. Astron.*, **52**, pp. 551–562.
39. LaMer, V. K., and Dinegar, R. H. (1950). Theory, production and mechanism of formation of monodispersed hydrosols, *J. Am. Chem. Soc.*, **72**, pp. 4847–4854.
40. Park, J., Joo, J., Kwon, S. G., Jang, Y., and Hyeon, T. (2007). Synthesis of monodisperse spherical nanocrystals, *Angew. Chem., Int. Ed.*, **46**, pp. 4630–4660.
41. Tompsett, G. A., Conner, W. C., and Yngvesson, K. S. (2006). Microwave synthesis of nanoporous materials. *ChemPhysChem*, **7**, pp. 296–319.
42. Gerbec, J. A., Magana, D., Washington, A., and Strouse, G. F. (2005). Microwave-enhanced reaction rates for nanoparticle synthesis, *J. Am. Chem. Soc.*, **127**, pp. 15791–15800.
43. Roxworthy, B. J., Bhuiya, A. M., Vanka, S. P., and Toussaint, K. C. (2014). Understanding and controlling plasmon-induced convection, *Nat. Commun.*, **5**, p. 3173.
44. Pineda, A., Gomez, L., Balu, A. M., Sebastian, V., Ojeda, M., Arruebo, M., Romero, A. A., Santamaria, J., and Luque, R. (2013). Laser-driven heterogeneous catalysis: efficient amide formation catalysed by Au/SiO_2 systems, *Green Chem.*, **15**, pp. 2043–2049.
45. Khanal, S., Bhattarai, N., Velazquez-Salazar, J. J., Bahena, D., Soldano, G., Ponce, A., Mariscal, M. M., Mejía-Rosales, S., and José-Yacamán,

M. (2013). Trimetallic nanostructures: the case of AgPd–Pt multiply twinned nanoparticles, *Nanoscale*, **5**, pp. 12456–12463.

46. Nadagouda, M. N., Speth, T. S., and Varma, R. S. (2011). Microwave-assisted green synthesis of silver nanostructures, *Acc. Chem. Res.*, **44**, pp. 469–478.

47. Galema, S. A. (1997). Microwave chemistry, *Chem. Soc. Rev.*, **26**, pp. 233–238.

48. Acharya, S. A., Dani, A., Sayyed, S. K., and Gaikawad, V. M. (2013). Understanding the role of mode of heating on phase formation of Fe-Pt nanoparticles. In *Advanced Nanomaterials and Nanotechnology*, Giri, P. K., Goswani, D. K., and Perumal, A., eds. (Springer-Verlag, Berlin, Heidelberg), pp. 51–63.

49. Kappe, C. O., Dallinger, D., and Murphree, S. S. (2009). *Practical Microwave Synthesis for Organic Chemists* (Wiley-VCH, Weinheim).

50. Kappe, C. O. (2013). Unraveling the mysteries of microwave chemistry using silicon carbide reactor technology, *Acc. Chem. Res.*, **46**, pp. 1579–1587.

51. Mingos, D. M. P., and Baghurst, D. R. (1991). Application of microwave dielectric heating effects to synthetic problems in chemistry, *Chem. Soc. Rev.*, **20**, pp. 1–47.

52. Jezequel, D., Guenot, J., Jouini, N., and Fievet, F. (1995). Submicrometer zinc oxide particles: elaboration in polyol medium and morphological characteristics, *J. Mater. Res.*, **10**, pp. 77–83.

53. Abdelsayed, V., Aljarash, A., El-Shall, M. S., Al Othman, Z. A., and Alghamdi, A. H. (2009). Microwave synthesis of bimetallic nanoalloys and CO oxidation on ceria-supported nanoalloys, *Chem. Mater.*, **21**, pp. 2825–2834.

54. Herring, N. P., AbouZeid, K., Mohamed, M. B., Pinsk, J., and El-Shall, M. S. (2011). Formation mechanism of gold-zinc oxide hexagonal nanopyramids by heterogeneous nucleation using microwave synthesis, *Langmuir*, **27**, pp. 15146–15154.

55. Panda, A. B., Glaspell, G., and El-Shall, M. S. (2006). Microwave synthesis of highly aligned ultra narrow semiconductor rods and wires, *J. Am. Chem. Soc.*, **128**, pp. 2790–2791.

56. Kappe, O. C. (2004). Controlled microwave heating in modern organic chemistry, *Angew. Chem., Int. Ed.*, **43**, pp. 6250–6284.

57. Abdelsayed, V., Panda, A. B., Glaspell, G. P., and El-Shall, M. S. (2008). Synthesis, passivation and stabilization of nanoparticles, nanorods, and nanowires by microwave irradiation. In *Nanoparticles: Synthesis,*

Stabilization, Passivation, and Functionalization, Nagarajan, M. B., and Hatton, T. A., eds. (American Chemical Society, Washington DC), pp. 225–247.

58. Mohamed, M. B., AbouZeid, K. M., Abdelsayed, V., Aljarash, A. A., and El-Shall, M. S. (2010). Growth mechanism of anisotropic gold nanocrystals *via* microwave synthesis: formation of dioleamide by gold nanocatalysis, *ACS Nano,* **4**, pp. 2766–2772.
59. Feldmann, C., and Jungk, H. O. (2001). Polyol-mediated preparation of nanoscale oxide particles, *Angew. Chem., Int. Ed.,* **40**, pp. 359–362.
60. Feldmann, C. (2003). Polyol-mediated synthesis of nanoscale functional materials, *Adv. Funct. Mater.,* **13**, pp. 101–107.
61. Zahn, M., Ohki, Y., Fenneman, D. B., Gripshover, R. J., and Gehman, V. H. (1986). Dielectric properties of water/ethylene glycol mixtures for use in pulsed power system design, *Proc. IEEE,* **74**, pp. 1182–1221.
62. Chen, W., Zhao, J., Lee, J. Y., and Liu, Z. (2005). Microwave heated polyol synthesis of carbon nanotubes supported Pt nanoparticles for methanol electrooxidation, *Mater. Chem. Phys.,* **91**, pp. 124–129.
63. Komarneni, S., Pidugu, R., Li, Q. H., and Roy, R. (1995). Microwave-hydrothermal processing of metal powders, *J. Mater. Res.,* **10**, pp. 1687–1692.
64. Suryawanshi, Y. R., Chakraborty, M., Jauhari, S., Mukhopadhyay, S., Shenoy, K. T., and Shridharkrishna, R. (2013). Microwave irradiation solvothermal technique: an optimized protocol for size-control synthesis of Ru nanoparticles, *Cryst. Res. Technol.,* **48**, pp. 69–74.
65. Li, D. S., and Komarneni, S. (2006). Microwave-assisted polyol process for synthesis of Ni nanoparticles, *J. Am. Ceram. Soc.,* **89**, pp. 1510–1517.
66. Bonet, F., Delmas, V., Grugeon, S., Urbina, R. H., Silvert, P. Y., and Tekaia-Elhsissen, K. (1999). Synthesis of monodisperse Au, Pt, Pd, Ru and Ir nanoparticles in ethylene glycol, *Nanostruct. Mater.,* **11**, pp. 1277–1284.
67. Ren, L. Z., Meng, L. J., Lu, Q. H., Fei, Z. F., and Dyson, P. J. (2008). Fabrication of gold nano- and microstructures in ionic liquids–a remarkable anion effect, *J. Colloid Interface Sci.,* **323**, pp. 260–266.
68. de la Hoz, A., Diaz-Ortiz, A., and Moreno, A. (2005). Microwaves in organic synthesis. Thermal and non-thermal effects, *Chem. Soc. Rev.,* **34**, pp. 164–178.
69. Berlan, J. (1995). Microwaves in chemistry: another way of heating reaction mixtures, *Radiat. Phys. Chem.,* **45**, pp. 581–589.

70. Santos, T., Valente, M. A., Monteiro, J., Sousa, J., and Costa, L. C. (2011). Electromagnetic and thermal history during microwave heating, *Appl. Therm. Eng.*, **31**, pp. 3255–3261.
71. Veerakumar, P., Velayudham, M., Lu, K.-L., and Rajagopal, S. (2013). Silica-supported PEI capped nanopalladium as potential catalyst in Suzuki, Heck, and Sonogashira coupling reactions, *Appl. Catal., A*, **455**, pp. 247–260.
72. Blavo, S. O., Qayyam, E., Baldyga, L. M., Castillo, V. A., Sanchez, M. D., Warrington, K., Barakat, M. A., and Kuhn, J. N. (2013). Verification of organic capping agent removal from supported colloidal synthesized Pt nanoparticle catalysts, *Top. Catal.*, **56**, pp. 1835–1842.
73. Zhu, Z., Gao, F., Lei, J., Dong, H., and Ju, H. (2012). A competitive strategy coupled with endonuclease-assisted target recycling for DNA detection using silver-nanoparticle-tagged carbon nanospheres as labels, *Chem. Eur. J.*, **18**, pp. 13871–13876.
74. Lin, L., Weng, S., Zhao, C., Liu, Q., Liu, A., and Lin, X. (2013). Hairpin LNA biosensor with enzyme tagged AuNPs as tracer for amperometric detection of K-ras mutation gene, *Electrochim. Acta*, **108**, pp. 808–813.
75. Adeyemi, O. S., and Whiteley, C. G. (2013). Interaction of nanoparticles with argenine kinase from *Trypanosoma brucei*: kinetic and mechanistic evaluation, *Int. J. Biol. Macromol.*, **62**, pp. 450–456.
76. Cuenya, B. R. (2010). Synthetic and catalytic properties of metal nanoparticles: size, shape, support, composition, and oxidation state effects, *Thin Solid Films*, **518**, pp. 3127–3150.
77. Roucoux, A., Schulz, J., and Patin, H. (2002). Reduced transition metal colloids: a novel family of reusable catalysts?, *Chem. Rev.*, **102**, pp. 3757–3778.
78. Kundu, S., Peng, L. H., and Liang, H. (2008). A new route to obtain high-yield multiple-shaped gold nanoparticles in aqueous solution using microwave irradiation, *Inorg. Chem.*, **47**, pp. 6344–6352.
79. Liu, F. K., Ker, C. J., Chang, Y. C., Ko, F. H., Chu, T. C., and Dai, B. T. (2003). Microwave heating for the preparation of nanometer gold particles, *Jpn. J. Appl. Phys.*, **42**, pp. 4152–4158.
80. Hu, B., Wang, S. B., Wang, K., Zhang, M., and Yu, S. H. (2008). Microwave-assisted rapid facile "green" synthesis of uniform silver nanoparticles: self-assembly into multilayered films and their optical properties, *J. Phys. Chem. C*, **112**, pp. 11169–11174.

81. Chen, J., Wang, J., Zhang, X., and Jin, Y. L. (2008). Microwave-assisted green synthesis of silver nanoparticles by carboxymethyl cellulose sodium and silver nitrate, *Mater. Chem. Phys.*, **108**, pp. 421–424.
82. Sharma, V. K., Yngard, R. A., and Lin, Y. (2009). Silver nanoparticles: green synthesis and their antimicrobial activities, *Adv. Colloid Interface Sci.*, **145**, pp. 83–96.
83. Pastoriza-Santos, I., and Liz-Marzán, L. (2002). Formation of PVP-protected metal nanoparticles in DMF, *Langmuir*, **18**, pp. 2888–2894.
84. Yu, W., Tu, W., and Liu, H. (1999). Synthesis of nanoscale platinum colloids by microwave dielectric heating, *Langmuir*, **15**, pp. 6–9.
85. Tu, W., and Liu, H. (2000). Rapid synthesis of nanoscale colloidal metal clusters by microwave irradiation, *J. Mater. Chem.*, **10**, pp. 2207–2211.
86. Luo, Y. L. (2007). A simple microwave-based route for size-controlled preparation of colloidal Pt nanoparticles, *Mater. Lett.*, **61**, pp. 1873–1875.
87. Soares, P. I. P., Alves, A. M. R., Pereira, L. C. J., Coutinho, J. T., Ferreira, I. M. M., Novo, C. M. M., and Borges, J. P. M. R. (2014). Effects of surfactants on the magnetic properties of iron oxide colloids, *J. Colloid Interface Sci.*, **419**, pp. 46–51.
88. Tsuji, M., Hashimoto, M., Nishizawa, Y., Kubokawa, M., and Tsuji, T. (2005). Microwave-assisted synthesis of metallic nanostructures in solution, *Chem. Eur. J.*, **11**, pp. 440–452.
89. Barret, C. S. (1943). *Structure of Metals: Crystallographic Methods, Principles and Data* (McGraw-Hill Book, New York).
90. Kobayashi, H., Morita, H., Yamauchi, M., Ikeda, R., Kitagawa, H., Kubota, Y., Kato, K., Takata, M., Toh, S., and Matsumura, S. (2012). Nanosize-induced drastic drop in equilibrium hydrogen pressure for hydride formation and structural stabilization in Pd-Rh solid-solution alloys, *J. Am. Chem. Soc.*, **134**, pp. 12390–12393.
91. Pearson, W. B. (1972). *The Crystal Chemistry and Physics of Metals and Alloys* (John Wiley and Sons, New York).
92. Mavrikakis, M., Hammer, B., and Nørskov, J. K. (1998). Effect of strain on the reactivity of metal surfaces, *Phys. Rev. Lett.*, **81**, pp. 2819–2822.
93. Strasser, P., Koh, S., Anniyev, T., Greeley, J., More, K., Yu, C., Liu, Z., Kaya, S., Nordlund, D., Ogasawara, H., Toney, M. F., and Nilsson, A. (2010). Lattice-strain control of the activity in dealloyed core-shell fuel cell catalysts, *Nat. Chem.*, **2**, pp. 454–460.
94. Chen, Z., Mochizuki, D., Maitani, M. M., and Wada, Y. (2013). Facile synthesis of bimetallic Cu-Ag nanoparticles under microwave

irradiation and their oxidation resistance, *Nanotechnology*, **24**, p. 265602 (9 pages).

95. Patel, K., Kapoor, S., Dave, D. P., and Mukherjee, T. (2006). Synthesis of Au, Au/Ag, Au/Pt and Au/Pd nanoparticles using the microwave-polyol method, *Res. Chem. Intermed.*, **32**, pp. 103–113.
96. Skriver, H. L., and Rosengaard, N. M. (1992). Surface energy and work function of elemental metals, *Phys. Rev. B*, **46**, p. 7157.
97. García, S., Anderson, R. M., Celio, H., Dahal, N., Dolocan, A., Zhou, J., and Humphrey, S. M. (2013). Microwave synthesis of Au–Rh core–shell nanoparticles and implications of the shell thickness in hydrogenation catalysis, *Chem. Commun.*, **49**, pp. 4241–4243.
98. Harpeness, R., and Gedanken, A. (2004). Microwave synthesis of core-shell gold/palladium bimetallic nanoparticles, *Langmuir*, **20**, pp. 3431–3434.
99. Patel, K., Kapoor, S., Dave, D. P., and Mukherjee, T. (2005). Synthesis of Pt, Pd, Pt/Ag and Pd/Ag nanoparticles by microwave-polyol method, *J. Chem. Sci.*, **117**, pp. 311–316.
100. Tang, W., and Henkelman, G. (2009). Charge redistribution in core-shell nanoparticles to promote oxygen reduction, *J. Chem. Phys.*, **130**, p. 194505.
101. Bensebaa, F., Patrito, N., Page, Y. L., L'Ecuyer, P., and Wang, D. (2004). Tunable platinum-ruthenium nanoparticle properties using microwave synthesis, *J. Mater. Chem.*, **14**, pp. 3378–3384.
102. Taufany, F., Pan, C.-J., Lai, F.- J., Chou, H.- L., Sarma, L. S., Rick, J., Lin, J.-M., Lee, J.-F., Tang, M.-T., and Hwang, B.-J. (2013). Relating the composition of Pt_xRu_{100-x}/C nanoparticles to their structural aspects and electrocatalytic activities in the methanol oxidation reaction, *Chem. Eur. J.*, **19**, pp. 905–915.
103. Mehta, S. K., and Gupta, S. (2013). Synthesis of Au-Pd alloy nanoparticles and their catalytic activity in the electrooxidation of formic acid and lower alcohols in alkaline media, *Sci. Adv. Mater.*, **5**, pp. 1377–1383.
104. Du, J.-Q., Zhang, Y., Tian, T., Yan, S.-C., and Wang, H.-T. (2009). Microwave irradiation assisted rapid synthesis of Fe-Ru bimetallic nanoparticles and their catalytic properties in water-gas shift reaction, *Mater. Res. Bull.*, **44**, pp. 1347–1351.
105. Ugalde, M., Chavira, E., Ochoa-Lara, M. T., Figueroa, I. A., Quintanar, C., and Tejeda, A. (2013). Synthesis by microwaves of bimetallic nano-rhodium-palladium, *J. Nanotechnol.*, **2013**, Article ID 578684 (9 pages).

106. Pal, A., Sha, S., and Devi, S. (2007). Synthesis of Au, Ag and Au–Ag alloy nanoparticles in aqueous polymer solution, *Colloids Surf., A,* **302**, pp. 51–57.

107. Shafii, S., Lihua, W., Nordin, R., and Yong, L. K. (2012). Synthesis of palladium-platinum bimetallic nanoparticles and their catalytic activity towards the hydrogenation reaction of palm olein, *J. Chem. Eng. Process Technol.,* **3**, pp. 1–8.

108. Gu, G., Xu, J., Wu, Y., Chen, M., and Wu, L. (2011). Synthesis and antibacterial property of hollow SiO_2/Ag nanocomposite sphere, *J. Colloid Interface Sci.,* **359**, pp. 327–333.

109. Quang, D. V., Sarawade, P. B., Hilonga, A., Kim, J.-K., Shim, Y. H., Shao, G. N., and Kim, H. T. (2012). Synthesis of silver nanoparticles within the pores of functionalized-free silica beads: the effect of pore size and porous structure, *Mater. Lett.,* **68**, pp. 350–353.

110. Ortiz-Ibarra, H., Casillas, N., Soto, V., Barcena-Soto, M., Vitela, R. T., Cruz, W. D., and Salazar, S. G. (2007). Surface characterization of electrodeposited on activated carbon for bactericidal purporses, *J. Colloid Interface Sci.,* **314**, pp. 562–571.

111. Pasini, T., Piccinini, M., Blosi, M., Bonelli, R., Albonetti, S., Dimitratos, N., Lopez-Sanchez, J. A., Sankar, M., He, Q., Kiely, C. J., Jutchings, G. J., and Cavani, F. (2011). Selective oxidation of 5-hydroxymethyl-2-furfural using supported gold-copper nanoparticles, *Green Chem.,* **13**, pp. 2091–2099.

112. Kakade, B. A., Sahoo, S., Halligudi, S. B., and Pillai, V, K. (2008). Highly selective catalytic hydrogenation of arenes using rhodium nanoparticles supported on multiwalled carbon nanotubes, *J. Phys. Chem. C,* **112**, pp. 13317–13319.

113. Bhattacharyya, S., and Gedanken, A. (2008). Microwave-assisted insertion of silver nanoparticles into 3-D mesoporous zinc oxide nanocomposites and nanorods, *J. Phys. Chem. C,* **112**, pp. 659–665.

114. Rogemond, E., Essayem, N., Frety, R., Perrichon, V., Primet, M., and Mathis, F. (1997). Characterization of model three-way catalysts, *J. Catal.,* **166**, pp. 229–235.

115. Sugiura, Y., Mukai, D., Murai, Y., Tochiya, S., and Sekine, Y. (2013). Oxidation resistance of $Ni/La_{0.7}Sr_{0.3}AlO_{3-\delta}$ catalyst for steam reforming of model aromatic hydrocarbon, *Int. J. Hydrogen Energy,* **38**, p. 7822.

116. Zhang, P., Huang, S.-Y., and Popov, B. N. (2010). Mesoporous tin oxide as an oxidation-resistant catalyst support for proton exchange membrane fuel cells, *J. Electrochem. Soc.,* **157**, pp. B1163–B1172.

117. Campelo, J. M., Luna, D., Luque, R., Marinas, J. M., and Romero, A. A. (2009). Sustainable preparation of supported metal nanoparticles and their applications in catalysis, *ChemSusChem,* **2**, pp. 18–45.

118. Zhao, J., Wang, P., Chen, W., Liu, R., Li, X., and Nie, Q. (2006). Microwave synthesis and characterization of acetate-stabilized Pt nanoparticles supported on carbon for methanol electro-oxidation, *J. Power Sources,* **160**, pp. 563–569.

119. Li, X., Chan, W.-X., Zhao, J., Xing, W., and Xu, Z-D. (2005). Microwave polyol synthesis of Pt/CNTs catalysts: effects of pH on particle size and electrocatalytic activity for methanol electrooxidation, *Carbon,* **43**, pp. 2168–2174.

120. Lordi, V., Yao, N., and Wei, J. (2001). Method for supporting platinum on single-walled carbon nanotubes for a selective hydrogenation catalyst, *Chem. Mater.,* **13**, pp. 733–737.

121. Li, W. Z., Liang, C. H., Qiu, J. S., Zhou, W. J., Han, H. M., and Wei, Z. B. (2002). Carbon nanotubes as support for cathode catalyst of a direct methanol fuel cell, *Carbon,* **40**, pp. 791–794.

122. Campelo, J. M., Conesa, T. D., Gracia, M. J., Jurado, M. J., Luque, R., Marinas, J. M., and Romero, A. A. (2008). Microwave facile preparation of highly active and dispersed SBA-12 supported metal nanoparticles, *Green Chem.*, **10**, pp. 853–858.

123. González-Arellano, C., Campelo, J. M., Macquarrie, D. J., Marinas, J. M., Romero, A. A., and Luque, R. (2008). Efficient microwave oxidation of alcohols using low-loaded supported metallic iron nanoparticles, *ChemSusChem,* **1**, pp. 746–750.

124. Glaspell, G., Hassan, H. M. A., Elzatahry, A., Abdalsayed, V., and El-Shall, M. S. (2008). Nanocatalysis on supported oxides for CO oxidation, *Top Catal.,* **47**, pp. 22–31.

125. Glaspell, G., Fuoco, L., and El-Shall, M. S. (2005). Microwave synthesis of supported Au and Pd nanoparticle catalysts for CO oxidation, *J. Phys. Chem. B,* **109**, pp. 17350–17355.

126. Liang, Y., Zhang, H., Zhong, H., Zhu, X., Tian, Z., Xu, D., and Yi, B. (2006). Preparation and characterization of carbon-supported PtRuIr catalyst with excellent CO-tolerant performance for proton-exchange membrane fuel cells, *J. Catal.,* **238**, pp. 468–476.

127. Boxall, D. L., Deluga, G. A., Kenik, E. A., King, W. D., and Lukehart, C. M. (2001). Rapid synthesis of a Pt_1Ru_1/Carbon nanocomposite using microwave irradiation: a DMFC anode catalyst of high relative performance, *Chem. Mater.*, **13**, pp. 891–900.

128. Siamaki, A. R., Khder, A. E. R. S., Abdelsayed, V., El-Shall, M. S., and Gupton, B. F. (2011). Microwave-assisted synthesis of palladium nanoparticles supported on grapheme: a highly active and recyclable catalyst for carbon-carbon cross-coupling reactions, *J. Catal.*, **279**, pp. 1–11.

129. Zhu, J., and Zhu, Y. (2006). Microwave-assisted one-step synthesis of polyacrylamide-metal (M = Ag, Pt, Cu) nanocomposites in ethylene glycol, *J. Phys. Chem. B*, **100**, pp. 8593–8597.

130. Wada, Y., Kobayashi, T., Tamasaki, H., Sakata, T., Hasegawa, N., Mori, H., and Tsukahara, Y. (2007). Nanohybrid polymer prepared by successive polymerization of methacrylate monomer containing silver nanoparticles in situ prepared under microwave irradiation, *Polymer*, **48**, pp. 1441–1444.

131. Mukherjee, P., Patra, C. R., Ghosh, A., Kumar, R., and Sastry, M. (2002). Characterization and catalytic activity of gold nanoparticles synthesized by autoreduction of aqueous chloroaurate ions with fumed silica, *Chem. Mater.*, **14**, pp. 1678–1684.

132. Tuval, T., and Gedanken, A. (2007). A microwave-assisted polyol method for the deposition of silver nanoparticles on silica spheres, *Nanotechnology*, **18**, 255601 (7 pp).

133. Harpeness, R., Peng, Z., Liu, X., Pol, V. G., Koltypin, Y., and Gedanken, A. (2005). Controlling the agglomeration of anisotropic Ru nanoparticles by the microwave-polyol process, *J. Colliod Interface Sci.*, **287**, pp. 678–684.

134. Schanche, J.-S. (2003). Microwave synthesis solutions from personal chemistry, *Mol. Diversity*, **7**, pp. 291–298.

Chapter 10

Microwave-Assisted Synthesis of Semiconductor Nanomaterials for Energy Conversion

Xinjuan Liu,[a] Likun Pan,[b] and Chang Q. Sun[a]

[a] *Institute of Coordination Bond Metrology and Engineering, College of Materials Science and Engineering, China Jiliang University, Hangzhou 310018, China*
[b] *Engineering Research Center for Nanophotonics & Advanced Instrument, Ministry of Education, Department of Physics, East China Normal University, Shanghai 200062, China*
lkpan@phy.ecnu.edu.cn

10.1 Introduction

The increasing demands for energy and growing concerns about air pollution and global warming have called for intense research on energy conversion from alternative energy sources. Sustainable and renewable energy resources are being intensively pursued owing to the diminishing supply of fossil fuels and climate change. Sunlight, being an abundant and easily available energy resource, possesses great potential in driving environmentally benign organic transformations. Solar energy conversion systems such as solar cells

Microwave Engineering of Materials: From Mesoscale to Nanoscale
Edited by Erwann Guénin

ISBN 978-981-4669-42-9 (Hardcover), 978-981-4669-43-6 (eBook)
www.panstanford.com

and photocatalysis are considered promising methods for effective and practical technologies.

Up to now, various methods such as hydrothermal or solvothermal synthesis have been employed to synthesize semiconductor nanostructures for solar cells and photocatalysis. However, conventional heating in hydrothermal or solvothermal processes is inevitably confronted with several disadvantages, such as sharp thermal gradients throughout the bulk solution, slow reaction kinetics, and nonuniform reaction conditions [1, 2]. Since the first reports in 1986 [3], the use of microwave ovens became an indispensable tool in modern organic synthesis. However, experimental parameters such as irradiation power, reaction temperature, and pressure inside the vessel are not known precisely enough in domestic microwave ovens. Nowadays, several manufacturers offer a large variety of reactor systems with different degrees of sophistication with respect to automation, database capabilities, safety features, temperature and pressure monitoring, and vessel design [4]. Microwave irradiation induces interaction of the dipole moment of polar molecules or molecular ionic aggregates with alternating electronic and magnetic fields, causing molecular-level heating, which leads to homogeneous and quick thermal reactions with low thermal gradient [5–14]. Compared to other solution-phase synthetic methods in conventional heating, microwave hydrothermal or solvothermal synthesis has been accepted as a promising method for rapid heating, higher reaction rate and selectivity, lower reaction temperature, less reaction time, homogeneous thermal transmission, and phase purity with better yield [15–19].

Recently, microwave-assisted synthesis of semiconductor nanostructures for energy conversion has become a hot research topic. The synthesis of nanoparticles and nanostructures, whose growth is highly sensitive to the reaction conditions, could benefit from the efficient and controlled heating as provided by microwave irradiation. Microwave heating does not only reduce the chemical reaction times by several orders of magnitude but also suppresses side reactions and thus improves the yield and reproducibility of a specific synthesis protocol. A variety of nanomaterials with different morphologies have been synthesized through the microwave heating approach, such as nanorods, nanowires, nanofibers, nanotubes, nanosheets, and nanoparticles [20–24]. The microwave heating technique is potentially competitive for scaling up industrial production of high-quality nanomaterials [25–27].

In this chapter, we highlight the recent progress in exploring various semiconductor nanomaterials via microwave-assisted synthesis. Application of these semiconductor nanomaterials in the solar energy conversion for solar cells and photocatalysis is also demonstrated.

10.2 Microwave-Assisted Synthesis of Semiconductor Nanomaterials for Photocatalysis

Currently considerable attention has been paid to semiconductor photocatalysis for water treatment due to its good destruction ability of pollutants and broad compound applicability [28–34], since the research of photocatalytic water split on TiO_2 electrodes was conducted in 1972 [35]. Photocatalysis is applied in many areas, including the elimination of contaminants from water and air, odor control, bacterial inactivation, water splitting to produce H_2, the inactivation of cancer cells, and many others. Organic compounds that are introduced into the environment as contaminants are of major concern, and the contaminants can be degraded photocatalytically, ideally to CO_2 and water, with no additional waste or by-products [36–40]. Nanostructured semiconductor materials have emerged as novel photocatalysts to harvest solar energy because of their unique structural, optical, and chemical properties. Diverse structures such as nanoparticles, porous nanospheres, nanotubes, nanorods, and nanowires exhibit efficient and tunable optical absorption, large surface areas, short charge carrier diffusion lengths, and low reflectivity for effective photocatalysis [41–48].

A photocatalytic reaction can be defined as a reaction induced by photoabsorption of a solid material, the photocatalyst, which remains unchanged during the reaction. Photocatalysis involves a complicated sequence of multiple synergistic or competing steps, including light harvesting, charge generation, separation, transportation, recombination, and utilization. Photocatalysis is normally based on the light absorption of the semiconductor oxide photocatalyst to excite the electrons from the valence band (VB) to the conduction band (CB) and create electron–hole pairs. The energy of the incident light should be greater than the band gap of the photocatalyst. The activation equation can be written as $TiO_2 + h\nu \rightarrow h^+ + e^-$.

These electrons and holes can migrate and initiate redox reactions with water and oxygen and then degrade organic molecules or reduce metal ions absorbed on the surface of the photocatalyst. For metal ion reduction, the CB level should be more negative than the metal ion potential. The photogenerated electrons and holes can recombine in bulk or on the surface of the semiconductor within a very short time, releasing energy in the form of heat or photons. The recombination of photogenerated electrons and holes is believed to be one of the main reasons for the low efficiency of photocatalysis. As revealed by time-resolved spectroscopic studies, electron–hole trapping or recombination rates are extremely fast, about the order of 10^{-6}–10^{-15}s, and can substantially lower the photocatalytic activity. There are many factors that can exert significant influence on photocatalytic performance, including the size, specific surface area, pore volume, pore structure, crystalline phase, and exposed surface facets of the photocatalyst. Thus, the development of performance improvements by adjusting these factors remains a focus of photocatalysis research.

In contrast to other semiconductors (i.e., WO_3, ZnS, Fe_2O_3, CdS, and $SrTiO_3$) [49–52], ZnO and TiO_2 have been proven to be promising photocatalysts for widespread environmental applications because of their intriguing optical and electric properties, low cost, and ease of availability [53, 54]. However, they can absorb only ultraviolet (UV) light because of their wide band gap (~3.2 eV) and are not responsive to visible light ($\lambda > 400$ nm), which accounts for around 46% of the total solar energy. To make full use of solar energy, the exploitation of a visible light photocatalyst becomes one of the hottest spots in the environmental photocatalytic field and is significant for the development of photocatalysis technology. It is a challenge of great importance to identify and design new semiconductor materials that are efficient, stable, and abundant. In this chapter, we summarize various semiconductor nanomaterials via microwave-assisted synthesis for photocatalysis. They show improved photocatalytic activities compared to the ones synthesized via conventional methods.

10.2.1 *Photocatalysts*

10.2.1.1 Semiconductor oxide

10.2.1.1.1 Bi_2O_3

Bi_2O_3 can crystallize in many structures, α-, β-, ω-, γ-, and σ-Bi_2O_3, among which, the α phase is stable at low temperature and σ at high temperature. Others are all metastable phases. Bi_2O_3 is an important metal oxide semiconductor with a direct band gap ranging from 2 to 3.9 eV and can be used to decompose dyes in water under UV and visible light irradiation [55]. Furthermore, Bi_2O_3 is rather inert in neutral water, which is a fundamental precondition for application as a photocatalyst for wastewater purification. The intrinsic polarizability induced by the Bi $6s^2$ lone pairs of electron favors the separation of photogenerated electron–hole pairs and the transfer of these charge carriers [56]. It is well known that photocatalytic activity depends on the morphologies and microstructures of photocatalysts, which are determined by the preparation method [57, 58]. Up to now, Bi_2O_3 with various morphologies, including nanoparticles [59], mesh-like [60], peanut-like [61], nanorods [62], hexagonal nanorings [62], and nanowires [63], have been fabricated by conventional heating in hydrothermal or solvothermal processes.

Huang et al. [64] synthesized directly amorphous β-Bi_2O_3 nanoparticles via a liquid-phase microwave reaction using $Bi(NO_3)_3{\cdot}5H_2O$ and urea as starting materials. The synthesized sample changed gradually into well-crystallized sheet-like nanoparticles of β-Bi_2O_3 or α-Bi_2O_3 during the following calcining at a lower (300°C) or higher (350°C) temperature. The reaction system was heated under microwave irradiation with 500 W of power for 90 min. The photocatalytic activity of the samples (1 g/L) under simulated sunlight (incandescent lamp with the power of 200 W) was investigated in the degradation of rhodamine B (RhB, 10^{-5} M) as a model reaction. β-Bi_2O_3 showed lower band-gap energy and higher absorbance in a wider visible light region than α-Bi_2O_3

did, resulting in its higher photocatalytic activity (91%). It was also found that higher crystallinity can improve the photocatalytic activity. Microwave irradiation can control the reaction kinetics to accelerate reactions, to control the morphology of the products, and even to promote the formation of metastable phases [65]. β-Bi_2O_3 is metastable and hard to prepare directly without calcining. Herein, for the first time, β-Bi_2O_3, though with low crystallinity, was prepared directly from a solution under microwave irradiation. By contrast, when the system was heated conventionally in an oil bath, little precipitate was obtained, which could not be indexed to β-Bi_2O_3. Microwave irradiation plays a very important role in the synthesis of β-Bi_2O_3, which is interesting, considering the higher photocatalytic activity than α-Bi_2O_3.

Liu et al. [8] successfully fabricated Bi_2O_3 via a microwave-assisted reaction using $Bi(NO_3)_3{\cdot}5H_2O$ and NaOH as starting materials in a microwave system (Explorer-48, CEM Co.). The reaction was treated at 150°C with a microwave irradiation power of 150 W for 10 min. The influence of the pH value on the morphology and photocatalytic activity of Bi_2O_3 samples was investigated. The as-synthesized Bi_2O_3 samples at pH values of 5, 7, 9, and 12 were named as B-5, B-7, B-9, and B-12. The morphology of Bi_2O_3 is notably affected by the pH value of the precursor solution. At a lower pH value of 5, the Bi_2O_3 sample displays the sheet nanostructure. With an increase in the pH value, the morphology changes drastically. The sheets nanostructure is evolved to rods at a higher pH value, and the diameter of the rods increases obviously at a pH value of 12. B-7 was identified by energy-dispersive X-ray spectroscopy (EDS) linked to FESEM. The peaks of Bi and O in the EDS profile proved the existence of Bi_2O_3. Bi_2O_3 rods with a diameter in the range of 20–40 nm are relatively uniform and dispersive. Photocatalytic degradation of methyl blue (MB) by B-5, B-7, B-9, and B-12 was performed under visible light irradiation. It was observed that the concentration of MB is hardly reduced under visible light irradiation in the absence of the photocatalyst and that the photocatalytic performance of Bi_2O_3 is dependent on the pH value of the precursor solution. The degradation rate of MB (5 mg/L) for Bi_2O_3 (2 g/L) sheets synthesized at a pH value of 5 is 9%. When the pH value of the precursor solution is increased to 7, the degradation rate reaches a maximum value of 76% for B-7, which is related to its higher visible light absorption and comparatively low electron–hole pair recombination. However, when the pH value of the precursor solution is further increased, the degradation rate decreases, which is ascribed to the much lower absorption of B_2O_3 obtained in the visible light range.

10.2.1.1.2 SnO_2

SnO_2 is recognized to be one of the most suitable candidates for widespread environmental applications due to its intriguing optical and electric properties and high stability [66–70]. Unfortunately, SnO_2 normally exhibits high photocatalytic activity under UV light irradiation due to its wide band gap (~3.6 eV), which significantly limits its practical applications [71]. Recently, investigations have been carried out to study the synthesis of a SnO_2 photocatalyst to improve its visible light photocatalytic performance [72, 73]

Liu et al. [74] successfully fabricated SnO_2 quantum dots (QDs) via microwave-assisted reaction of a $SnCl_4$ precursor solution in an aqueous solution using a microwave system in an automated focused microwave system (Explorer-48, CEM Co.) and treated at 150°C with a microwave irradiation power of 150 W for 10 min. Figure 10.1a,b shows the low-magnification and high-magnification FESEM images of S-5. It can be clearly observed that SnO_2 is constructed from a random agglomeration of particles and small SnO_2 QDs aggregate into large particles. Figure 10.1c,d shows the low-magnification and high-magnification HRTEM images of S-5. The lattice fringes with an interplanar distance of 0.334 nm can be assigned to the (110) plane of SnO_2 (JCPDS 41-1445). The SnO_2 QDs are relatively uniform and dispersive with diameters of 3–5 nm. The photocatalytic performance of the as-prepared samples (2 g/L) was evaluated through the photocatalytic degradation of MB (20 mg/L) under visible light irradiation produced by a 400 W metal halogen lamp with a cutoff filter ($\lambda > 400$ nm). As shown in Fig. 10.2, the photocatalytic performance of the SnO_2 QDs is related to the pH value of the precursor solution. The SnO_2 QDs synthesized at a pH value of 5 exhibits an optimal photocatalytic performance with a degradation rate of 90% at 240 min under visible light irradiation, which should be ascribed to the easier adsorption of pollutants,

Figure 10.1 (a and b) FESEM images of S-5; (c) TEM and (d) HRTEM images of S-5 [74].

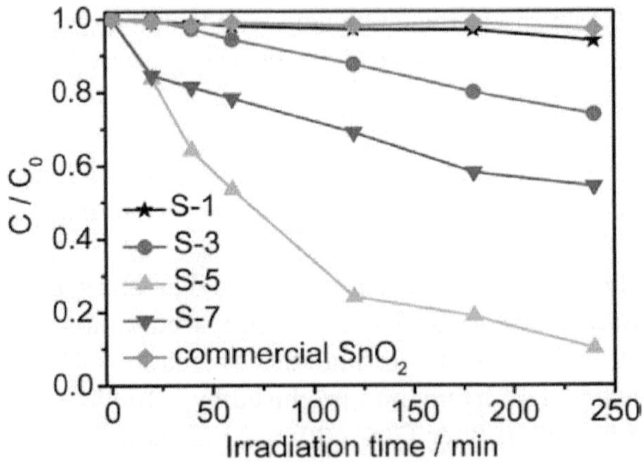

Figure 10.2 Photocatalytic degradation of MB by S-1, S-3, S-5, and S-7 under visible light irradiation [74]. (The SnO_2 samples synthesized at pH values of 1, 3, 5, and 7 were named S-1,S-3, S-5, and S-7.)

higher visible light absorption intensity, and lower electron–hole pair recombination.

10.2.1.1.3 ZnO

ZnO, a semiconductor with a direct band gap of 3.37 eV and a large exciton binding energy of 60 meV, possesses unique electrical, optoelectronic, and luminescent properties. Because of these attractive properties, nanostructured ZnO, including particles, rods, wires, belts, tubes, and disks, has many important practical applications in catalysis, photoluminescence, and functional devices (e.g., solar cells, resonators, field-effect transistors, and gas sensors) [75, 76].

Shinde et al. [77] demonstrated the shape- and size-selective growth of ZnO nanostructures on indium tin oxide–coated glass substrates using a microwave-assisted chemical bath deposition (MACBD) method by systematically controlling the deposition parameters such as pH and temperature of the solution. $Zn(NO_3)_3$ and aqueous NH_3 were used as starting reagents, which were heated under microwave irradiation (Milestone Multisynthetic Labstation) to a specific temperature (from 90°C to 120°C) for 1 h. It is found that the morphology (bundled, tapered tip, and flat tip) and size of ZnO can be controlled by changing the pH and temperature of the reaction bath. The diameter of the ZnO rods decreased with increasing temperature at a constant pH value. An increase in the pH of the solution promoted the nucleation rate, resulting in a decrease in the size of the rods and modification of the growth rates along different planes to give varying shapes and sizes of rods. The prepared rods were active catalysts in the degradation of MB (10^{-4} M) under UV radiation (Xe lamp) and exhibited size-dependent photocatalytic activity. Larger ZnO nanorods (pH = 10, T = 120°C) photocatalytically degraded 85% of MB, whereas smaller ZnO nanorods (pH = 12, T = 120°C) degraded 92% of MB.

Ivanov et al. [78] synthesized highly dispersed ZnO powders based on microwave heating of a mixture of $Zn(NO_3)_2$ and NaCl and low-temperature microwave-assisted hydrothermal treatment of $Zn(OH)_2$ in a domestic microwave oven (650 W) at 75°C–170°C for 15–240 min. ZnO powders consisted of rod-like particles (250 ×

100 nm). The obtained powders (3 g/L) posed high photocatalytic activity in the degradation of methyl orange (MO) (13 mg/L).

Ivanov and coworkers [79] also studied the effect of synthesis temperature on the morphology of fine ZnO powder prepared by microwave-assisted hydrothermal processing of $Zn(OH)_2$ suspensions using a Volta MS-6 microwave oven in pulsed mode at temperatures of 75°C–170°C for 15 min. Upon an increase in the temperature of microwave-assisted hydrothermal treatment, the mean size of ZnO particles gradually decreases, while an increase in the temperature of conventional hydrothermal treatment does not result in a significant change in the particle size. Probably the effect is due to the increase in the nucleation rate against the particle growth rate under microwave heating. The photocatalytic activity of ZnO powders (~3 g/L) in degradation of MO (13 mg/L) under UV light irradiation has been assessed as a function of synthesis temperature. Microwave processing, compared with conventional hydrothermal processes, was found to decrease the amorphous content of the sample and increase the specific surface area of the material and, hence, enhance its photocatalytic activity.

Parida and Parija [80] prepared ZnO by microwave-assisted synthesis and conventional sol–gel and precipitation methods, respectively. During microwave-assisted synthesis, the mixture of $NaHCO_3$ and $Zn(OAc)_2$ was put into the microwave oven for a particular time in the range of 5–30 min. The photocatalytic degradation of phenol (10 mg/L) using ZnO (0.6 g/L) was performed under sunlight and UV light irradiation using a low-pressure (6 W) mercury lamp, which emits radiation at 258 nm. It is found that microwave irradiation decreases the crystallite size and increases the surface area and acid sites of ZnO compared to the conventional method. The 15 min microwave-irradiated sample showed 88%, 82%, and 80% conversion at pH 5.0 under sunlight, visible light, and UV light irradiation, respectively.

Ghule et al. [81] synthesized nanosized ZnO rods using a microwave-assisted aqueous method in a microwave oven with a power output of 695 W by following the on–off cycle (1 min on and 2 min off) for 10 min using $Zn(NO_3)_3$ as a precursor and high-molecular-weight polyvinyl alcohol (PVA) as a stabilizing agent. They showed that ZnO nanorods had diameters of about 50 nm and lengths of a few micrometers and that ZnO nanorods were of hexagonal wurtzite structure. The effects of catalyst loading, pH value, and initial concentration of MO on the photocatalytic degradation efficiency using ZnO nanorods as a photocatalyst under UV irradiation (125 W high-pressure mercury lamp with maximum emission at about 365 nm) were discussed. The results revealed that ZnO nanorods with a diameter of 50 nm showed the highest photocatalytic activity at pH $= 7$ and 0.2 g/dm^3 loading of catalyst within 80 min.

10.2.1.1.4 TiO_2

TiO_2 photocatalysis has attracted increasing attention due to its biological and chemical inertness, strong photo-oxidization power, cost-effectiveness, and long-term stability against photo- and chemical corrosion [82–84]. However, the photocatalytic efficiency of TiO_2 still needs to be further improved for practical applications.

Zhang et al. [85] synthesized a microsheet anatase TiO_2 single-crystal photocatalyst with a remarkable 80% level of reactive facets via a microwave-assisted hydrothermal route involving titanium tetrafluoride and a tetrafluoroborate-based ionic liquid. The microwave reaction was carried out using a microwave digestion system (ETHOS TC, Milestone) at 210°C for 90 min. The high reactivity of facets makes these single crystals (1.33 g/L) highly photocatalytically active in degradation of 4-chlorophenol (10 ppm) in aqueous solution under UV light irradiation (300 W high-pressure mercury vapor lamp). They are easily recyclable and thermally stable up to 800°C.

Zheng et al. [86] synthesized rapidly the anatase TiO_2 nanocrystals with tunable percentage of reactive {001} facets by a microwave-assisted hydrothermal treatment of the mixed solution of tetrabutyl titanate, HF solution, and additional water at 200°C for 30 min using a microwave digestion system (MDS-6, Sineo, Shanghai, China). With increasing the amount of additional water from 0 to 21 mL, the shapes of TiO_2 nanocrystals evolved from nanosheets to truncated octahedral bipyramids, resulting in a steady decrease in the percentage of exposed {001} facets (from 71% to 23%). The photocatalytic activity of TiO_2 (1 g/L) was evaluated in the degradation of brilliant-red X3B (10^{-4} M) under visible light irradiation by a light source (375 W, Shanghai Yamin) emitted mainly at 365 nm. The photocatalytic activity of the resulted TiO_2 nanocrystals increased first and then decreased with an increase in the amount of water, and the sample with 60% of {001} facets exposed showed the highest photocatalytic activity.

Zhang et al. [87] employed a high-energy-efficient microwave-assisted hydrothermal method to prepare TiO_2 from commercially available $K_2Ti_4O_9$ particles without using any organic species at 190°C for a desired length of time. After the microwave reaction, regardless of the reaction time, all the obtained products showed a single anatase phase of TiO_2 and consisted of small particles around 5 nm in diameter. The specific surface area of TiO_2 was above 200 m^2/g. The photocatalytic activities were evaluated by the decomposition of acetaldehyde under 352 nm UV light irradiation. The TiO_2 prepared through this process showed high photocatalytic activities (89.3%) equal to the commercial one P25 (87.7%).

10.2.1.2 Metal tungstate

Metal tungstate is an important family of inorganic materials that have great application potential in the areas of photoluminescence, optics, humidity sensors, magnetics, and catalysis [88, 89]. A family of metal tungstate nanostructures has been synthesized over the past two decades, mainly focusing on zero-dimensional, one-dimensional, and complex structures [90].

10.2.1.2.1 Bi_2WO_6

Bi_2WO_6, one of the simplest Aurivillius phase oxides, has a layered structure in which the perovskite layer $(WO_4)^{2-}$ lies between $(Bi_2O_2)^{2+}$ layers. Bi_2WO_6 has attracted a great deal of attention because of its excellent physical and chemical properties such as ferroelectric piezoelectricity, pyroelectricity, catalytic behavior, oxide anion conduction, nonlinear dielectric susceptibility, and luminescence [91]. Since the discovery of visible-light-driven photocatalytic activities of Bi_2WO_6 for O_2 evolution [92] and the degradation of organic compounds [93], a great deal of research has been conducted to study its photocatalytic performance under the visible light irradiation [94–98]. Many research groups have carried out studies of Bi_2WO_6 nano-/microstructures, fabricating many kinds of Bi_2WO_6 with various morphologies, such as nanoplates, nanoflake films, porous thin films, nanocages, and nanofibrous mat-, flower-, tire-, and helix-like shapes [99–103].

Huang et al. [104] synthesized pure orthorhombic-phase Bi_2WO_6 powders by a microwave hydrothermal method in the absence of surfactants and templates using $Bi(NO_3)_3 \cdot 5H_2O$ and Na_2WO_4 as raw materials and examined the effects of different temperatures on the photocatalytic activity of the flower-like Bi_2WO_6. The results indicated that different microwave temperatures had an important effect on the photocatalytic performance. The flower-like Bi_2WO_6 (1 g/L) self-assembling into the uniform size with a diameter of 2 μm synthesized at 180°C had the maximal catalytic reaction rate constant about 0.04167 min^{-1}, and the degradation rate of RhB (5 mg/L) reached more than 96% under visible light irradiation by a 350 W Xe lamp for 70 min.

10.2.1.2.2 $ZnWO_4$

$ZnWO_4$ is a kind of tungstate that has received considerable attention and is widely applied in X-ray and γ-scintillator, photoanodes, solid-state laser hosts, electrooptics, and acoustic and optical fibers [105]. Until now, several methods have been developed to synthesize nanocrystalline $ZnWO_4$, including the Czochralski technique [106], the sol–gel technique [107], the hydrothermal process [108], the polymerized complex method [109], the hydrothermal-combined-with-annealing treatment [110], the template method [111], the solid-state metathetic approach [112], the self-propagating combustion method [113], etc. Besides these established synthesis methods, finding a simple and cost-effective route to synthesize nanocrystalline $ZnWO_4$ is still a challenge.

Garadkar et al. [105] synthesized $ZnWO_4$ nanoparticles by a simple microwave-assisted technique using zinc nitrate and Na_2WO_4 as starting reagents in a microwave oven (input power 800 W) with the on–off cycle (20 s on and 40 s off) for about 45 min. The synthesized nanoparticles exhibited only the wolframite structure. The size of $ZnWO_4$ was found to be 10 nm, and its band gap was about 3.4 eV. The photocatalytic activities of $ZnWO_4$ nanoparticles in the degradation of RhB and MB (10^{-5} M) were investigated under

UV irradiation using a 115 W low-pressure mercury lamp with an average light intensity of 0.9 mW/cm^2 ($\lambda = 254$ nm), and they exhibited highly enhanced photocatalytic activity toward RhB than MB.

Bi et al. [114] synthesized $ZnWO_4$ nanoparticles by a facile microwave-assisted solvothermal process using a microwave synthesizer (ETHOS TC, Milestone) at 160°C for different times. $ZnSO_4 \cdot 7H_2O$, $5(NH_4)_2O \cdot 12WO_3 \cdot 5H_2O$, and ethylene glycol were used as the starting materials. The as-synthesized nanoparticles exhibited the wolframite structure without impurities. Compared with the conventional solvothermal route, the irradiation under the microwave produces better crystallinity in a shorter treatment time. This indicates that the introduction of microwaves can save energy and time with faster kinetics of crystallization. The photocatalytic activities were evaluated by the decomposition of salicylic acid (2.5×10^{-4} M) and RhB (10^{-5} M) under UV light irradiation. Four 4 W UV lamps with a wavelength centered at 254 nm (Philips, TUV4W/G4 T5) were used as the illumination source. The results showed that the samples prepared by the microwave solvothermal process exhibited higher photocatalytic activities than those prepared by the conventional solvothermal process. The sample prepared by the microwave-assisted solvothermal process for 3 h exhibited the highest photocatalytic activity in the degradation of salicylic acid and RhB because of its higher crystallinity and more adsorbed oxygen.

Yan et al. [43] successfully synthesized highly crystalline $ZnWO_4$ nanoparticles via a highly effective and environmentally friendly microwave route using a commercial multimode microwave synthesizer coupled with an automation system (Initiator 8 Exp) by controlling the reaction time and temperature. Na_2WO_4 and $ZnCl_2$ were the staring materials. The crystallinity was enhanced with an increase in the reaction temperature and time. The photocatalytic activity of the $ZnWO_4$ catalyst was evaluated by the degradation of RhB (10 mg/L) under 300 W UV irradiation. The photocatalytic activity of samples prepared at a reaction temperature of 140°C and a reaction time of 5 min was the best. Furthermore, the $ZnWO_4$ powder showed stable photocatalytic activity in the degradation of RhB under UV light irradiation.

Figure 10.3 (a) SEM and (b) TEM images of $AgIn(WO_4)_2$ mesocrystals prepared by microwave-assisted synthesis at 180°C for 20 min.

10.2.1.2.3 $AgIn(WO_4)_2$

$AgIn(WO_4)_2$, as a semiconductor with a band gap (E_g) around 3.12 eV, has high photocatalytic activity under UV light because of the transfer of photogenerated electrons from the hybridized band of Ag $4d$ and O $2p$ to the hybridized band of Ag $5s$ and In $5s$.

Hu et al. [115] synthesized a new kind of $AgIn(WO_4)_2$ mesocrystal by a microwave-assisted approach from $AgNO_3$, $In(NO_3)_3$, and Na_2WO_4. These mesocrystals show a high hierarchy of the caterpillar-like architectures, as shown in the scanning electron microscopy (SEM) image (Fig. 10.3a). It can be seen from Fig. 10.3b that these architectures include the initially formed olive-like core and a striated pattern of outgrowths that were oriented vertically to the core. All structures are uniform with diameters of 350–500 nm at the middle part and lengths of 800–1000 nm. The $AgIn(WO_4)_2$ mesocrystals show high and selective photocatalytic activity in the degradation of different organic dyes (Eosin Y, RhB, and MO) under UV and visible light irradiation.

10.2.1.2.4 $M_3PW_{12}O_{40}$($M = NH_4^+$, Cs^+, Ag^+, and Cu^{2+})

Nanosize and bimodal porous polyoxotungstate anatase can photocatalytically decompose the organophosphorus pesticide under visible-light excitation.

Deng et al. [116] prepared $M_3PW_{12}O_{40}$ ($M = NH_4^+$, Cs^+, Ag^+, and Cu^{2+}) photocatalysts by the microwave radiation solid-phase synthesis method using $H_3PW_{12}O_{40}{\cdot}19H_2O$, $(NH_4)_2C_2O_4{\cdot}2H_2O$,

$Cs_3PW_{12}O_{40}$, $Ag_3PW_{12}O_{40}$, and $Cu_3(PW_{12}O_{40})_2$ as raw materials in the microwave oven for 5 min. The average particle size of $M_3PW_{12}O_{40}$ was 15–80 nm. The absorption band of $M_3PW_{12}O_{40}$ shifted from 260 to 270–350 nm and became broad in comparison with those of $H_3PW_{12}O_{40}$. The photocatalytic degradation of formaldehyde was performed over $H_3PW_{12}O_{40}$ (0.1 g) under the irradiation of an 18 W UV lamp (λ = 253.7 nm). The results proved that the shift of UV-visible (UV-Vis) absorption band was proportionally related to the activity of the catalyst. The $Cu_3(PW_{12}O_{40})_2$ catalyst that had the maximal red shift to 350 nm in its UV-Vis absorption band was found to be the most efficient catalyst. When the concentration of formaldehyde is 32.79 mg/m^3, the conversion of the formaldehyde is basically completed under a flow rate of 40 mL/min over 0.1 g $M_3PW_{12}O_{40}$ catalyst.

10.2.1.3 Metal vanadates

10.2.1.3.1 Ag_3VO_4

The use of the hydrothermal method for the preparation of Ag_3VO_4 was first reported by Hu et al. [117]. They studied the effects of the ratio of silver to vanadium on the formation of Ag_3VO_4 and found that Ag_3VO_4 prepared in excessive vanadium at 160°C for 48 h exhibited higher visible-light-driven activity than that of samples prepared in a stoichiometric ratio. Huang et al. [118] investigated the effects of time on the hydrothermal synthesis of Ag_3VO_4 and found that Ag_3VO_4 synthesized at 140°C for 4 h exhibited the highest photocatalytic activity. However, the processing time of conventional hydrothermal methods always requires a few hours to a few days.

Pan et al. [119] synthesized a variety of visible-light-driven silver vanadates, including α-$AgVO_3$, β-$AgVO_3$, and α-Ag_3VO_4, using $AgNO_3$ and NH_4VO_3 as precursors, by the microwave-assised hydrothermal synthesis method in a microwave oven equuipped with a 3 GHz microwave generator (CEM Discover). The microwave power was set at 50 W, and the reaction was held at 140°C for 60 min. In the conventional hydrothermal method, the reaction was carried out in an autoclave at 140°C for 240 min. Each of the silver vanadate particles had strong visible light absorption associated with band gaps in the range of 2.2–2.5 eV. The α-Ag_3VO_4 crystalline sample with rich hydroxyl functional groups on the surface exhibited the highest degree of photocatalytic activity. The reaction rates of the photodegradation of isopropanol and benzene vapors were approximately eight times higher than those of P25 under visible light irradiation. The photocatalytic activities of microwave-assisted hydrothermal samples in the degradation of isopropanol and benzene vapors irradiated with a white fluorescent lamp (the main emission band was at 545 nm) were higher than those of samples produced by conventional hydrothermal techniques. This was due to an increase in the specific surface area and additional hydroxyl functional groups on the surface.

10.2.1.3.2 $BiVO_4$

$BiVO_4$ micro-/nanostructures are of special attention as one of the most promising visible-light-driven photocatalysts by virtue of their fascinating photophysical and photochemical properties. The various $BiVO_4$ nanostructures reported exhibited outstanding photocatalytic activities in the degradation of different organic pollutants, water splitting for O_2 and H_2 evolution, and destruction of microorganisms under visible light irradiation [120–124]. $BiVO_4$ appears in three crystal structures: tetragonal zircon, monoclinic scheelite, and tetragonal scheelite. It was reported that tetragonal $BiVO_4$ with a 2.9 eV band gap mainly possessed a UV absorption band, while monoclinic $BiVO_4$ with a 2.4 eV band gap had a characteristic visible light absorption band besides the UV band [125]. Among the three crystal types of $BiVO_4$, monoclinic $BiVO_4$ exhibited much higher photocatalytic activity than the other two tetragonal phases [120].

Tan et al. [120] prepared $BiVO_4$ powders with hierarchical structures by the microwave hydrothermal method in an MDS-8 closed-vessel microwave chemistry workstation at different pH values and 200°C for 40 min, using $Bi(NO_3)_3$ and NH_4VO_3 as raw materials. It is found that $BiVO_4$ with different crystal phases and morphologies can be prepared by varying the pH values of the precursors. The results showed that when the pH value of the precursor was 0.59, the as-prepared powders were monoclinic $BiVO_4$ crystals with octahedron and decahedron morphologies. Spherical and polyhedral $BiVO_4$ with particle sizes in the range of 2–4 μm can be prepared under strong acid conditions (pH = 0.70–1.21) and posed a mixed crystal consisting of tetragonal and monoclinic phases, whereas rod-like and dendritic $BiVO_4$ with a pure monoclinic phase can be obtained within a very wide pH range (pH = 4.26–9.76). The phase transformation from the tetragonal phase to the monoclinic phase occurs at pH 3.65. At pH >9.76, the powders were nonstoichiometric crystals between the mixed-phase $BiVO_4$ and non-$BiVO_4$. Photocatalytic activities of samples (2 g/L) were determined by decolorization of RhB aqueous solution (2×10^{-5} mol/L) under simulated sunlight and UV light irradiation. A 350 W Xe lamp was used as a simulated sunlight source, and a 300 W mercury lamp was used as a UV light source. Spherical and polyhedral $BiVO_4$ was able to completely degrade the RhB after 2 h of UV light irradiation. The octahedron and decahedron monoclinic scheelite structure $BiVO_4$ with a high specific surface area (5.16 m^2/g) exhibited the best visible light photocatalytic activity, indicating that visible light photocatalytic activity of $BiVO_4$ was related not only to its crystal structure but also greatly to its morphology and specific surface area. At pH >9.76, the photocatalytic activities of the powders begin to decrease.

10.2.1.3.3 $InVO_4$

As an important ternary oxide semiconductor with a narrow band gap (E_g = 2.0 eV), $InVO_4$ had huge potential in visible-light-driven photocatalytic hydrogen production from water splitting [126–128]. $InVO_4$ nanostructures with fairly small size and high surface area have already been synthesized through various methods, including the solid-state high-temperature reaction, the sol–gel method, the aerosol flow synthetic method, and the hydrothermal method [129].

Yan et al. [130] proposed a simple and versatile microwave-assisted synthesis of $InVO_4$ nanocrystal photocatalysts using $In(NO_3)_3 \cdot 4.5H_2O$ and NH_4VO_3 as raw materials in a microwave reactor (MCR-3, BeiLun, China) with an operating power of 800 W and a working temperature of 100°C for 5 h. The as-prepared products showed high crystallinity and ultrasmall size (20 nm). The 20-nm-sized nanocrystal samples showed reduced visible light response and enlarged reductive energy compared to large-sized $InVO_4$ samples. The photocatalytic hydrogen production experiments were carried out in a Lab-H_2 photocatalytic hydrogen production system under visible light irradiation. A 305 W Xe lamp with a UV cutoff filter (420 nm) was used as a visible light source. The hydrogen production rate of $InVO_4$ nanocrystal photocatalysts was 14.16 μmol/hg, which was higher than that of large-sized $InVO_4$ microspheres (9 μmol/hg).

10.2.1.3.4 YVO_4

Tetragonal-phased YVO_4 has been extensively used in powder form as red phosphor with several rare-earth metal ions as a dopant in cathode ray tubes and color television. As an attractive host material, YVO_4 can be excited under UV light irradiation and the phonon energy can transfer efficiently from the exited vanadate to the luminescent center, rare-earth metal ions. Recently, some new synthetic methods, including the solution combustion process, the hydrolyzed colloid reaction technique, the microemulsion-mediated method, hydrothermal processing, and the wet-chemical method [131, 132], have been developed to synthesize YVO_4 nanopowders in view of their important applications in the optoelectronic field. However, these methods need longer reaction times and more complicated processing.

Xu et al. [131, 132] prepared YVO_4 photocatalysts with different grain sizes by annealing the YVO_4 nanopowders synthesized via microwave irradiation in a microwave oven (700 W, 37% power) for 10 min from a $NaVO_3$ and $Y(NO_3)_3$ solution. Photocatalytic measurements showed that YVO_4 nanopowders (2 g/L) with a grain size of about 5 nm exhibited superior photocatalytic properties (90%) in the decolorization of MO (10 mg/L) under UV light irradiation using a 400 W high-pressure Hg lamp with a maximum emission at about 360 nm.

10.2.1.3.5 $LnVO_4$(Ln = Ce, Pr, and Nd)

Lanthanide orthovanadates belong to ABO_4-type structures. These compounds exhibit unique optical, catalytic, electrical, magnetic, and luminescent properties and are used as sensors and electrodes in electrochromic devices. Recently, lanthanide orthovanadates have been examined as potential photocatalysts in the degradation of dyes and organics [133, 134].

Mahapatra et al. [135] synthesized lanthanide orthovanadates, $CeVO_4$, $PrVO_4$, and $NdVO_4$, of nanodimensions under microwave exposure using $Ce(NO_3)_3 \cdot 6H_2O$, $Pr(NO_3)_3 \cdot 6H_2O$, $Nd(NO_3)_3 \cdot 6H_2O$, $NaVO_3$, and NH_4VO_3 as the starting materials in a domestic microwave oven (model MS2342AE4 of LG) at different power levels from 320 to 800 W. The degradation rates of dyes and organics in the presence of microwave-synthesized orthovanadates were higher under UV irradiation with a high-pressure mercury vapor lamp of 125 W (Philips, India) compared to that observed with orthovanadates synthesized by the solid-state technique. The microwave-synthesized orthovandates also showed chloro-specificity, with the rate of degradation of 2,4-dichlorophenol significantly higher than that observed in the presence of P25.

10.2.1.4 Metal molybdates

Over the past few years, metal molybdates have attracted much attention because of their luminescent behavior, structural determined properties, and potential applications [136–138].

10.2.1.4.1 Bi_2MoO_6

Bismuth molybdate, ϒ-Bi_2MoO_6, is the simplest member of the Aurivillius family of layered perovskites, which structurally comprise alternating perovskite-like and fluorite-like blocks, with a general formula $[Bi_2O_2][A_{n-1}B_nO_{3n+1}]$. For Bi_2MoO_6, $n = 1$ [139]. It has been found that Bi_2MoO_6 can be used for some applications, such as a gas sensor of alcohol and acetone, an ionic conductor, a catalyst for CO conversion, a catalyst for the selective oxidation and ammoxidation of lower olefins, a photocatalyst for water splitting under visible light irradiation, etc. Among all these potentials, using Bi_2MoO_6 as a visible light responsive photocatalyst interests us the most [140–142]. Various synthesis methods of Bi_2MoO_6 have been reported, including solid-state reactions, spray drying, reflux, ultrasonic-assisted synthesis, conventional hydrothermal synthesis, etc. [143]. Different synthetic approaches may give rise to particular properties due to possible variation of surface morphology and structure. The conventional hydrothermal method is an efficient low-temperature method for the synthesis of nanosized compounds, which is also environmental friendly. However, this method requires prolonged reaction time.

Bi et al. [144] successfully synthesized nanocrystalline Bi_2MoO_6 photocatalysts by microwave solvothermal routes using $Bi(NO_3)_3{\cdot}5H_2O$ and $(NH_4)_6Mo_7O_{24}{\cdot}4H_2O$ as starting materials. The samples presented nanorods with some nanosheets. The crystallite sizes were in the range of 16–35 nm. The absorption edge of the samples was at about 491 nm, corresponding to a band-gap energy of about 2.53 eV. The photocatalytic activities of Bi_2MoO_6 (1 g/L) were evaluated by the decomposition of RhB (10^{-5} M) in an aqueous solution under visible light irradiation with a 300 W halogen lamp (Philips Electronics). The sample obtained at 160°C for 2 h showed the highest photocatalytic activity.

10.2.1.5 Perovskite-type materials

Perovskite-type samples, with the common formula ABO_3 (the A site is a rare-earth metal cation and the B site is a transition metal), have been studied extensively as promising photocatalysts [145, 146]. Due to the excellent photocatalytic property, a variety of methods, such as solid-state reaction, sol–gel, hydrothermal technology, polymerized complex technique, etc. [147, 148], have been applied to synthesize perovskite-type samples. Conventional processes of synthesis involve many defects, for instance, high calcination temperature for a long time, no control of particle size, and no phase purity. To solve these problems, the microwave-assisted method can be applied for the preparation of perovskite-type samples.

10.2.1.5.1 $CdSnO_3{\cdot}3H_2O$

Among the perovskite-type stannates, $CdSnO_3{\cdot}3H_2O$ has been recently considered to be a new semiconductor material for gas sensors and dye-sensitized solar cells (DSSCs) and as an anode material for lithium-ion batteries [149]. At present, synthesis methods for $CdSnO_3{\cdot}3H_2O$ have focused on precipitation reactions or conventional hydrothermal processes [150, 151].

Liu et al. [149] successfully synthesized a $CdSnO_3{\cdot}3H_2O$ hollow-nanocuboid photocatalyst via a facile template-free microwave hydrothermal method using $CdCl_2{\cdot}2.5H_2O$ and $Na_2SnO_3{\cdot}4H_2O$ as precursors in a microwave instrument (ETHOS A, Milestone) for 2 h at different temperatures. The sample prepared at 160°C for 2 h at pH = 11 exhibited the best photocatalytic activity and stability in the degradation of gaseous benzene under UV irradiation as compared to the other samples. Four 4 W UV lamps (λ = 254 nm, Philips Co., TUV 4 W/G4 T5) were used as the light source. The conversion and mineralization rates of benzene were about 12.5% and 67%, respectively. The mineralization rate was twice as high as that of commercial TiO_2 (Degussa Co.). Furthermore, the excellent stability in benzene decomposition was attributed to the positive charge on the catalyst surface, which was analyzed both by electron spin resonance and zeta potentials.

10.2.1.5.2 $LaFeO_3$

As is well known, perovskite $LaFeO_3$ has promising applications in gas sensors, fuel cells, solid electrolytes, and superconductors [152, 153]. In addition, it has been found that $LaFeO_3$ is visible light photocatalytic active due to its unique optoelectronic properties and narrow band gap [154, 155]. Currently, many approaches have been developed to prepare perovskite $LaFeO_3$, such as coprecipitation, combustion, solid-state reaction, sol–gel, and hydrothermal methods [156–160].

Tang et al. [161] prepared $LaFeO_3$ nanoparticles by a one-step microwave-assisted route without additional high-temperature calcination process with $La(NO_3)_3{\cdot}6H_2O$ and $Fe(NO_3)_2{\cdot}6H_2O$ as raw materials. The prepared $LaFeO_3$ was single-phase perovskite and sphere-like shape. The prepared $LaFeO_3$ had strong visible light absorption and an optical absorption onset of 525 nm, corresponding to an optical band gap of 2.36 eV. The photocatalytic activity of $LaFeO_3$ (2 g/L) was investigated by the degradation of MB (10 mg/L) under the visible light irradiation. The visible light was obtained from a 150 W metal halogen lamp with a JB400 filter cutoff. The photocatalytic experiment indicated that the as-prepared $LaFeO_3$ was much more visible-light photocatalytic active than that of commercial P25 photocatalyst. It is suggested that the narrow optical band gap and strong visible light absorption are responsible for the high visible light photocatalytic activity.

10.2.1.5.3 $NaTaO_3$

A fast and facile process for the preparation of perovskite $NaTaO_3$ nanocrystals with Ta_2O_5 and NaOH as starting materials by a microwave-assisted hydrothermal technique in a MARS-5 microwave digestion system operated at 2.45 GHz (XP-1500, CEM Co.) with a maximum power output of 1600 W was reported [162]. The band gap of the as-prepared samples was estimated to be 4.0 eV from the onset of the absorption. The $NaTaO_3$ photocatalyst obtained by the microwave-assisted hydrothermal technique showed high activity in overall water splitting under UV irradiation with a 450 W high-pressure mercury lamp (Ushio; UM453B-A), which was ascribed to the good crystallinity and large surface area achieved in a short reaction time.

10.2.1.5.4 $YFeO_3$

Using nitrate precursors, a novel microwave-assisted progress for the preparation of a nanocrystalline pure $YFeO_3$ phase has been developed using a microwave digestion system (NN-S570MFS) at 2450 MHz frequency and 550 W power [163]. Photocatalytic activity in the degradation of orange II (10 mg/L) under visible light irradiation with a 450 W blended-light mercury fluorescent lamp ($\lambda \geq 410$ nm) showed that perovskite $YFeO_3$ was superior to P25.

10.2.1.5.5 $SrTiO_3$

Recently, perovskite-type strontium titanate ($SrTiO_3$) has attracted considerable attention due to its unique photocatalytic activity, high stability, nontoxicity, etc. [164].

Sulaeman et al. [165] synthesized $SrTiO_3$ nanoparticles by microwave-assisted solvothermal reaction of $SrCl_2 \cdot 6H_2O$ and $Ti(OC_3H_7)_4$ in KOH methanol-oleic acid solution at 200°C for 3 h. The photocatalytic activity for NO oxidation was studied using light-emitting diode (LED) lamps with various wavelengths such as 627 nm (red), 530 nm (green), 445 nm (blue), and 390 nm (UV). The photocatalytic activity of $SrTiO_3$ under visible light ($\lambda = 530$ nm) irradiation could be generated by modification of the surface with the carboxyl group (–COO) from oleic acid, which enabled the absorption of visible light.

Sulaeman et al. [166] also synthesized $SrTiO_3$ powders with various Sr/Ti atomic ratios by microwave-assisted solvothermal reactions of $SrCl_2$ and $Ti(OC_3H_7)_4$ in KOH aqueous solutions at 200°C for 3 h. The photocatalytic activity was determined by deNOx ability using LED lamps of various wavelengths such as 627 nm (red), 530 nm (green), 445 nm (blue), and 390 nm (UV) with 2 mW/cm^2 of light intensity. The photocatalytic activity was dependent on the Sr/Ti atomic ratio, that is, the strontium-rich sample (Sr/Ti atomic ratio >1) showed excellent visible-light-responsive photocatalytic activity in the oxidative destruction of NO.

10.2.1.5.6 $ReFeO_3$(Re: La, Sm, Eu, and Gd)

Ding et al. [167] applied the microwave-assisted method to synthesize perovskite-type samples $ReFeO_3$ (Re: La, Sm, Eu, and Gd). Rare-earth nitrates were blended with $Fe(NO_3)_3$ to form the precursors, activate carbon was used to enhance microwave absorption, and PVA and urea were introduced as the inhibitor and the homogeneous precipitator, respectively. The microwave processing was carried out for a period until complete combustion, accompanying the liberation of various gases (H_2O, CO_2, CO, etc.). $ReFeO_3$ with a pure perovskite phase has well-crystallized nanosized particles and homogeneous aggregates with porous aspects. Photocatalytic property was evaluated by the degradation of RhB aqueous solution (10 mg/L) under the irradiation of a 150 W visible light lamp ($\lambda > 400$ nm). All these samples exhibited good photocatalytic activity due to their perovskite structure.

10.2.1.6 Hydroxy salts

10.2.1.6.1 GaOOH

GaOOH, another wide-band-gap p-block metal oxyhydroxide, has been synthesized in various morphologies using a protein filament template process, sonochemical reaction, forced hydrolysis, hydrothermal treatment, and sol–gel method [168].

Sun et al. [168] synthesized GaOOH nanorods from $Ga(NO_3)_3$ via a facile microwave hydrothermal method in a single-mode CEM Discover system (Explorer-48, CEM Co.) operating at 200 W and 180°C for 20 min. It was found that the as-synthesized sample consisted of rod-like particles and possessed a surface area of 14.3 m^2/g and a band gap of 4.75 eV. The photocatalytic property of GaOOH nanorods was evaluated by the degradation of aromatic compounds (such as benzene and toluene) in an oxygen gas stream under UV light illumination. The results demonstrated that GaOOH nanorods exhibited superior photocatalytic activity and stability as compared to commercial TiO_2 (P25, Degussa Co.) in both benzene and toluene degradation under 254 nm UV light irradiation. In the extended (35 h) reaction test toward benzene, GaOOH maintained high photocatalytic activity, and no obvious deactivation was observed.

10.2.1.6.2 $Cd(OH)_2$

$Cd(OH)_2$ is a wide-band-gap semiconductor with many possible applications, such as in solar cells, catalysts, transparent electrodes, and gas sensors [169].

Novel pancake-like $Cd(OH)_2$ superstructures have been prepared via a facile and rapid microwave-assisted hydrothermal method in a microwave system (2.45 GHz, Discover S-Class, CEM Co.) employing cadmium acetate and hydrazine hydrate as the starting materials in a mixed medium of water and glycerol at 150°C for 10 min [170]. Low-magnification SEM observations show that the as-obtained product is composed of a great deal of pancake-like $Cd(OH)_2$ microstructures with diameters of 5–10 mm (Fig. 10.4a). The high-magnification SEM image in Fig. 10.4b clearly reveals that the entire structure of these architectures is constructed by many ultrathin nanoflakes via a layer-by-layer self-assembly mode. In addition, there are many nanoparticles on the surface of the sheet-like building units. Figure 10.4c shows a typical transmission electron microscopy (TEM) image of a layered $Cd(OH)_2$ superstructure with a diameter of ~5 mm, which further confirms the results of the SEM. The selected-area electron diffraction (SAED) pattern displays strongly ordered electron diffraction spots and confirms the high crystallinity of the product (Fig. 10.4d). The formation process of the layered $Cd(OH)_2$ superstructures is given in Scheme 10.1. In this formation process nanoflakes are first quickly formed under microwave-assisted hydrothermal conditions. With a prolonged reaction time, the layer-by-layer self-assembly of these building nanoflakes can occur along

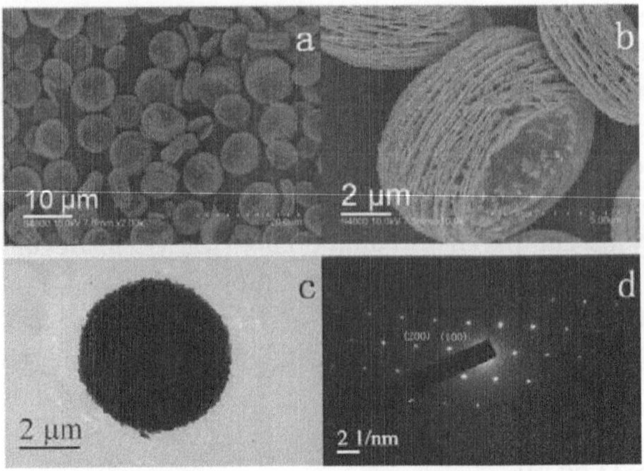

Figure 10.4 (a and b) SEM images, (c) TEM images, and (d) SAED of the $Cd(OH)_2$ sample prepared at 150°C by 10 min microwave irradiation.

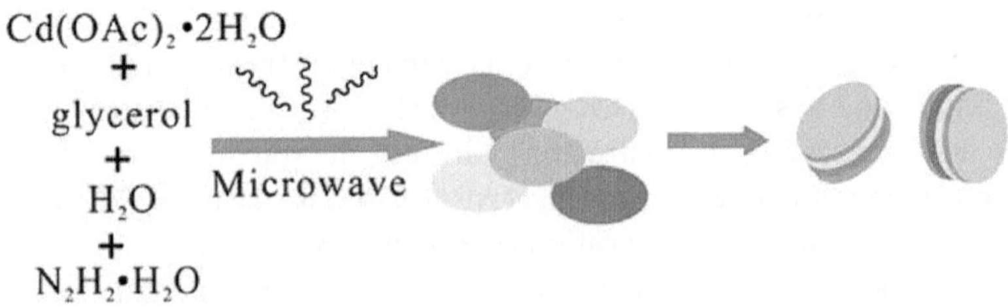

Scheme 10.1 A possible formation process to form layered $Cd(OH)_2$ superstructures.

a certain direction and lead to nanoflake-assembled structures. Finally, the number of layers of the superstructures gradually increases and fully self-assembled layered $Cd(OH)_2$ superstructures are produced. The as-obtained $Cd(OH)_2$ superstructures (1 g/L) showed enhanced photocatalytic activity in the decomposition of RhB (10^{-5} M) under basic conditions and UV light irradiation (500 W high-pressure mercury lamp with main wavelength 365 nm).

10.2.1.6.3 $CaSb_2O_5(OH)_2$

Sun et al. [171] synthesized a nanocrystalline $CaSb_2O_5(OH)_2$ photocatalyst from $CaCl_2$ and $K_2H_2Sb_2O_7$ and used it to degrade benzene in the gas phase. The reaction was performed under microwave irradiation using a single-mode CEM Discover system (Explorer-48, CEM Co.) at 180°C for 20 min. Microwave radiation can rapidly heat the system to 180°C within 2 min, whereas it would take ~100 min for the conventional hydrothermal method. The $CaSb_2O_5(OH)_2$ sample had an average particle size of approximately 8 nm, a specific surface area of 101.8 m^2/g, and a band gap of 4.6 eV. Photocatalytic activity of the sample was evaluated by the degradation of benzene in an O_2 gas stream under UV light irradiation (4 W UV lamps with a wavelength centered at 254 nm). The results demonstrated that the photocatalytic activity of $CaSb_2O_5(OH)_2$ was higher than that of commercial P25. In the photocatalytic degradation of benzene, it reached a steady conversion ratio of 29%. $CaSb_2O_5(OH)_2$ also exhibited photocatalytic activity toward other aromatic organic compounds.

10.2.1.6.4 $Bi_6O_6(OH)_3(NO_3)_3 \cdot 1.5H_2O$

Xie et al. [172] reported the synthesis of basic $Bi_6O_6(OH)_3(NO_3)_3 \cdot 1.5H_2O$ via a microwave hydrothermal process in a microwave synthesizer (ETHOS A, Milestone) at 160°C for 1 h. $Bi(NO_3)_3 \cdot 5H_2O$ and NaOH were used as precursors for the synthesis. Comparing with this basic bismuth nitrate synthesized under conventional hydrothermal method, the sample from the microwave hydrothermal method had a much higher specific surface area and smaller particles. Photocatalytic evaluations via the decomposition of MO and malachite green (20 ppm) under UV light irradiation (three 4 W UV lamps with a wavelength centered at 254 nm [Philips, TUV 4 W/G4T5]) revealed that samples obtained by the microwave hydrothermal method (1 g/L) exhibited higher photocatalytic efficiency as well as mineralization efficiency due to a higher surface area, which was relevant to the size and sheet-like morphology.

10.2.1.7 Bismuth oxyhalides

Bismuth oxyhalides, BiOX (X = Cl, Br, and I), are of great importance because of their optical properties and applications as catalysts, ferroelectric materials, pigments, and photocatalysts [173]. Recently, as a new family of promising photocatalysts, BiOXs

(X = Cl, Br, and I) have demonstrated remarkable photocatalytic activities due to their uniquely layered structures with an internal static electric field perpendicular to each layer, which can induce effective separation of photogenerated electron–hole pairs [174]. Other complex bismuth oxyhalides, such as Bi_3O_4Cl, Bi_4NbO_8Cl, $PbBiO_2Cl$, $Bi_4O_5I_2$, and Bi_5O_7I, have been investigated and show good photocatalytic ability under visible light irradiation for the degradation of various organic contaminants [175].

10.2.1.7.1 BiOBr

Recently, hierarchical-structured BiOBr microspheres have attracted much attention and have already been successfully prepared via various synthetic routes [176–178]. Zhang et al. [178] prepared two kinds of BiOBr nanosheet-assembled microspheres via a facile, rapid, and reliable microwave solvothermal route employing $Bi(NO_3)_3{\cdot}5H_2O$ and cetyltrimethylammonium bromide as starting reagents in the absence or presence of oleic acid in a microwave system (200 W, Discover S-Class, CEM Co.) at 180°C for 10 min. It was found that the formation of these building blocks of microspheres could be ascribed to the self-assembly of nanoparticles according to the mesocrystal growth mode. The microstructure and morphology of the product were investigated by SEM. As shown in Fig. 10.5a, the sample comprises many microspheres with diameters ranging from 3 to 5 μm. A high-magnification SEM image of an individual microsphere shows that the as-prepared product is composed of many nanosheets (Fig. 10.5b). Careful observation reveals that there is a great deal of nanopores on the surface of microspheres, which are formed by stacking of nanosheets. If oleic acid is introduced, BiOBr nanosheet-assembled microspheres with relatively good dispersion and uniform diameters of ~4 μm are obtained (Fig. 10.5c,d). It should be noted that the presence of oleic acid did not change the morphology of these building blocks. However, it has an important influence on the stacking mode of constructing units and results in different pore structures on the surface of microspheres. Both samples (1 g/L) exhibited not only strong adsorption abilities but also excellent photocatalytic activities for MO (60 mg/L, 96% and 85%), RhB (60 mg/L), and phenol (20 mg/L, 99% and 94%) under UV light irradiation (500 W high-pressure mercury lamp). The

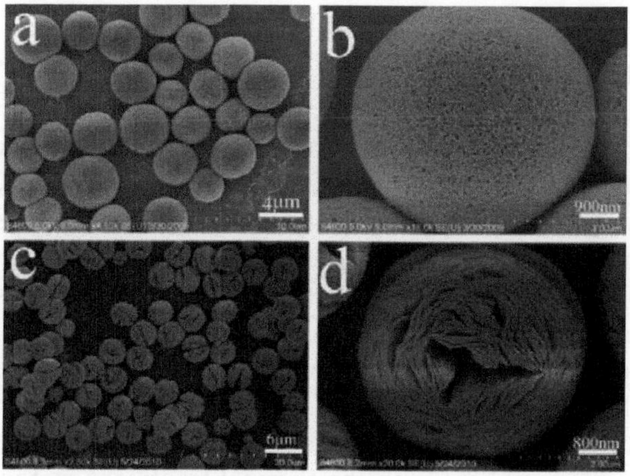

Figure 10.5 SEM images of BiOBr hierarchical microspheres prepared in the (a and b) absence and (c and d) presence of oleic acid.

sample prepared in the absence of oleic acid had a larger Brunauer–Emmett–Teller (BET) surface area, which contributed to its higher photocatalytic activity.

10.2.1.7.2 $Bi_7O_9I_3$

Xiao et al. [179] demonstrated a simple and energy-saving microwave heating route for rapid synthesis of $Bi_7O_9I_3$ sheet-like hierarchical architectures in a microwave chemical reactor (MCR-3, Beijing Rui Cheng Wei Industry Equipment Co. Ltd., China) at a fixed frequency of 2.45 GHz and a maximum output of 800 W under different microwave irradiation powers and reaction times. $Bi(NO_3)_3{\cdot}5H_2O$ and KI were used as starting materials. The as-synthesized materials demonstrated excellent visible light photocatalytic activity, high mineralization capacity, and good stability in the degradation of bisphenol-A (BPA) in aqueous solution under visible light irradiation of a 1000 W Xe lamp combined with a 420 nm cutoff filter. Using a $Bi_7O_9I_3$ photocatalyst synthesized by 400 W microwaves heating for 180 s and with a catalyst dosage of 1 g/L in a 20 mg/L aqueous solution of BPA, a degradation percentage of 99% was obtained under visible light irradiation for 60 min.

10.2.1.8 Other semiconductors

10.2.1.8.1 $Y_4Al_2O_9$

Y_2O_3–Al_2O_3 compound systems are important functional materials, including yttrium aluminum garnet, $Y_3Al_5O_{12}$, (YAG) and $YAlO_3$ (YAP), and both compounds have been intensively studied, focusing on their synthesis, structures, and luminescence properties because of their use in laser and fluorescence areas. Another compound, $Y_4Al_2O_9$, is considered as an intermediate crystal phase in the process of producing YAG.

Xu et al. [180] developed a microwave solvothermal synthesis (MDS-10, Sineo, China) using $Y(NO_3)_3$ and $Al(NO_3)_3$ in mixed solvents of water and *N*, *N*-dimethylformamide (DMF) at 200°C for 60 min combined with the thermal treatment method for the preparation of $Y_4Al_2O_9$ hierarchically nanostructured microspheres assembled with nanosheets. $Y_4Al_2O_9$ consisted of hierarchically nanostructured microspheres with an average diameter of about 1 μm, and the microspheres were constructed by a three-dimensional assembly of nanosheets with a thickness of about 30–40 nm. The photocatalytic activity of the as-prepared $Y_4Al_2O_9$ hierarchical microspheres (1 g/L) over phenol (20 mg/L) under UV light irradiation was investigated, and the degradation rate of phenol was up to 91.2% in a period of 240 min.

10.2.1.8.2 $SnNb_2O_6$

Layered niobates are one of the most attractive functional materials due to their wide variety of structural and electronic properties. Foordite $SnNb_2O_6$ is a typical layered niobate, whose structure consists of a two-octahedron-thick sheet built by edge-sharing octahedral NbO_6 units and Sn^{2+} ions inserted between two sheets. In a NbO_6 unit, the metallic atom is surrounded by six oxygen atoms placed at the apex of an octahedron. For its electronic band structure, the hybrid orbitals of Sn 5*s* and O 2*p* will form a new VB above the VB consisting of O 2*p* orbitals. The unique band structure makes photoabsorption possible for $SnNb_2O_6$ in the visible light region [181, 182].

Liang et al. [181] prepared visible-light-activated $SnNb_2O_6$ nanosheets with a high surface area and small crystallites by a microwave-assisted template-free hydrothermal method without exfoliation using $Nb_2O_5 \cdot nH_2O$ ($n = 5$–8) as a precursor in a microwave synthesizer (ETHOS A/260, 800 W, Milestone) at 200°C for various times. A hypsochromic shift of the photoabsorption edge was observed, which reflected an obvious quantum size effect. $SnNb_2O_6$ was composed of numerous irregular nanosheets with a thickness of 1–4 nm versus several hundred nanometers in lateral size. Due to the unique nanosheets structure, larger surface area, smaller crystallites, and stronger redox ability of the photogenerated hole–electron pair, these photocatalysts (1 g/L) showed much higher photocatalytic activity in the degradation of RhB (10 μmol/L) under visible light irradiation (300 W Xe lamp, $420 < \lambda < 760$ nm) compared to their counterparts prepared by the traditional solid-state reaction.

10.2.1.8.2 $ZnGa_2O_4$

$ZnGa_2O_4$ with spinel-type structure has received much attention due to its potential applications, such as a low-voltage field emission display or a vacuum fluorescent display [183]. The synthesis of $ZnGa_2O_4$ has been examined mostly by the traditional solid-state reaction, the combustion method, the template method, the soft-chemical method, and the coprecipitation method [184]. Sol–gel and sputtering processes have also been tried for the preparation of $ZnGa_2O_4$ thin films [185, 186].

Sun et al. [184] synthesized $ZnGa_2O_4$ from $Ga(NO_3)_3$ and $ZnCl_2$ via a rapid and facile microwave-assisted hydrothermal method in a single-mode CEM Discover system (Explorer-48, CEM Co.) operating at 200 W, 2.45 GHz at 160°C for 20 min. The as-synthesized $ZnGa_2O_4$ sample had an average diameter of 10 nm, and the band gap was approximately 4.5 eV, corresponding to an optical absorption edge of 275 nm. The BET specific surface area of $ZnGa_2O_4$ was 101.8 m^2/g. The photocatalytic property of the as-prepared $ZnGa_2O_4$ was evaluated by the degradation of pollutants in air and aqueous solution under UV light illumination. Four 4 W UV lamps with a wavelength centered at 254 nm (Philips, TUV 4 W/G4 T5) were used as the light source (light intensity: 4.8×10^{16} quants cm^{-2} s^{-1}). The results demonstrated that $ZnGa_2O_4$ exhibited efficient photocatalytic activity higher than that of commercial P25 (Degussa

Co.) in the degradation of benzene, toluene, and ethylbenzene. In the liquid-phase degradation of dyes (MO, RhB, and methylene blue), $ZnGa_2O_4$ also exhibited remarkable activity higher than that of P25. After 32 min of UV light irradiation, the decomposition ratio of MO (10 ppm, 150 mL) over $ZnGa_2O_4$ (0.06 g) was up to 99%. The total organic carbon test revealed that the mineralization ratio of MO (10 ppm, 150 mL) was 88.1% after 90 min.

10.2.1.8.3 ZnGaNO

Yang et al. [187] proposed a facile microwave route to synthesize a porous ZnGaNO photocatalyst using urea as a nitridation source in an ordinary domestic microwave oven (model EG823EA6-NS, Midea, China) at a maximum power of 800 W and an operating frequency of 2.45 GHz for 10 min by using CuO as a heated medium material. ZnGaNO sample had a porous microstructure aggregated by nanoparticles in about 30 nm. The photocatalytic activity of porous ZnGaNO (0.02 g) in the degradation of isopropyl alcohol (IPA, 230 mL, 5000 ppm) under visible light irradiation of a 300 W Xe lamp with cutoff filter L42 was much higher than that of the ZnGaNO solid solution prepared by a solid-state reaction. The as-prepared porous ZnGaNO photocatalyst exhibited improved photocatalytic activity due to the strong gas adsorption by the porous structure and the enhancement of extended light absorption by the higher Zn content.

10.2.2 *Summary*

Semiconductor photocatalytic technology using either UV light or solar energy has become more prominent owing to its advantages of the use of vast additive chemicals or disinfectants and its mineralization aspects. Microwave-assisted synthesis has shown great potential in the preparation of semiconductor photocatalysts compared to other traditional technologies. Increasing the energy-harvesting efficiency for endless solar energy under ambient conditions will continue to be one of the most important objectives in designing photocatalysts. The applications of such photocatalytic systems not only convert inexhaustible light energy into chemical energy but also protect our environment. Searching for visible-light-active semiconductor photocatalytic materials synthesized by the microwave-assisted method will receive rapidly growing research attention. Despite the promising developments, there is a long way ahead before photocatalysis can find widespread applications in environmental preservation and renewable energy supplies.

10.3 Microwave-Assisted Synthesis of Semiconductor Nanomaterials for Sensitized Solar Cells

With increasing concern over the twin crises of fossil fuel depletion and environmental degradation, sensitized solar cells (SSCs) have attracted considerable attention and represent a key class of cell architecture that has emerged as a promising candidate for the development of the next generation of solar cells, due to their acceptable power conversion efficiency and low production cost [188, 189]. SSCs are based on the photosensitization of nanocrystalline TiO_2 or ZnO semiconductor photoanodes by absorbed dyes or QD sensitizers. As a key component of SSCs, the structure, size, and characteristics of photoanodes have been studied to improve SSC efficiency. Many studies have also devoted to explore more sophisticated structures to optimize the performance of photoanodes.

10.3.1 *QD-Sensitized Solar Cells*

As sensitizers for SSCs, inorganic semiconductor QDs, such as PbS, CdS, CdSe, and InAs, have been suggested along with organometallic or organic dyes because QDs have advantages of high extinction coefficient, spectral tunability by particle size, and good stability, which are known to increase the overall power conversion efficiency of solar cells. Quantum dot–sensitized solar cells (QDSSCs) are based on the photosensitization of semiconductor photoanodes, typically TiO_2, by absorbed QD sensitizers, typically CdS or CdSe. The working mechanism of QDSSCs, which is similar to DSSCs, is depicted in Fig. 10.6, involving the key processes 1–7 [190, 191]. In brief, electrons, excited from the VB to the CB of the QDs by absorbing light (path 1), are rapidly injected into the CB of the

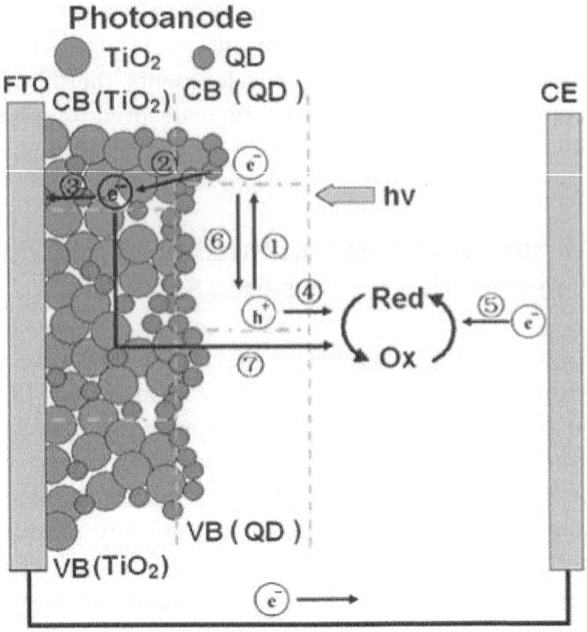

Figure 10.6 Schematic diagram of the working mechanism of QDSSCs. Reprinted from Ref. [190], Copyright (2011), with permission from Elsevier.

TiO_2 particles (path 2) and then transported to the F-doped SnO_2 (FTO) (path 3). The oxidized QDs are regenerated by accepting the electrons from a counterelectrode (CE) (path 5) via a redox pair (path 4). Furthermore, the charge recombination (paths 6 and 7) will happen in the meantime and deteriorate the cell performance. Pursuing high efficiency is always a core task for solar cells, and one of the current key issues is to enhance the light harvesting in a visible light region. In QDSSCs, the performance of QDs and their interconnectivity with the TiO_2 substrate is the key for device performance. Many studies have been devoted to explore different fabrication techniques to attach QDs onto TiO_2. So far, the QDs have been successfully sensitized on the surface of TiO_2 by self-assembled monolayers via linker assistance or direct adsorption, chemical bath deposition (CBD), electrochemical deposition, and photodeposition techniques.

10.3.1.1 Semiconductor sulfides

10.3.1.1.1 CdS

Among the semiconductor QDs, CdS QDs have been paid much more attention because of their high potential in light harvesting in a visible light region [192].

Zhu et al. [193] fabricated SSCs based on TiO_2 photoanodes and CdS QDs as sensitizers, in which CdS QDs are prepared using the MACBD method in a focused microwave system (Explorer-48, CEM Co.) at 150°C with microwave irradiation power of 100 W for 30 min. The photovoltaic performances of the cells fabricated using CdS precursor solutions with different concentrations were investigated. Compared with those methods, the MACBD technique can rapidly synthesize CdS QDs and precisely control their sizes with a narrow distribution, improve the wettability of TiO_2 surface, and form a good contact between CdS QDs and the TiO_2 layer due to rapidly elevated temperature during microwave irradiation. Furthermore, this technique can offer easy control over all experimental parameters without the requirement of a repetitive immersing operation, an organic linker, or high-temperature heating. An optimized thickness of a CdS layer was obtained at a precursor concentration of 0.05 M, which resulted in the highest short-circuit current density (J_{SC}) of 7.20 mA/cm^2 and a conversion efficiency (η) of 1.18% under one sun illumination (AM 1.5G, 100 mW/cm^2), providing a good interfacial structure between TiO_2 and CdS films and reducing the recombination of the injected electrons from TiO_2 to the electrolyte because of a well-covered CdS layer on the TiO_2 surface. Higher J_{SC} and η values were achieved via the MACBD method compared to the cells fabricated using CBD and spray pyrolysis deposition methods, which should be ascribed to the following reasons:

(1) Good contact between CdS QDs and the TiO_2 layer is formed due to rapidly elevated temperature during microwave irradiation.
(2) Microwave irradiation can heat up the aqueous solution homogeneously and fast because of the penetration characteristic of microwaves and high utilization factor of microwave energy. Therefore, the nucleation and growth of CdS QDs can be finished in an extremely short period of time, which is extraordinarily beneficial for reducing the concentration of surface defects of

QDs. The carrier recombination at surface defects of QDs is correspondingly suppressed and thus the cell performance is increased.

Furthermore, the photovoltaic performances of the cells fabricated using CdS precursor solutions with different deposition times also were investigated [190]. An optimized thickness of a CdS layer was obtained at a deposition time of 30 min, which resulted in the highest J_{SC} of 6.69 mA/cm^2 and η of 1.03% under one sun illumination (AM 1.5G, 100 mW/cm^2).

10.3.1.1.2 PdS

Zhu et al. [194] explored the application of the MNCBD method in the fabrication of deposited PdS QDs on the surface of a TiO_2 film as a photoanode for QDSSCs using a microwave system (Explorer-48, CEM Co.) at 150°C with a microwave irradiation power of 150 W for 20 min. Compared with the absorption spectrum of a pure TiO_2 film, the absorption edge, obtained from the intersection of the sharply decreasing region of the spectrum with its baseline, was about 750 nm for the TiO_2/PbS electrode. The band gap corresponding to the absorption edge was about 1.65 eV, which was higher than the value of PbS in the bulk (0.42 eV), indicating that the size of the PbS particles is within the scale of QDs. A J_{SC} of 0.42 mW/cm^2 and η of 0.15% under one sun illumination (AM 1.5 G, 100 mW/cm^2) had been achieved for prepared PbS QDSSCs.

10.3.1.2 Semiconductor selenide

10.3.1.2.1 CdSe

As compared with CdS, CdSe with a smaller band gap (1.7 eV) can absorb a wider range of wavelengths of light, which is advantageous for sensitizers [195, 196].

Zhu et al. [191] introduced a simple, rapid, and effective approach to deposit CdSe QDs on the surface of TiO_2 films for QDSSCs using the MACBD method in a microwave system (Explorer-48, CEM Co.) at 150°C with a microwave irradiation power of 100 W for different time periods. This technique allowed direct and rapid deposition and good contact between the CdSe and TiO_2 films. The photovoltaic performances of the cells with CdSe deposited at different times were investigated. The cell based on MACBD-deposited TiO_2/CdSe electrodes at a deposition time of 20 min achieved a maximum J_{SC} of 12.1 mA/cm^2 and η of 1.75% at one sun illumination (AM 1.5 G, 100 mW/cm^2), which is comparable with those fabricated using conventional techniques. The thickness of the CdSe layer increased with an increase of deposition time. Such an increment of CdSe loading led to more excited electrons under the illumination of light and reduction of charge recombination, which is advantageous for the photocurrent.

10.3.1.2.2 CdS–CdSe

Compared with single (CdS and CdSe) QDs, their cosensitized structure can provide superior ability owing to extension of the light absorption range and effective charge injection from QDs to TiO_2. Attempts to combine CdS and CdSe QDs to assemble on TiO_2 nanoparticles, nanowires, or nanotubes have been reported in efforts to obtain efficient QDSSCs with η of 2%–3% [197, 198]. It should be noticed that, to date, these QDSSCs based on a cosensitized porous TiO_2 electrode have been almost fabricated by a successive ionic layer adsorption and reaction method, which requires the repetitive immersing operation of a TiO_2 film in precursor solutions containing Cd, S, or Se.

Zhu et al. [199] fabricated a CdS/CdSe QD cosensitized TiO_2 film using the MACBD technique in a microwave system (Explorer-48, CEM Co.) at 150°C with a microwave irradiation power of 100 W for 20 min. The CdS/CdSe QD cosensitized TiO_2 film was used as a photoanode for QDSSCs. The cosensitization of CdS and CdSe QDs can extend the absorption range and increase the absorbance. The cascade energy level structure in the order of TiO_2<CdS <CdSe is formed via the combination of CdS and CdSe. The favorable alignment of the Fermi levels at TiO_2/CdS/CdSe interfaces facilitates electron injection and hole recovery for both inner CdS and outer CdSe layers. The performance of the CdS/CdSe cosensitized cell achieved a J_{SC} of 16.1 mW/cm^2 and η of 3.06% at one sun illumination (AM 1.5 G, 100 mW/cm^2), which was comparable to the one fabricated using the conventional successive ionic layer adsorption and reaction technique. The incident photon-to-current conversion efficiency peak value reached 73% for the CdS/CdSe

cosensitized cell, which was ascribed to the joint contributions from CdS and CdSe QDs in the light harvesting.

10.3.2 *Dye-Sensitized Solar Cells*

Since the first report of DSSCs by O'Regan and Grfitzeli in 1991 [188], DSSCs have attracted considerable attention [200, 201]. A typical DSSC is composed of a photoanode and a CE separated by an electrolyte containing an iodide–triiodide (I^-–I_3^-) redox couple. Until recently, the highest DSSCs power conversion efficiency that has been achieved is more than 12%, which is lower than that achieved with bulk silicon solar cells [202]. Much research has been carried out to improve the performance of DSSCs and has mainly focused on the design of dyes with a high absorption coefficient, the development of various semiconductor nanostructures with a high surface area to increase dye loading, and the use of ionic liquid solvent–based electrolytes for better temperature stability.

10.3.2.1 Semiconductor oxide

10.3.2.1.1 SnO_2

SnO_2 is a stable, n-type, and wide-band-gap semiconductor (~3.6 eV at 300 K) with excellent optical and electrical properties, such as fast electron conductivity (which is orders of magnitude higher than TiO_2) and high stability even under long-term UV radiation. Nanostructured SnO_2 materials have been investigated extensively for application as photoelectrodes in DSSCs [203, 204].

Birkel et al. [205] fabricated highly efficient DSSCs with excellent long-term stability based on SnO_2 nanocrystals with tunable morphologies and band energy levels. SnO_2 nanocrystals were prepared by a facile, fast, and energy-saving microwave-assisted solvothermal reaction in a CEM XP-1500 microwave vessel for 2 h at an internal pressure of 46 bar corresponding to a temperature of 240°C. The impact of the nanocrystal morphology on the device performance was explored. The different morphologies of the SnO_2 nanocrystals depended on precursor bases. Figure 10.7 shows the TEM micrographs of various SnO_2 nanocrystals obtained with different precursor bases. With NaOH, homogeneous bundles of

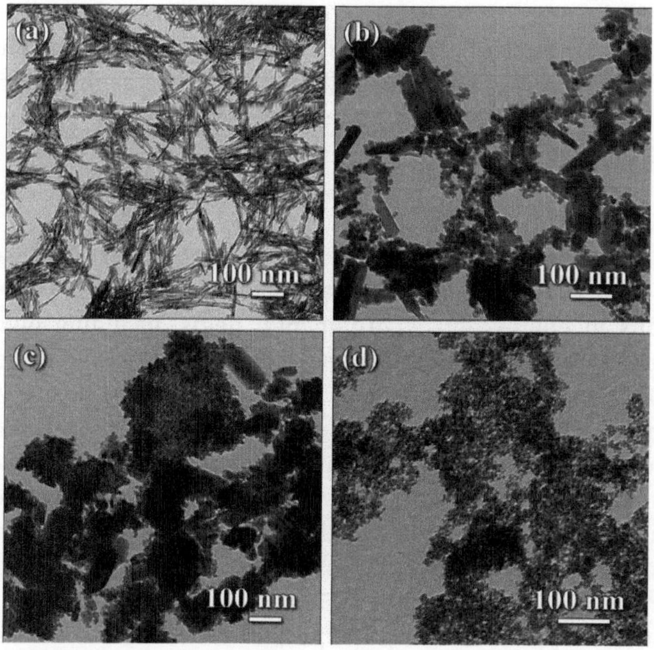

Figure 10.7 TEM micrographs of as-prepared SnO_2 particles with various morphologies obtained from microwave-assisted reactions.

nanorods around 150 nm in length and 5–10 nm in diameter, as shown in Fig. 10.7a, are obtained. SnO_2 nanocrystals obtained with KOH are less anisotropic: In addition to polydisperse rods (from 10 nm up to 200–300 nm in length and 100 nm in diameter), the sample contains spherical particles (with diameters ranging from 5 to 10 nm) and larger plate-like structures (Fig. 10.7b). Tetramethylammonium hydroxide (TMAH) leads to the formation of SnO_2 with a mixture of spherical-, plate-, and rod-like particles (Fig. 10.7c). The use of NH_4OH as a precursor base in the synthesis of SnO_2 nanocrystals leads to the formation of small and spherical particles ~8 nm in diameter and nearly monodisperse in size (Fig. 10.7d). The SnO_2 photoanodes prepared with NH_4OH (small spheres) posed a large surface area (149 m^2/g) that is larger than the surface area of the SnO_2 nanorod films prepared with NaOH (51 m^2/g), KOH (66 m^2/g), and TMAH (69 m^2/g) as precursor bases. A very high η of up to 3.2%, based on a very high open-circuit voltage (V_{oc}) and comparable J_{sc} and fill factor under one sun condition (AM 1.5, 100 mW/cm^2, with shading masks), was achieved with the cell based on SnO_2 with a mixture of different nanostructures as photoanodes.

10.3.2.1.2 TiO_2

TiO_2 including anatase, rutile, and brookite phases are among the most interesting nanostructures for a wide variety of applications in energy and environmental fields, including pigments, photocatalysts, catalyst supports, sensors, and DSSCs, because of their unique photo-induced electron transfer efficiency, high chemical stability, nontoxicity, and electrical conductivity [206, 207]. Various morphologies of TiO_2, including nanoparticles, nanorods, nanotubes, and ordered structures, have been fabricated to develop high-efficiency and stable DSSCs [208, 209]. The performance of nanocrystalline films for DSSCs is profoundly dependent on the microstructures of mesoporous TiO_2, such as the crystalline phase, crystal size, pore texture, optical properties, and electron transport.

Wang et al. [210] synthesized porous anatase TiO_2 spheres with sizes ranging from 150 to 250 nm by a rapid microwave treatment of spherical titanium glycolate precursors preformed via an ethylene glycol-mediated sol–gel process using a microwave system (Discover S-Class, CEM Co.) at 150°C for 10 min for DSSCs. The photoanode film made of porous TiO_2 spheres with a thickness of about 10 μm. A 5% η was achieved, which represented an almost 40% enhancement over that made of the standard commercial Degussa P25 TiO_2 nanopowders. The porous TiO_2 spheres exhibited a higher specific surface area (60.2 m^2/g) than that of the commercial P25 nanopowders (~45 m^2/g), which possessed higher dye adsorption ability than the one made of P25 nanopowders at a similar particle packing density and film thickness. The photoanode film made of porous spheres with a larger particle size (150–250 nm) provided better light-scattering properties, which can thus increase the light-traveling path and enhance the interaction between TiO_2 and dye molecules. This resulted in higher light-harvesting efficiencies and a corresponding higher photocurrent. The charge transport through the photoanode films made of porous TiO_2 spheres might be improved due to better interparticle connectivity, which contributes to better charge collection and higher J_{sc}.

Uchida et al. [211, 212] prepared the TiO_2 nanocrystalline electrode via a microwave method using a multimode microwave heating system operating at a frequency of 28 GHz for DSSCs. Photoelectron energies, η, of 5.51% and 2.16% were achieved in an electrode prepared by 28 GHz microwave irradiation at 0.7 and 1.0 kW for 5 min.

Huang et al. [213] developed a novel and facile method for rapid synthesis of mesoporous TiO_2 via a microwave-assisted hydrothermal route using stable and water-soluble titanium citrate complexes as the precursor in a microwave reaction system (Model MARS-X, CEM Co.) at 400 W. The synthesized TiO_2 contained mainly the anatase phase with a crystallite size of 5.0–8.6 nm at various temperatures ranging from 150°C to 180°C and from 30 to 120 min, respectively. The specific surface areas of the TiO_2 nanocrystals were 217–323 m^2/g with a pore diameter of 5.8–6.9 nm, clearly showing the mesoporous characteristics of the TiO_2 nanocrystals. The mesoporous nanocrystals synthesized at 180°C were used to prepare the TiO_2 photoelectrode using yjr screen-printing deposition method for DSSC application. The mesoporous TiO_2 photoelectrodes exhibited an excellent η of 4.8%–7.1%, a J_{sc} of 11–13.9 mA/cm^2, and a V_{oc} of 0.76–0.77 V, respectively, depending on active area (0.25 and 1.0 cm^2) and film thickness (6.1 and 7.4 μm) under illumination of AM 1.5G. The photovoltaic η and durability of mesoporous TiO_2-based DSSCs decreased slightly from 4.8% and 4.3% after 490 h of storage, clearly depicting the long-term stability of the DSSCs. The thick TiO_2 film would adsorb more dye molecules, leading to enhancement in the J_{sc} and η of DSSCs. The pure anatase phase with a large crystalline size and thick film enhanced the photocurrent and contributed the effective electron transport to the DSSCs.

A single-crystalline rutile nanorod was grown directly on top of a fluorine-doped tin oxide (FTO) substrate via a microwave-assisted hydrothermal reaction using a CEM microwave accelerated reactor system (MARS) for DSSCs [214]. An increase in the reaction temperature and time results in an increase in the length and diameter of the nanorods. The growth rate of the nanorods in the microwave-assisted hydrothermal reaction was much faster than in the conventional hydrothermal reaction. When the surface of the nanorods was treated with $TiCl_4$, the η of DSSC using a 2.5 μm thick film increased to 3.7%, which has been explained in several viewpoints such as enhanced dye adsorption, enhanced charge transport due to dye bonding or shift of the CB edge, the scattering effect, and reduced surface recombination.

Menzies et al. [215] produced the TiO_2 and ZrO_2-coated TiO_2 working electrodes and calcined at 450°C in a 2.45 GHz microwave furnace (MPC, Model 941) prior to incorporation in DSSCs. Microwave sintering of both bare TiO_2 and ZrO_2-coated TiO_2 working electrodes was found to have a dramatic improvement in the performance of the DSSCs. The microwave-processed TiO_2 electrodes were found to be 2.4% efficient, whereas the ZrO_2-coated TiO_2 electrodes performed at 3.1% efficiency. The increased photovoltaic performance was due to improvements in the V_{oc} from 692 to 719 mV and the J_{sc} from 6.24 to 6.56 mA/cm^2.

10.3.2.1.3 ZnO

Min et al. [216] prepared ultralong ZnO nanowires on a large scale by a microwave-assisted aqueous route without using any surfactant or template at a relatively low temperature of 120°C in a microwave oven with power 900 W. The mean diameter of the nanowires was about 80 nm, with lengths up to tens of micrometers. DSSCs were prepared using the ZnO nanowires and commercial ZnO powder with an average particle size of 2 μm. The V_{oc} was the same for the two cells, whereas the J_{sc}, FF, and η for cell-a was much higher than for cell-b, which might be attributed to the nanocrystalline nature and unique synthetic route of the nanowires.

10.3.2.1.3 NiO_x

Awais et al. [217] prepared p-type NiO_x coatings to be used as a photocathode in DSSCs using a microwave plasma sintering treatment. The microwave-assisted plasma sintering process was carried out using a circumferential antenna plasma (CAP) microwave system at a pressure of 5 mbar in an argon and oxygen (10:1) atmosphere with input powers of 2.4 kW and a frequency of 2.45 GHz. The mesoporous structure of the p-type electrode favored dye adsorption and charge injection from the sensitizer to the photocathode with a resulting increase in the light-to-photocurrent conversion efficiency. A J_{sc} of 0.53 mW/cm^2 and η of 0.0118% using 870 mW/cm^2 AM 1.5 solar illumination have been achieved for prepared NiO_x DSSCs after 5 min sintered coatings.

10.3.3 *Summary*

Nanomaterials provide the promise of being able to meet the global challenge of supplying clean energy. The recent advances in utilizing QDs or semiconductor nanocrystals synthesized via a microwave-assisted method as light energy harvesters provide unique opportunities for the development of SSCs. Further improvement in the performance of SSCs is necessary to fulfill the purpose of large-scale applications and replacement of traditional fossil fuels. To further enhance the η of SSCs, it is critical to improve the V_{oc} by reducing the charge recombination and increasing the electron injection efficiency. With the recent surge in interest of renewable energy we can expect major breakthroughs in developing economically viable solar energy conversion devices in the near future.

10.4 Summary and Outlook

The use of microwave opens up unique opportunities for semiconductor nanomaterials' synthesis in energy storage and conversion. In future research, microwave-assisted methods will be expected to synthesize core–shell particles, thin films on substrates, and ceramic and polymeric composites. Microwave heating effects have a major impact on many areas of preparative science, particularly in the area of synthetic chemistry. Microwave synthesis offers several advantages such as rapid heating, higher reaction rate and selectivity, lower reaction temperature, shorter reaction time, homogeneous thermal transmission, and phase purity with better yields. These advantages are mainly due to the thermal and non-thermal effects of microwaves, that is, superheating, polarization, dielectric properties, hotspot formation, nuclear spin rotation, and spin alignment [218]. Further major impacts on sample preparation for biology, materials science, nanotechnology, and polymer science are all likely to result in the next 5–10 years. One of the major drawbacks of the relatively new technology is the equipment cost. This fact has severely limited the penetration of microwave synthesis in academic laboratories around the world. Another crucial challenge is to manipulate the shape and size of the as-synthesized nanomaterials. Attempts such as adding catalysts, using small molecules as surfactants, and changing reaction parameters such as temperature, pressure, etc., can be adapted in the microwave system [219]. As with any new technology, the current situation is bound to change over the next several years and less expensive equipment should become available.

References

1. Feng, S. H., and Xu, R. R. (2000). New materials in hydrothermal synthesis, *Acc. Chem. Res.*, **34**, pp. 239–247.
2. Siskin, M., and Katritzky, A. R. (1991). Reactivity of organic compounds in hot water: geochemical and technological implications, *Science*, **254**, pp. 231–237.
3. Gedye, R., Smith, F., Westaway, K., Ali, H., Baldisera, L., Laberge, L., and Rousell, J. (1986). The use of microwave ovens for rapid organic synthesis, *Tetrahedron Lett.*, **27**, pp. 279–282.
4. Kappe, C. O. (2006). The use of microwave irradiation in organic synthesis. From laboratory curiosity to standard practice in twenty years, *Chin. Int. J. Chem.*, **60**, pp. 308–312.
5. Baghbanzadeh, M., Carbone, L., Cozzoli, P. D., and Kappe, C. O. (2011). Microwave-assisted synthesis of colloidal inorganic nanocrystals, *Angew. Chem., Int. Ed.*, **50**, pp. 11312–11359.
6. Gerbec, J. A., Magana, D., Washington, A., and Strouse, G. F. (2005). Microwave-enhanced reaction rates for nanoparticle synthesis, *J. Am. Chem. Soc.*, **127**, pp. 15791–15800.
7. Hayes, B. L. (2004). Recent advances in microwave-assisted synthesis, *Aldrichim. Acta*, **37**, pp. 66–77.
8. Liu, X. J., Pan, L. K., Li, J. L., Yu, K., and Sun, Z. (2013). Visible Light-induced photocatalytic activity of Bi_2O_3 prepared via microwave-assisted method, *J. Nanosci. Nanotechnol.*, **13**, pp. 5044–5047.

9. Liu, X. J., Pan, L. K., Li, J. L., Yu, K., Sun, Z., and Sun, C. Q. (2013). Light down-converting characteristics of ZnO-Y_2O_2S:Eu^{3+} for visible light photocatalysis, *J. Colloid Interface Sci.*, **404**, pp. 150–154.
10. Liu, X. J., Pan, L. K., Lv, T., Lu, T., Zhu, G., Sun, Z., and Sun, C. Q. (2011). Microwave-assisted synthesis of ZnO-graphene nanocomposite for photocatalytic reduction of Cr(VI), *Catal. Sci. Technol.*, **1**, pp. 1189–1193.
11. Liu, X. J., Pan, L. K., Lv, T., Sun, Z., and Sun, C. Q. (2012). Enhanced photocatalytic reduction of Cr(VI) by ZnO-TiO_2-CNTs composites synthesized via microwave-assisted reaction, *J. Mol. Catal. A: Chem.*, **363–364**, pp. 417–422.
12. Liu, X. J., Pan, L. K., Lv, T., Zhu, G., Lu, T., Sun, Z., and Sun, C. Q. (2011). Microwave-assisted synthesis of TiO_2-reduced graphene oxide composites for photocatalytic reduction of Cr(VI), *RSC Adv.*, **1**, pp. 1245–1249.
13. Liu, X. J., Pan, L. K., Lv, T., Zhu, G., Sun, Z., and Sun, C. Q. (2011). Microwave-assisted synthesis of CdS-reduced graphene oxide composites for photocatalytic reduction of Cr(VI), *Chem. Commun.*, **47**, pp. 11984–11986.
14. Liu, X. J., Wang, X. J., Li, H. L., Pan, L. K., Lv, T., Sun, Z., and Sun, C. Q. (2012). Microwave-assisted synthesis of ZnO-$Y_3Al_5O_{12}$: Ce^{3+} composites with enhanced visible light photocatalysis, *J. Mater. Chem.*, **22**, pp. 16293–16298.
15. Kappe, C. O. (2008). Microwave dielectric heating in synthetic organic chemistry, *Chem. Soc. Rev.*, **37**, pp. 1127–1139.
16. Lin, Z., Wragg, D. S., and Morris, R. E. (2006). Microwave-assisted synthesis of anionic metal–organic frameworks under ionothermal conditions, *Chem. Commun.*, pp. 2021–2023.
17. Lidstrom, P., Tierney, J., Wathey, B., and Westman, J. (2001). Microwave assisted organic synthesis: a review, *Tetrahedron*, **57**, pp. 9225–9283.
18. Roberts, B. A., and Strauss, C. R. (2005). Toward rapid, "green", predictable microwave-assisted synthesis, *Acc. Chem. Res.*, **38**, pp. 653–661.
19. Nuchter, M., Ondruschka, B., Bonrath, W., and Gum, A. (2004). Microwave assisted synthesis–a critical technology overview, *Green Chem.*, **6**, pp. 128–141.
20. Vijayakumar, S., Ponnalagi, A. K., Nagamuthu, S., and Muralidharan, G. (2013). Microwave assisted synthesis of Co_3O_4 nanoparticles for high-performance supercapacitors, *Electrochim. Acta*, **106**, pp. 500–505.

21. Jena, A., Vinu, R., Shivashankar, S. A., and Madras, G. (2010). Microwave assisted synthesis of nanostructured titanium dioxide with high photocatalytic activity, *Ind. Eng. Chem. Res.*, **49**, pp. 9636–9643.
22. Wang, J., Niu, B., Du, G., Zeng, R., Chen, Z., Guo, Z., and Dou, S. (2011). Microwave homogeneous synthesis of porous nanowire Co_3O_4 arrays with high capacity and rate capability for lithium ion batteries, *Mater. Chem. Phys.*, **126**, pp. 747–754.
23. Zhu, Z., Yang, D., and Liu, H. (2011). Microwave-assisted hydrothermal synthesis of ZnO rod-assembled microspheres and their photocatalytic performances, *Adv. Powder Technol.*, **22**, pp. 493–497.
24. Liu, X. J., Lv, T., Pan, L. K., Sun, Z., and Sun, C. Q. (2012). Microwave-assisted synthesis of ZnO for photocatalytic reduction of Cr (VI) in aqueous solution, *Desalin. Water Treat.*, **42**, pp. 216–221.
25. Tan, G. Q., Zheng, Y. Q., Miao, H. Y., Xia, A., Ren, H. J., and Hintzen, H. (2012). Controllable microwave hydrothermal synthesis of bismuth ferrites and photocatalytic characterization, *J. Am. Ceram. Soc.*, **95**, pp. 280–289.
26. Ai, Z. H., Deng, K. J., Wan, Q. F., Zhang, L. Z., and Lee, S. C. (2010). Facile microwave-assisted synthesis and magnetic and gas sensing properties of Fe_3O_4 nanoroses, *J. Phys. Chem. C*, **114**, pp. 6237–6242.
27. Larhed, M., Moberg, C., and Hallberg, A. (2002). Microwave-accelerated homogeneous catalysis in organic chemistry, *Acc. Chem. Res.*, **35**, pp. 717–727.
28. Wang, W., Ding, M., Lu, C., Ni, Y., and Xu, Z. (2014). A study on upconversion UV-vis-NIR responsive photocatalytic activity and mechanisms of hexagonal phase $NaYF_4$: Yb^{3+}, Tm^{3+}@TiO_2 core-shell structured photocatalyst, *Appl. Catal. B: Environ.*, **144**, pp. 379–385.
29. Bai, X. J., Wang, L., Wang, Y. J., Yao, W. Q., and Zhu, Y. F. (2014). Enhanced oxidation ability of g-C_3N_4 photocatalyst via C_{60} modification, *Appl. Catal. B: Environ.*, **152–153**, pp. 262–270.
30. Zhang, N., Zhang, Y. H., Yang, M. Q., Tang, Z. R., and Xu, Y. J. (2013). A critical and benchmark comparison on graphene-, carbon nanotube-, and fullerene-semiconductor nanocomposites as visible light photocatalysts for selective oxidation, *J. Catal.*, **299**, pp. 210–221.
31. Zhang, J., Huang, Z., Xu, Y., and Kang, F. (2013). Hydrothermal synthesis of graphene/Bi_2WO_6 composite with high adsorptivity and photoactivity for azo dyes, *J. Am. Ceram. Soc.*, **96**, pp. 1562–1569.

32. Yang, X., Cui, H., Li, Y., Qin, J., Zhang, R., and Tang, H. (2013). Fabrication of Ag_3PO_4-graphene composites with highly efficient and stable visible light photocatalytic performance, *ACS Catal.*, **3**, pp. 363–369.

33. Yang, M., Weng, B., and Xu, Y. (2013). Improving the visible light photoactivity of in_2s_3-graphene nanocomposite via a simple surface charge modification approach, *Langmuir*, **29**, pp. 10549–10558.

34. Yang, J., Wang, D., Han, H., and Li, C. (2013). Roles of cocatalysts in photocatalysis and photoelectrocatalysis, *Acc. Chem. Res.*, **46**, pp. 1900–1909.

35. Fujishima, A., and Honda, K. (1972). Electrochemical photolysis of water at a semiconductor electrode, *Nature*, **238**, pp. 37–38.

36. Lang, X., Ma, W., Chen, C., Ji, H., and Zhao, J. (2013). Selective aerobic oxidation mediated by TiO_2 photocatalysis, *Acc. Chem. Res.*, **47**, pp. 355–363.

37. Xu, X., Randorn, C., Efstathiou, P., and Irvine, J. T. S. (2012). A red metallic oxide photocatalyst, *Nat. Mater.*, **11**, pp. 595–598.

38. Xiang, Q. J., Yu, J. G., and Jaroniec, M. (2012). Graphene-based semiconductor photocatalysts, *Chem. Soc. Rev.*, **41**, pp. 782–796.

39. Tong, H., Ouyang, S. X., Bi, Y. P., Umezawa, N., Oshikiri, M., and Ye, J. H. (2012). Nano-photocatalytic materials: possibilities and challenges, *Adv. Mater.*, **24**, pp. 229–251.

40. Yang, F., Takahashi, Y., Sakai, N., and Tatsuma, T. (2011). Visible light driven photocatalysts with oxidative energy storage abilities, *J. Mater. Chem.*, **21**, pp. 2288–2293.

41. Low, J., Yu, J., Li, Q., and Cheng, B. (2014). Enhanced visible-light photocatalytic activity of plasmonic Ag and graphene co-modified Bi_2WO_6 nanosheets, *Phys. Chem. Chem. Phys.*, **16**, pp. 1111–1120.

42. Gui, M. S., Wang, P. F., Yuan, J. G., Li, Y. Q., and Zeng, W. G. (2014). Preparation and visible light photocatalytic activity of Bi_2WO_6 hollow microspheres, *Adv. Mater. Res.*, **864**, pp. 1323–1326.

43. Yan, J., Shen, Y. H., Li, F., and Li, T. H. (2013). Synthesis and photocatalytic properties of $ZnWO_4$ nanocrystals via a fast microwave-assisted method, *Sci. World J.*, **2013**, Article ID 458106 (8 pp).

44. Wang, H., Li, S., Zhang, L., Chen, Z., Hu, J., Zou, R., Xu, K., Song, G., Zhao, H., and Yang, J. (2013). Surface decoration of Bi_2WO_6 superstructures with Bi_2O_3 nanoparticles: an efficient method to improve visible-light-driven photocatalytic activity, *CrystEngComm*, **15**, pp. 9011–9019.

45. Tu, W., Zhou, Y., Liu, Q., Yan, S., Bao, S., Wang, X., Xiao, M., and Zou, Z. (2013). An in situ simultaneous reduction-hydrolysis technique for fabrication of TiO_2-graphene 2D sandwich-like hybrid nanosheets: graphene-promoted selectivity of photocatalytic-driven hydrogenation and coupling of CO_2 into methane and ethane, *Adv. Funct. Mater.*, **23**, pp. 1743–1749.

46. Tang, Y., Di, W., Zhai, X., Yang, R., and Qin, W. (2013). NIR-responsive photocatalytic activity and mechanism of $NaYF_4$: Yb, Tm@ TiO_2 core-shell nanoparticles, *ACS Catal.*, **3**, pp. 405–412.

47. Shi, Y., Hua, C., Li, B., Fang, X., Yao, C., Zhang, Y., Hu, Y. S., Wang, Z., Chen, L., and Zhao, D. (2013). Highly ordered mesoporous crystalline $MoSe_2$ material with efficient visible-light-driven photocatalytic activity and enhanced lithium storage performance, *Adv. Funct. Mater.*, **23**, pp. 1832–1838.

48. Zhao, G., Liu, S. W., Lu, Q. F., and Song, L. J. (2012). Controllable synthesis of Bi_2WO_6 nanofibrous mat by electrospinning and enhanced visible photocatalytic degradation performances, *Ind. Eng. Chem. Res.*, **51**, pp. 10307–10312.

49. Guo, X., Di, W., Chen, C., Liu, C., Wang, X., and Qin, W. (2014). Enhanced near-infrared photocatalysis of $NaYF_4$: Yb, Tm/CdS/TiO_2 composites, *Dalton Trans.*, **43**, pp. 1048–1054.

50. Li, H., Yin, S., Wang, Y., Sekino, T., Lee, S. W., and Sato, T. (2012). Roles of Cr^{3+} doping and oxygen vacancies in $SrTiO_3$ photocatalysts with high visible light activity for NO removal, *J. Catal.*, **297**, pp. 65–69.

51. Bi, D. Q., and Xu, Y. M. (2011). Improved photocatalytic activity of WO3 through clustered $Fe_2O_3$for organic degradation in the presence of H_2O_2, *Langmuir*, **27**, pp. 9359–9366.

52. Zhang, Y. H., Zhang, N., Tang, Z. R., and Xu, Y. J. (2012). Graphene transforms wide band gap ZnS to visible light photocatalyst. The new role of graphene as a macromolecular photosensitizer, *ACS Nano*, **6**, pp. 9777–9789.

53. Long, R., and English, N. J. (2011). Electronic structure of cation-codoped TiO_2 for visible-light photocatalyst applications from hybrid density functional theory calculations, *Appl. Phys. Lett.*, **98**, p. 142103.

54. Zhao, T. Y., Liu, Z. Y., Nakata, K., Nishimoto, S., Murakami, T., Zhao, Y., Jiang, L., and Fujishima, A. (2010). Multichannel TiO_2 hollow fibers with enhanced photocatalytic activity, *J. Mater. Chem.*, **20**, pp. 5095–5099.

55. Hameed, A., Montini, T., Gombac, V., and Fornasiero, P. (2008). Surface phases and photocatalytic activity correlation of Bi_2O_3/Bi_2O_{4-x} nanocomposite, *J. Am. Chem. Soc.*, **130**, pp. 9658–9659.
56. Lin, X. P., Huang, F. Q., Wang, W. D., and Shi, J. L. (2007). Photocatalytic activity of $Bi_{24}Ga_2O_{39}$ for degrading methylene blue, *Scr. Mater.*, **56**, pp. 189–192.
57. Shang, M., Wang, W. Z., Ren, J., Sun, S. M., and Zhang, L. (2010). A novel $BiVO_4$ hierarchical nanostructure: controllable synthesis, growth mechanism, and application in photocatalysis, *CrystEngComm*, **12**, pp. 1754–1758.
58. Liu, Y. X., Dai, H. X., Deng, J. G., Zhang, L., and Au, C. T. (2012). Three-dimensional ordered macroporous bismuth vanadates: PMMA-templating fabrication and excellent visible light-driven photocatalytic performance for phenol degradation, *Nanoscale*, **4**, pp. 2317–2325.
59. Gui, M. S., and Zhang, W. D. (2012). A facile preparation strategy for hollow-structured Bi_2O_3/Bi_2WO_6 heterojunction with high visible light photocatalytic activity, *J. Phys. Chem. Solids*, **73**, pp. 1342–1349.
60. Chen, R., Shen, Z. R., Wang, H., Zhou, H. J., Liu, Y. P., Ding, D. T., and Chen, T. H. (2011). Fabrication of mesh-like bismuth oxide single crystalline nanoflakes and their visible light photocatalytic activity, *J. Alloys Compd.*, **509**, pp. 2588–2596.
61. Ai, Z. H., Huang, Y. H., Lee, S. C., and Zhang, L. Z. (2011). Monoclinic Bi_2O_3 photocatalyst for efficient removal of gaseous NO and HCHO under visible light irradiation, *J. Alloys Compd.*, **509**, pp. 2044–2049.
62. Lu, H. B., Wang, S. M., Zhao, L., Dong, B. H., Xu, Z. X., and Li, J. C. (2012). Surfactant-assisted hydrothermal synthesis of Bi_2O_3 nano/microstructures with tunable size, *RSC Adv.*, **2**, pp. 3374–3378.
63. Wu, C., Shen, L., Huang, Q., and Zhang, Y. (2011). Hydrothermal synthesis and characterization of Bi_2O_3 nanowires, *Mater. Lett.*, **65**, pp. 1134–1136.
64. Huang, Q. Q., Zhang, S. N., Cai, C. X., and Zhou, B. (2011). Bi_2O_3 nanoparticles synthesized via microwave-assisted method and their photocatalytic activity towards the degradation of rhodamine B, *Mater. Lett.*, **65**, pp. 988–990.
65. Bilecka, I., and Niederberger, M. (2010). Microwave chemistry for inorganic nanomaterials synthesis, *Nanoscale*, **2**, pp. 1358–1374.
66. Jiang, C., Zhang, G., Wu, Y., Li, L., and Shi, K. Y. (2012). Facile synthesis of SnO_2 nanocrystalline tubes by electrospinning and their fast response

and high sensitivity to NO_x at room temperature, *CrystEngComm*, **14**, pp. 2739–2747.

67. Wang, X., Fan, H. Q., and Ren, P. R. (2013). Electrospinning derived hollow SnO_2 microtubes with highly photocatalytic property, *Catal. Commun.*, **31**, pp. 37–41.
68. Wu, W., Zhang, S. F., Zhou, J., Xiao, X. H., Ren, F., and Jiang, C. Z. (2011). Controlled synthesis of monodisperse sub-100 nm hollow SnO_2 nanospheres: a template-and surfactant-free solution-phase route, the growth mechanism, optical properties, and application as a photocatalyst, *Chem. Eur. J.*, **17**, pp. 9708–9719.
69. Manjula, P., Boppella, R., and Manorama, S. V. (2012). A facile and green approach for the controlled synthesis of porous SnO_2 nanospheres: application as an efficient photocatalyst and an excellent gas sensing material, *ACS Appl. Mater. Interfaces*, **4**, pp. 6252–6260.
70. Wang, W. W., Zhu, Y. J., and Yang, L. X. (2007). ZnO-SnO_2 hollow spheres and hierarchical nanosheets: hydrothermal preparation, formation mechanism, and photocatalytic properties, *Adv. Funct. Mater.*, **17**, pp. 59–64.
71. Du, Y. L., Zhang, N., and Wang, C. M. (2010). Photo-catalytic degradation of trifluralin by SnO_2-doped Cu_2O crystals, *Catal. Commun.*, **11**, pp. 670–674.
72. Wang, H. J., Sun, F. Q., Zhang, Y., Li, L. S., Chen, H. Y., Wu, Q. S., and Yu, J. C. (2010). Photochemical growth of nanoporous SnO_2 at the air-water interface and its high photocatalytic activity, *J. Mater. Chem.*, **20**, pp. 5641–5645.
73. Wang, G., Lu, W., Li, J. H., Choi, J. Y., Jeong, Y., Choi, S. Y., Park, J. B., Ryu, M. K., and Lee, K. (2006). V-shaped tin oxide nanostructures featuring a broad photocurrent signal: an effective visible-light-driven photocatalyst, *Small*, **2**, pp. 1436–1439.
74. Liu, X. J., Pan, L. K., Chen, T. Q., Li, J. L., Yu, K., Sun, Z., and Sun, C. Q. (2013). Visible light photocatalytic degradation of methylene blue by SnO_2 quantum dots prepared via microwave-assisted method, *Catal. Sci. Technol.*, **3**, pp. 1805–1809.
75. Repins, I., Contreras, M. A., Egaas, B., DeHart, C., Scharf, J., Perkins, C. L., To, B., and Noufi, R. (2008). 19.9%-efficient ZnO/CdS/$CuInGaSe_2$ solar cell with 81.2% fill factor, *Prog. Photovoltaics Res. Appl.*, **16**, pp. 235–239.

76. Fan, Z., Wang, D., Chang, P., Tseng, W.-Y., and Lu, J. G. (2004). ZnO nanowire field-effect transistor and oxygen sensing property, *Appl. Phys. Lett.*, **85**, pp. 5923–5925.

77. Shinde, V. R., Gujar, T. P., Noda, T., Fujita, D., Vinu, A., Grandcolas, M., and Ye, J. (2010). Growth of shape-and size-selective zinc oxide nanorods by a microwave-assisted chemical bath deposition method: effect on photocatalysis properties, *Chem. Eur. J.*, **16**, pp. 10569–10575.

78. Ivanov, V. K., Shaporev, A. S., Sharikov, F. Y., and Baranchikov, A. Y. (2007). Hydrothermal and microwave-assisted synthesis of nanocrystalline ZnO photocatalysts, *Superlatt. Microstruct.*, **42**, pp. 421–424.

79. Shaporev, A. S., Ivanov, V. K., Baranchikov, A. E., and Tret'yakov, Y. D. (2007). Microwave-assisted hydrothermal synthesis and photocatalytic activity of ZnO, *Inorg. Mater.*, **43**, pp. 35–39.

80. Parida, K. M., and Parija, S. (2006). Photocatalytic degradation of phenol under solar radiation using microwave irradiated zinc oxide, *Sol. Energy*, **80**, pp. 1048–1054.

81. Ghule, L. A., Patil, A. A., Sapnar, K. B., Dhole, S. D., and Garadkar, K. M. (2011). Photocatalytic degradation of methyl orange using ZnO nanorods, *Toxicol. Environ. Chem.*, **93**, pp. 623–634.

82. Zhang, W., Zou, L. D., and Wang, L. Z. (2011). A novel charge-driven self-assembly method to prepare visible-light sensitive TiO_2/activated carbon composites for dissolved organic compound removal, *Chem. Eng. J.*, **168**, pp. 485–492.

83. Zhu, Y. F., Du, R. G., Chen, W., Qi, H. Q., and Lin, C. J. (2010). Photocathodic protection properties of three-dimensional titanate nanowire network films prepared by a combined sol-gel and hydrothermal method, *Electrochem. Commun.*, **12**, pp. 1626–1629.

84. Xiong, Z., Dou, H., Pan, J., Ma, J., Xu, C., and Zhao, X. S. (2010). Synthesis of mesoporous anatase TiO_2 with a combined template method and photocatalysis, *CrystEngComm*, **12**, pp. 3455–3457.

85. Zhang, D., Li, G., Yang, X., and Jimmy, C. Y. (2009). A micrometer-size TiO2 single-crystal photocatalyst with remarkable 80% level of reactive facets, *Chem. Commun.*, pp. 4381–4383.

86. Zheng, Y., Lv, K., Wang, Z., Deng, K., and Li, M. (2012). Microwave-assisted rapid synthesis of anatase TiO_2 nanocrystals with exposed {001} facets, *J. Mol. Catal. A: Chem.*, **356**, pp. 137–143.

87. Zhang, P., Yin, S., and Sato, T. (2009). Synthesis of high-activity TiO_2 photocatalyst via environmentally friendly and novel microwave assisted hydrothermal process, *Appl. Catal. B: Environ.*, **89**, pp. 118–122.

88. Kozma, P., Bajgar, R., and Kozma Jr, P. (2002). Radiation damage of $PbWO_4$ crystals due to irradiation by ^{60}Co gamma rays, *Radiat. Phys. Chem.*, **65**, pp. 127–130.

89. Song, S. Y., Zhang, Y., Xing, Y., Wang, C., Feng, J., Shi, W. D., Zheng, G. L., and Zhang, H. J. (2008). Rectangular AgIn $(WO_4)_2$ nanotubes: a promising photoelectric material, *Adv. Funct. Mater.*, **18**, pp. 2328–2334.

90. Yu, S. H., Liu, B., Mo, M. S., Huang, J. H., Liu, X. M., and Qian, Y. T. (2003). General synthesis of single-crystal tungstate nanorods/nanowires: a facile, low-temperature solution approach, *Adv. Funct. Mater.*, **13**, pp. 639–647.

91. Li, Y., Liu, J., Huang, X., and Li, G. (2007). Hydrothermal synthesis of Bi_2WO_6 uniform hierarchical microspheres, *Cryst. Growth Des.*, **7**, pp. 1350–1355.

92. Kudo, A., and Hijii, S. (1999). H_2 or O_2 Evolution from aqueous solutions on layered oxide photocatalysts consisting of Bi^{3+} with $6s^2$ configuration and d^0 transition metal ions, *Chem. Lett.*, **1999**, pp. 1103–1104.

93. Tang, J. W., Zou, Z. G., and Ye, J. H. (2004). Photocatalytic decomposition of organic contaminants by Bi_2WO_6 under visible light irradiation, *Catal. Lett.*, **92**, pp. 53–56.

94. Zhang, Y. H., Zhang, N., Tang, Z. R., and Xu, Y. J. (2013). Identification of Bi_2WO_6 as a highly selective visible-light photocatalyst toward oxidation of glycerol to dihydroxyacetone in water, *Chem. Sci.*, **4**, pp. 1820–1824.

95. Tian, J., Sang, Y., Yu, G., Jiang, H., Mu, X., and Liu, H. (2013). A Bi_2WO_6-based hybrid photocatalyst with broad spectrum photocatalytic properties under UV, visible, and near-infrared irradiation, *Adv. Mater.*, **25**, pp. 5075–5080.

96. Ma, H. W., Shen, J. F., Shi, M., Lu, X., Li, Z. Q., Long, Y., Li, N., and Ye, M. X. (2012). Significant enhanced performance for Rhodamine B, phenol and Cr(VI) removal by Bi_2WO_6 nancomposites via reduced graphene oxide modification, *Appl. Catal. B: Environ.*, **121–122**, pp. 198–205.

97. Zhang, Z. J., Wang, W. Z., Yin, W. Z., Shang, M., Wang, L., and Sun, S. M. (2010). Inducing photocatalysis by visible light beyond the absorption

edge: effect of upconversion agent on the photocatalytic activity of Bi_2WO_6, *Appl. Catal. B: Environ.*, **101**, pp. 68–73.

98. Zhu, S. B., Xu, T. G., Fu, H. B., Zhao, J. C., and Zhu, Y. F. (2007). Synergetic effect of Bi_2WO_6 photocatalyst with C_{60} and enhanced photoactivity under visible irradiation, *Environ. Sci. Technol.*, **41**, pp. 6234–6239.

99. Sun, S., Wang, W., and Zhang, L. (2013). Bi_2WO_6 quantum dots decorated reduced graphene oxide: improved charge separation and enhanced photoconversion efficiency, *J. Phys. Chem. C*, **117**, pp. 9113–9120.

100. Cao, X. F., Zhang, L., Chen, X. T., and Xue, Z. L. (2011). Microwave-assisted solution-phase preparation of flower-like Bi_2WO_6 and its visible-light-driven photocatalytic properties, *CrystEngComm*, **13**, pp. 306–311.

101. Shang, M., Wang, W. Z., Ren, J., Sun, S. M., Wang, L., and Zhang, L. (2009). A practical visible-light-driven Bi_2WO_6 nanofibrous mat prepared by electrospinning, *J. Mater. Chem.*, **19**, pp. 6213–6218.

102. Shang, M., Wang, W. Z., and Xu, H. L. (2008). New Bi_2WO_6 nanocages with high visible-light-driven photocatalytic activities prepared in refluxing EG, *Cryst. Growth Des.*, **9**, pp. 991–996.

103. Wu, J., Duan, F., Zheng, Y., and Xie, Y. (2007). Synthesis of Bi_2WO_6 nanoplate-built hierarchical nest-like structures with visible-light-induced photocatalytic activity, *J. Phys. Chem. C*, **111**, pp. 12866–12871.

104. Huang, J., Tan, G. Q., Xia, A., Ren, H. J., and Zhang, L. L. (2014). Growth mechanism and photocatalytic properties of flower-like Bi_2WO_6 powders synthesized by a microwave hydrothermal method, *Res. Chem. Intermediat.*, **40**, pp. 903–911.

105. Garadkar, K. M., Ghule, L. A., Sapnar, K. B., and Dhole, S. D. (2013). A facile synthesis of $ZnWO_4$ nanoparticles by microwave assisted technique and its application in photocatalysis, *Mater. Res. Bull.*, **48**, pp. 1105–1109.

106. Yang, F., Tu, C., Wang, H., Wei, Y., You, Z., Jia, G., Li, J., Zhu, Z., Lu, X., and Wang, Y. (2008). Growth and spectroscopy of $ZnWO_4{:}Ho^{3+}$ crystal, *J. Alloys Compd.*, **455**, pp. 269–273.

107. Wu, Y., Zhang, S. C., Zhang, L. W., and Zhu, Y. F. (2007). Photocatalytic activity of nanosized $ZnWO_4$ prepared by the sol-gel method, *Chem. Res. Chin. Univ.*, **23**, pp. 465–468.

108. Fu, H., Lin, J., Zhang, L., and Zhu, Y. (2006). Photocatalytic activities of a novel $ZnWO_4$ catalyst prepared by a hydrothermal process, *Appl. Catal. A: Gen.*, **306**, pp. 58–67.

109. Ryu, J. H., Lim, C. S., and Auh, K. H. (2003). Synthesis of $ZnWO_4$ nanocrystalline powders, by the polymerized complex method, *Mater. Lett.*, **57**, pp. 1550–1554.

110. Huang, G., Zhang, C., and Zhu, Y. (2007). $ZnWO_4$ photocatalyst with high activity for degradation of organic contaminants, *J. Alloys Compd.*, **432**, pp. 269–276.

111. Zhao, W., Song, X., Chen, G., and Sun, S. (2009). One-step template-free synthesis of $ZnWO_4$ hollow clusters, *J. Mater. Sci.*, **44**, pp. 3082–3087.

112. Parhi, P., Karthik, T., and Manivannan, V. (2008). Synthesis and characterization of metal tungstates by novel solid-state metathetic approach, *J. Alloys Compd.*, **465**, pp. 380–386.

113. Dong, T., Li, Z., Ding, Z., Wu, L., Wang, X., and Fu, X. (2008). Characterizations and properties of Eu^{3+}-doped $ZnWO_4$ prepared via a facile self-propagating combustion method, *Mater. Res. Bull.*, **43**, pp. 1694–1701.

114. Bi, J. H., Wu, L., Li, Z. H., Ding, Z. X., Wang, X. X., and Fu, X. Z. (2009). A facile microwave solvothermal process to synthesize $ZnWO_4$ nanoparticles, *J. Alloys Compd.*, **480**, pp. 684–688.

115. Hu, B., Wu, L. H., Liu, S. J., Yao, H. B., Shi, H. Y., Li, G. P., and Yu, S. H. (2010). Microwave-assisted synthesis of silver indium tungsten oxide mesocrystals and their selective photocatalytic properties, *Chem. Commun.*, **46**, pp. 2277–2279.

116. Deng, Q., Zhou, W., Li, X., Peng, Z., Jiang, S., Yue, M., and Cai, T. (2007). Microwave radiation solid-phase synthesis of phosphotungstate nanoparticle catalysts and photocatalytic degradation of formaldehyde, *J. Mol. Catal. A: Chem.*, **262**, pp. 149–155.

117. Hu, X., Hu, C., and Qu, J. (2008). Preparation and visible-light activity of silver vanadate for the degradation of pollutants, *Mater. Res. Bull.*, **43**, pp. 2986–2997.

118. Huang, C. M., Pan, G. T., Li, Y. C. M., Li, M. H., and Yang, T. C. K. (2009). Crystalline phases and photocatalytic activities of hydrothermal synthesis Ag_3VO_4 and $Ag_4V_2O_7$ under visible light irradiation, *Appl. Catal. A: Gen.*, **358**, pp. 164–172.

119. Pan, G. T., Lai, M. H., Juang, R. C., Chung, T. W., and Yang, T. C. K. (2011). Preparation of visible-light-driven silver vanadates by a microwave-assisted hydrothermal method for the photodegradation of volatile organic vapors, *Ind. Eng. Chem. Res.*, **50**, pp. 2807–2814.

120. Tan, G., Zhang, L., Ren, H., Wei, S., Huang, J., and Xia, A. (2013). Effects of pH on the hierarchical structures and photocatalytic performance

of $BiVO_4$ powders prepared via the microwave hydrothermal method, *ACS Appl. Mater. Interfaces*, **5**, pp. 5186–5193.

121. Chen, L., Yin, S. F., Huang, R., Zhang, Q., Luo, S. L., and Au, C. T. (2012). Hollow peanut-like m-$BiVO_4$: facile synthesis and solar-light-induced photocatalytic property, *CrystEngComm*, **14**, pp. 4217–4222.
122. Wang, D. E., Jiang, H. F., Zong, X., Xu, Q. A., Ma, Y., Li, G. L., and Li, C. (2011). Crystal facet dependence of water oxidation on $BiVO_4$ sheets under visible light irradiation, *Chem. Eur. J.*, **17**, pp. 1275–1282.
123. Zhang, A. P., Zhang, J. Z., Cui, N. Y., Tie, X. Y., An, Y. W., and Li, L. J. (2009). Effects of pH on hydrothermal synthesis and characterization of visible-light-driven $BiVO_4$ photocatalyst, *J. Mol. Catal. A: Chem.*, **304**, pp. 28–32.
124. Ke, D. N., Peng, T. Y., Ma, L., Cai, P., and Jiang, P. (2008). Photocatalytic water splitting for O_2 production under visible-light irradiation on $BiVO_4$ nanoparticles in different sacrificial reagent solutions, *Appl. Catal. A: Gen.*, **350**, pp. 111–117.
125. Kudo, A., Omori, K., and Kato, H. (1999). A novel aqueous process for preparation of crystal form-controlled and highly crystalline $BiVO_4$ powder from layered vanadates at room temperature and its photocatalytic and photophysical properties, *J. Am. Chem. Soc.*, **121**, pp. 11459–11467.
126. Ye, J., Zou, Z., Oshikiri, M., Matsushita, A., Shimoda, M., Imai, M., and Shishido, T. (2002). A novel hydrogen-evolving photocatalyst $InVO_4$ active under visible light irradiation, *Chem. Phys. Lett.*, **356**, pp. 221–226.
127. Zhang, L., Fu, H., Zhang, C., and Zhu, Y. (2006). Synthesis, characterization, and photocatalytic properties of $InVO_4$ nanoparticles, *J. Solid State Chem.*, **179**, pp. 804–811.
128. Xiao, G., Wang, X., Li, D., and Fu, X. (2008). $InVO_4$-sensitized TiO_2 photocatalysts for efficient air purification with visible light, *J. Photochem. Photobiol. A: Chem.*, **193**, pp. 213–221.
129. Yan, Y., Liu, X., Fan, W., Lv, P., and Shi, W. (2012). $InVO_4$ microspheres: preparation, characterization and visible-light-driven photocatalytic activities, *Chem. Eng. J.*, **200**, pp. 310–316.
130. Yan, Y., Cai, F. P., Song, Y., and Shi, W. D. (2013). $InVO_4$ nanocrystal photocatalysts: microwave-assisted synthesis and size-dependent activities of hydrogen production from water splitting under visible light, *Chem. Eng. J.*, **233**, pp. 1–7.

131. Xu, H. Y., Wang, H., and Yan, H. (2007). Preparation and photocatalytic properties of YVO_4 nanopowders, *J. Hazard. Mater.*, **144**, pp. 82–85.

132. Xu, H. Y., Wang, H., Meng, Y. Q., and Yan, H. (2004). Rapid synthesis of size-controllable YVO_4 nanoparticles by microwave irradiation, *Solid State Commun.*, **130**, pp. 465–468.

133. Mahapatra, S., Madras, G., and Guru Row, T. (2007). Synthesis, characterization and photocatalytic activity of lanthanide (Ce, Pr and Nd) orthovanadates, *Ind. Eng. Chem. Res.*, **46**, pp. 1013–1017.

134. Mahapatra, S., Madras, G., and Row, T. G. (2007). Structural and photocatalytic activity of lanthanide (Ce, Pr, and Nd) molybdovanadates, *J. Phys. Chem. C*, **111**, pp. 6505–6511.

135. Mahapatra, S., Nayak, S. K., Madras, G., and Guru Row, T. N. (2008). Microwave synthesis and photocatalytic activity of nano lanthanide (Ce, Pr, and Nd) orthovanadates, *Ind. Eng. Chem. Res.*, **47**, pp. 6509–6516.

136. Zhang, L. W., Xu, T. G., Zhao, X., and Zhu, Y. F. (2010). Controllable synthesis of Bi_2MoO_6 and effect of morphology and variation in local structure on photocatalytic activities, *Appl. Catal. B: Environ.*, **98**, pp. 138–146.

137. Yin, W. Z., Wang, W. Z., and Sun, S. M. (2010). Photocatalytic degradation of phenol over cage-like Bi_2MoO_6 hollow spheres under visible-light irradiation, *Catal. Commun.*, **11**, pp. 647–650.

138. Martínez-de la Cruz, A., and Obregon Alfaro, S. (2010). Synthesis and characterization of γ-Bi_2MoO_6 prepared by co-precipitation: photoassisted degradation of organic dyes under vis-irradiation, *J. Mol. Catal. A: Chem.*, **320**, pp. 85–91.

139. Buttrey, D. J., Vogt, T., and White, B. D. (2000). High-temperature incommensurate-to-commensurate phase transition in the Bi_2MoO_6 catalyst, *J. Solid State Chem.*, **155**, pp. 206–215.

140. Zhang, M. Y., Shao, C. L., Mu, J. B., Zhang, Z. Y., Guo, Z. C., Zhang, P., and Liu, Y. C. (2012). One-dimensional Bi_2MoO_6/TiO_2 hierarchical heterostructures with enhanced photocatalytic activity, *CrystEngComm*, **14**, pp. 605–612.

141. Wang, P. F., Ao, Y. H., Wang, C., Hou, J., and Qian, J. (2012). A one-pot method for the preparation of graphene-Bi_2MoO_6 hybrid photocatalysts that are responsive to visible-light and have excellent photocatalytic activity in the degradation of organic pollutants, *Carbon*, **50**, pp. 5256–5264.

142. Xie, L. J., Ma, J. F., and Xu, G. J. (2008). Preparation of a novel Bi_2MoO_6 flake-like nanophotocatalyst by molten salt method and evaluation for photocatalytic decomposition of rhodamine B, *Mater. Chem. Phys.*, **110**, pp. 197–200.

143. Xie, H. D., Shen, D. Z., Wang, X. W., and Shen, G. Q. (2008). Microwave hydrothermal synthesis and visible-light photocatalytic activity of Bi_2MoO_6 nanoplates, *Mater. Chem. Phys.*, **110**, pp. 332–336.

144. Bi, J. H., Wu, L., Li, J., Li, Z. H., Wang, X. X., and Fu, X. Z., (2007). Simple solvothermal routes to synthesize nanocrystalline Bi_2MoO_6 photocatalysts with different morphologies, *Acta Mater.*, **55**, pp. 4699–4705.

145. Yoshimura, J., Ebina, Y., Kondo, J., Domen, K., and Tanaka, A. (1993). Visible light-induced photocatalytic behavior of a layered perovskite-type rubidium lead niobate, RbPb2Nb3O10, *J. Phys. Chem.*, **97**, pp. 1970–1973.

146. Kato, H., Kobayashi, H., and Kudo, A. (2002). Role of Ag+ in the band structures and photocatalytic properties of $AgMO_3$ (M: Ta and Nb) with the perovskite structure, *J. Phys. Chem. B*, **106**, pp. 12441–12447.

147. Choudhary, V., Banerjee, S., and Uphade, B. (2000). Activation by hydrothermal treatment of low surface area ABO_3-type perovskite oxide catalysts, *Appl. Catal. A: Gen.*, **197**, pp. L183–L186.

148. Zhong, Z., Chen, K., Ji, Y., and Yan, Q. (1997). Methane combustion over B-site partially substituted perovskite-type $LaFeO_3$ prepared by sol-gel method, *Appl. Catal. A: Gen.*, **156**, pp. 29–41.

149. Liu, G. D., Liang, S. J., Wu, W. M., Lin, R., Qing, N., Liang, R. W., and Wu, L. (2013). Template-free synthesis of a $CdSnO_3 \cdot 3H_2O$ hollow-nanocuboid photocatalyst via a facile microwave hydrothermal method, *Nanotechnology*, **24**, p. 255601.

150. Jia, X., Fan, H., Lou, X., and Xu, J. (2009). Synthesis and gas sensing properties of perovskite $CdSnO_3$ nanoparticles, *Appl. Phys. A*, **94**, pp. 837–841.

151. Zhao, X., Li, Z., Lou, X., Li, M., and Zhang, N. (2013). Room-temperature chlorine gas sensor based on $CdSnO_3$ synthesized by hydrothermal process, *J. Adv. Ceram.*, **2**, pp. 31–36.

152. Toan, N., Saukko, S., and Lantto, V. (2003). Gas sensing with semiconducting perovskite oxide $LaFeO_3$, *Phys. B: Condens. Matter*, **327**, pp. 279–282.

153. Xu, J. J., Wang, Z. L., Xu, D., Meng, F. Z., and Zhang, X. B. (2014). 3D ordered macroporous $LaFeO_3$ as efficient electrocatalyst for Li–O_2 batteries with enhanced rate capability and cyclic performance, *Energy Environ. Sci.*, **7**, pp. 2213–2219.

154. Su, H., Jing, L., Shi, K., Yao, C., and Fu, H. (2010). Synthesis of large surface area $LaFeO_3$ nanoparticles by SBA-16 template method as high active visible photocatalysts, *J. Nanopart. Res.*, **12**, pp. 967–974.

155. Parida, K., Reddy, K., Martha, S., Das, D., and Biswal, N. (2010). Fabrication of nanocrystalline $LaFeO_3$: an efficient sol–gel auto-combustion assisted visible light responsive photocatalyst for water decomposition, *Int. J. Hydrogen Energy*, **35**, pp. 12161–12168.

156. Nakayama, S. (2001). $LaFeO_3$ perovskite-type oxide prepared by oxide-mixing, co-precipitation and complex synthesis methods, *J. Mater. Sci.*, **36**, pp. 5643–5648.

157. Qi, X., Zhou, J., Yue, Z., Gui, Z., and Li, L. (2003). Auto-combustion synthesis of nanocrystalline $LaFeO_3$, *Mater. Chem. Phys.*, **78**, pp. 25–29.

158. Chu, X., Zhou, S., Zhang, W., and Shui, H. (2009). Trimethylamine sensing properties of nano-$LaFeO_3$ prepared using solid-state reaction in the presence of PEG400, *Mater. Sci. Eng. B*, **164**, pp. 65–69.

159. Yang, Z., Huang, Y., Dong, B., and Li, H.-L. (2006). Controlled synthesis of highly ordered $LaFeO_3$ nanowires using a citrate-based sol–gel route, *Mater. Res. Bull.*, **41**, pp. 274–281.

160. Zheng, W., Liu, R., Peng, D., and Meng, G. (2000). Hydrothermal synthesis of $LaFeO_3$ under carbonate-containing medium, *Mater. Lett.*, **43**, pp. 19–22.

161. Tang, P. S., Tong, Y., Chen, H. F., Cao, F., and Pan, G. X. (2013). Microwave-assisted synthesis of nanoparticulate perovskite $LaFeO_3$ as a high active visible-light photocatalyst, *Curr. Appl. Phys.*, **13**, pp. 340–343.

162. Shi, J. Y., Liu, G. J., Wang, N., and Li, C. (2012). Microwave-assisted hydrothermal synthesis of perovskite $NaTaO_3$ nanocrystals and their photocatalytic properties, *J. Mater. Chem.*, **22**, pp. 18808–18813.

163. Lü, X. M., Xie, J., Shu, H. M., Liu, J., Yin, C. Q., and Lin, J. M. (2007). Microwave-assisted synthesis of nanocrystalline $YFeO_3$ and study of its photoactivity, *Mater. Sci. Eng. B*, **138**, pp. 289–292.

164. Miyauchi, M., Takashio, M., and Tobimatsu, H. (2004). Photocatalytic activity of $SrTiO_3$ codoped with nitrogen and lanthanum under visible light illumination, *Langmuir*, **20**, pp. 232–236.

165. Sulaeman, U., Yin, S., and Sato, T. (2011). Visible light photocatalytic activity induced by the carboxyl group chemically bonded on the surface of $SrTiO_3$, *Appl. Catal. B: Environ.*, **102**, pp. 286–290.

166. Sulaeman, U., Yin, S., and Sato, T. (2010). Solvothermal synthesis of designed nonstoichiometric strontium titanate for efficient visible-light photocatalysis, *Appl. Phys. Lett.*, **97**, p. 103102.

167. Ding, J., Lu, X., Shu, H., Xie, J., and Zhang, H. (2010). Microwave-assisted synthesis of perovskite $ReFeO_3$(Re: La, Sm, Eu, Gd) photocatalyst, *Mater. Sci. Eng. B*, **171**, pp. 31–34.

168. Sun, M., Li, D., Zhang, W., Fu, X., Shao, Y., Li, W., Xiao, G., and He, Y. (2010). Rapid microwave hydrothermal synthesis of GaOOH nanorods with photocatalytic activity toward aromatic compounds, *Nanotechnology*, **21**, p. 355601.

169. Weinhardt, L., Gleim, T., Fuchs, O., Heske, C., Umbach, E., Bar, M., Muffler, H.-J., Fischer, C., Lux-Steiner, M. C., and Zubavichus, Y. (2003). CdS and Cd $(OH)_2$ formation during Cd treatments of Cu (In, Ga)$(S, Se)_2$ thin-film solar cell absorbers, *Appl. Phys. Lett.*, **82**, pp. 571–573.

170. Li, Z. Q., Lin, X. S., Zhang, L., Chen, X. T., and Xue, Z. L. (2012). Microwave-assisted hydrothermal synthesis, growth mechanism and photocatalytic properties of pancake-like $Cd(OH)_2$ superstructures, *CrystEngComm*, **14**, pp. 3495–3500.

171. Sun, M., Li, D., Zheng, Y., Zhang, W., Shao, Y., Chen, Y., Li, W., and Fu, X. (2009). Microwave hydrothermal synthesis of calcium antimony oxide hydroxide with high photocatalytic activity toward benzene, *Environ. Sci. Technol.*, **43**, pp. 7877–7882.

172. Xie, L., Wang, J., Hu, Y., Zheng, Z., Weng, S., Liu, P., Shi, X., and Wang, D. (2012). Template-free microwave-assisted hydrothermal synthesis and photocatalytic performance of $Bi_6O_6(OH)_3(NO_3)_3{\cdot}1.5\ H_2O$ nanosheets, *Mater. Chem. Phys.*, **136**, pp. 309–312.

173. Zhang, X., Ai, Z., Jia, F., and Zhang, L. (2008). Generalized one-pot synthesis, characterization, and photocatalytic activity of hierarchical BiOX (X= Cl, Br, I) nanoplate microspheres, *J. Phys. Chem. C*, **112**, pp. 747–753.

174. Chang, X. F., Huang, J., Cheng, C., Sui, Q., Sha, W., Ji, G. B., Deng, S. B., and Yu, G. (2010). BiOX (X=Cl, Br, I) photocatalysts prepared using $NaBiO_3$ as the Bi source: characterization and catalytic performance, *Catal. Commun.*, **11**, pp. 460–464.

175. Liu, Q. C., Ma, D. K., Hu, Y. Y., Zeng, Y. W., and Huang, S. M. (2013). Various bismuth oxyiodide hierarchical architectures: alcohothermal-controlled synthesis, photocatalytic activities, and adsorption capabilities for phosphate in water, *ACS Appl. Mater. Interfaces*, **5**, pp. 11927–11934.

176. Song, S. Y., Gao, W., Wang, X., Li, X. Y., Liu, D. P., Xing, Y., and Zhang, H. J. (2012). Microwave-assisted synthesis of BiOBr/graphene nanocomposites and their enhanced photocatalytic activity, *Dalton Trans.*, **41**, pp. 10472–10476.

177. Jiang, G. H., Wang, R. J., Wang, X. H., Xi, X. G., Hu, R. B., Zhou, Y., Wang, S., Wang, T., and Chen, W. X. (2012). Novel highly active visible-light-induced photocatalysts based on BiOBr with Ti doping and Ag decorating, *ACS Appl. Mater. Interfaces*, **4**, pp. 4440–4444.

178. Zhang, L., Cao, X. F., Chen, X. T., and Xue, Z. L. (2011). BiOBr hierarchical microspheres: microwave-assisted solvothermal synthesis, strong adsorption and excellent photocatalytic properties, *J. Colloid Interface Sci.*, **354**, pp. 630–636.

179. Xiao, X., Hao, R., Zuo, X., Nan, J., Li, L., and Zhang, W. (2012). Microwave-assisted synthesis of hierarchical $Bi_7O_9I_3$ microsheets for efficient photocatalytic degradation of bisphenol-A under visible light irradiation, *Chem. Eng. J.*, **209**, pp. 293–300.

180. Xu, J. S., and Zhu, Y. J. (2011). $Y_4Al_2O_9$ hierarchically nanostructured microspheres assembled with nanosheets: microwave-solvothermal synthesis combined with thermal treatment and photocatalytic property, *Mater. Lett.*, **65**, pp. 2793–2796.

181. Liang, S., Zhu, S., Chen, Y., Wu, W., Wang, X., and Wu, L. (2012). Rapid template-free synthesis and photocatalytic performance of visible light-activated $SnNb_2O_6$ nanosheets, *J. Mater. Chem.*, **22**, pp. 2670–2678.

182. Yuan, L., Yang, M.-Q., and Xu, Y.-J. (2014). Tuning the surface charge of graphene for self-assembly synthesis of $SnNb_2O_6$ nanosheets-graphene (2D-2D) nanocomposite with enhanced visible light photoactivity, *Nanoscale*, **6**, pp. 6335–6345.

183. Yu, M., Lin, J., Zhou, Y., and Wang, S. (2002). Citrate-gel synthesis and luminescent properties of $ZnGa_2O_4$ doped with Mn^{2+} and Eu^{3+}, *Mater. Lett.*, **56**, pp. 1007–1013.

184. Sun, M., Li, D. Z., Zhang, W. J., Chen, Z. X., Huang, H. J., Li, W. J., He, Y. H., and Fu, X. Z. (2012). Rapid microwave hydrothermal synthesis of $ZnGa_2O_4$ with high photocatalytic activity toward aromatic compounds in air and dyes in liquid water, *J. Solid State Chem.*, **190**, pp. 135–142.

185. Zhang, W., Zhang, J., Li, Y., Chen, Z., and Wang, T. (2010). Preparation and optical properties of $ZnGa_2O_4$: Cr^{3+} thin films derived by sol–gel process, *Appl. Surf. Sci.*, **256**, pp. 4702–4707.

186. Hsieh, I., Feng, M., Kuo, K., and Lin, P. (1994). Growth of $ZnGa_2O_4$ phosphor by radio frequency magnetron sputtering, *J. Electrochem. Soc.*, **141**, pp. 1617–1621.

187. Yang, M., Huang, Q. Q., and Jin, X. Q. (2012). Microwave synthesis of porous ZnGaNO solid solution for improved visible light photocatalytic performance, *Solid State Sci.*, **14**, pp. 465–470.

188. O'Regan, B., and Grfitzeli, M. (1991). A low-cost, high-efficiency solar cell based on dye-sensitized, *Nature*, **353**, pp. 737–739.

189. Gratzel, M. (2001). Photoelectrochemical cells, *Nature*, **414**, pp. 338–344.

190. Zhu, G., Pan, L. K., Xu, T., and Sun, Z. (2011). Microwave assisted chemical bath deposition of CdS on TiO_2 film for quantum dot-sensitized solar cells, *J. Electroanal. Chem.*, **659**, pp. 205–208.

191. Zhu, G., Pan, L. K., Xu, T., Zhao, Q., Lu, B., and Sun, Z. (2011). Microwave assisted CdSe quantum dot deposition on TiO_2 films for dye-sensitized solar cells, *Nanoscale*, **3**, pp. 2188–2193.

192. Wang, H., Bai, Y., Zhang, H., Zhang, Z., Li, J., and Guo, L. (2010). CdS quantum dots-sensitized TiO_2 nanorod array on transparent conductive glass photoelectrodes, *J. Phys. Chem. C*, **114**, pp. 16451–16455.

193. Zhu, G., Pan, L. K., Xu, T., and Sun, Z. (2011). One-step synthesis of CdS sensitized TiO_2 photoanodes for quantum dot-sensitized solar cells by microwave assisted chemical bath deposition method, *ACS Appl. Mater. Interfaces*, **3**, pp. 1472–1478.

194. Zhu, G., Pan, L. K., Sun, H. C., and Sun, Z. (2011). Microwave assisted chemical bath deposition of PbS quantum dots for quantum dot-sensitized solar cells, *Proc. Shanghai Int. Nanotechnol. Cooperation Symp.*, pp. 76–79.

195. Lee, Y.-L., Huang, B.-M., and Chien, H.-T. (2008). Highly efficient CdSe-sensitized TiO_2 photoelectrode for quantum-dot-sensitized solar cell applications, *Chem. Mater.*, **20**, pp. 6903–6905.

196. Robel, I., Subramanian, V., Kuno, M., and Kamat, P. V. (2006). Quantum dot solar cells. Harvesting light energy with CdSe nanocrystals molecularly linked to mesoscopic TiO_2 films, *J. Am. Chem. Soc.*, **128**, pp. 2385–2393.

197. Lee, Y. L., and Lo, Y. S. (2009). Highly efficient quantum-dot-sensitized solar cell based on Co-sensitization of CdS/CdSe, *Adv. Funct. Mater.*, **19**, pp. 604–609.

198. Huang, S., Zhang, Q., Huang, X., Guo, X., Deng, M., Li, D., Luo, Y., Shen, Q., Toyoda, T., and Meng, Q. (2010). Fibrous CdS/CdSe quantum dot co-sensitized solar cells based on ordered TiO_2 nanotube arrays, *Nanotechnology*, **21**, p. 375201.

199. Zhu, G., Pan, L. K., Xu, T., and Sun, Z. (2011). CdS/CdSe-cosensitized TiO_2 photoanode for quantum-dot-sensitized solar cells by a microwave-assisted chemical bath deposition method, *ACS Appl. Mater. Interfaces*, **3**, pp. 3146–3151.

200. Gratzel, M. (2003). Dye-sensitized solar cells, *J. Photochem. Photobiol. C: Photochem. Rev.*, **4**, pp. 145–153.

201. Hagfeldt, A., Boschloo, G., Sun, L., Kloo, L., and Pettersson, H. (2010). Dye-sensitized solar cells, *Chem. Rev.*, **110**, pp. 6595–6663.

202. Yella, A., Lee, H. W., Tsao, H. N., Yi, C., Chandiran, A. K., Nazeeruddin, M. K., Diau, E. W. G., Yeh, C. Y., Zakeeruddin, S. M., and Grätzel, M. (2011). Porphyrin-sensitized solar cells with cobalt (II/III)–based redox electrolyte exceed 12 percent efficiency, *Science*, **334**, pp. 629–634.

203. Sitaramanjaneya–Mouli, T. (2010). A novel coral-like porous SnO_2 hollow architecture: biomimetic swallowing growth mechanism and enhanced photovoltaic property for dye-sensitized solar cell application, *Chem. Commun.*, **46**, pp. 472–474.

204. Gubbala, S., Chakrapani, V., Kumar, V., and Sunkara, M. K. (2008). Band-edge engineered hybrid structures for dye-sensitized solar cells based on SnO_2 nanowires, *Adv. Funct. Mater.*, **18**, pp. 2411–2418.

205. Birkel, A., Lee, Y. G., Koll, D., Van Meerbeek, X., Frank, S., Choi, M. J., Kang, Y. S., Char, K., and Tremel, W. (2012). Highly efficient and stable dye-sensitized solar cells based on SnO_2 nanocrystals prepared by microwave-assisted synthesis, *Energy Environ. Sci.*, **5**, pp. 5392–5400.

206. Liu, S., Yu, J., and Jaroniec, M. (2011). Anatase TiO_2 with dominant high-energy {001} facets: synthesis, properties, and applications, *Chem. Mater.*, **23**, pp. 4085–4093.

207. Oh, J.-K., Lee, J.-K., Kim, H.-S., Han, S.-B., and Park, K.-W. (2009). TiO_2 branched nanostructure electrodes synthesized by seeding method for dye-sensitized solar cells, *Chem. Mater.*, **22**, pp. 1114–1118.

208. Chen, D., Huang, F., Cheng, Y., and Caruso, R. A. (2009). Mesoporous anatase TiO_2 beads with high surface areas and controllable pore sizes: a superior candidate for high-performance dye-sensitized solar cells, *Adv. Mater.*, **21**, pp. 2206–2210.

209. Zhang, H., Han, Y., Liu, X., Liu, P., Yu, H., Zhang, S., Yao, X., and Zhao, H. (2010). Anatase TiO_2 microspheres with exposed mirror-like plane {001} facets for high performance dye-sensitized solar cells (DSSCs), *Chem. Commun.*, **46**, pp. 8395–8397.

210. Wang, H. E., Zheng, L. X., Liu, C. P., Liu, Y. K., Luan, C. Y., Cheng, H., Li, Y. Y., Martinu, L., Zapien, J. A., and Bello, I. (2011). Rapid microwave synthesis of porous TiO_2 spheres and their applications in dye-sensitized solar cells, *J. Phys. Chem. C*, **115**, pp. 10419–10425.

211. Uchida, S., Tomiha, M., Takizawa, H., and Kawaraya, M. (2004). Flexible dye-sensitized solar cells by 28GHz microwave irradiation, *J. Photochem. Photobiol. A: Chem.*, **164**, pp. 93–96.

212. Uchida, S., Tomiha, M., Masaki, N., Miyazawa, A., and Takizawa, H. (2004). Preparation of TiO_2 nanocrystalline electrode for dye-sensitized solar cells by 28 GHz microwave irradiation, *Sol. Energy Mater. Sol. Cells*, **81**, pp. 135–139.

213. Huang, C. H., Yang, Y. T., and Doong, R. A. (2011). Microwave-assisted hydrothermal synthesis of mesoporous anatase TiO_2 via sol–gel process for dye-sensitized solar cells, *Microporous Mesoporous Mater.*, **142**, pp. 473–480.

214. Yang, M., Ding, B., Lee, S., and Lee, J. K. (2011). Carrier transport in dye-sensitized solar cells using single crystalline TiO_2 nanorods grown by a microwave-assisted hydrothermal reaction, *J. Phys. Chem. C*, **115**, pp. 14534–14541.

215. Menzies, D. B., Dai, Q., Cheng, Y. B., Simon, G. P., and Spiccia, L. (2006). One-step microwave calcination of ZrO_2-coated TiO_2 electrodes for use in dye-sensitized solar cells, *C. R. Chim.*, **9**, pp. 713–716.

216. Min, C., Shen, X., and Sheng, W. (2009). Microwave-assisted aqueous synthesis of ultralong ZnO nanowires: photoluminescence and photovoltaic performance for dye-sensitized solar cell, *Appl. Phys. A*, **96**, pp. 799–803.

217. Awais, M., Rahman, M., Don MacElroy, J. M., Dini, D., Vos, J. G., and Dowling, D. P. (2011). Application of a novel microwave plasma treatment for the sintering of nickel oxide coatings for use in dye-sensitized solar cells, *Surf. Coat. Technol.*, **205**, pp. S245–S249.

218. Remya, N., and Lin, J. G. (2011). Current status of microwave application in wastewater treatment: a review, *Chem. Eng. J.*, **166**, pp. 797–813.

219. Zhang, X. Y., and Liu, Z. (2012). Recent advances in microwave initiated synthesis of nanocarbon materials, *Nanoscale*, **4**, pp. 707–714.

Chapter 11

Microwave Nanosurface Engineering

S. Richard,[a] R. Aufaure,[b] E. Nehlig,[a] L. Motte,[b] and E. Guénin[b]

[a]*Laboratoire CSPBAT, CNRS UMR 7244, UFR SMBH, Université Paris 13, Sorbonne Paris Cité, F-93017 Bobigny, France*
[b]*Inserm, U1148, Laboratory for Vascular Translational Science, UFR SMBH, Université Paris 13, Sorbonne Paris Cité, F-93017 Bobigny, France*
guenin@univ-paris13.fr

11.1 Introduction

Since its first description by Gedye et al. [1], the use of microwave (MW) energy in chemistry rapidly spread, and MW ovens are now usual tools that can be found in most modern laboratories [2–6]. The unique features of MW heating are yield enhancement, a decrease in the reaction time, and in most cases better purity of obtained products. All these aspects explained that it was studied in all areas of organic chemistry [7–13]. But the use of MW energy has not been restricted to synthesis of organic compounds and its development went far beyond organic chemistry. The use of MWs gradually democratized itself in several scientific areas, and it is now a common methodology in domains as different as protein digestion as mining and metallurgy processes [14–19]. This progress was of course concomitant with the development of dedicated equipment specifically designed for MW reactions, and now the kitchen MW

Microwave Engineering of Materials: From Mesoscale to Nanoscale
Edited by Erwann Guénin

ISBN 978-981-4669-42-9 (Hardcover), 978-981-4669-43-6 (eBook)
www.panstanford.com

ovens used in the beginning have been replaced by proper materials that could allow for perfect control of reaction parameters and for scaling up [20].

Material chemistry is one of the fields where MW irradiation technologies are studied more and more. In recent years the development of nanotechnologies has increased the interest of material scientists for these new technologies. MW methodologies are now routinely used in several areas of materials science, and new advances are ongoing. They were used first in the development of new materials and especially in nanomaterial synthesis. Several recent reviews described the advances in this field that will not be treated in this chapter [21–25]. The other growing interest in MW methodologies in materials science was found in their use for the postsynthetic modifications of nanosurfaces. With the rapid increase in the number of applications in nanotechnology the ability to introduce or modify surface functionality is becoming a principal concern. The need for rapid and efficient methodologies to tailor surfaces is a rapidly growing field of interest. In this context, due to its unique features MW irradiation is gaining popularity in surface chemistry at the nanometric level.

In the present contribution we will give an overview of the recent advances in MW surface functionalization of nanomaterials to improve their properties or introduce new ones. We will rapidly go back to the fundamentals on MW technology and introduce MW methodologies dealing with nanomaterials. We will then focus on the use of MW irradiation for the surface modification of four main groups of nanomaterials: mesoporous silica and silicon, carbon-based materials, inorganic nanoparticles (NPs), and nanoplasmonic structures. The modifications of inorganic NPs will represent the most important part of this chapter as it is a wide area with several possibilities of surface modifications that we will classify into four major aspects: (i) silanization and formation of a TiO_2 shell, (ii) formation of a metallic shell, (iii) formation of an organic polymeric shell, and (iv) functionalization of small molecules at the NP's surface. For a broader picture on the uses of MWs in surface chemistry of various materials we invite readers to go back to our recent review [12].

11.2 Microwave Fundamentals

MW heating has become increasingly popular in the past two decades, owing to its successful applications in molecular chemistry, materials science, and nanotechnology. This heating method is known for being more efficient than conventional ones. The heat transfer on the reflux setup is based on convective currents and thermal conduction that is mainly absorbed on the reaction vessel. In contrast, MW heating allows a uniform increase of temperature throughout the whole liquid volume, called *in-core volumetric heating*. In fact, the absorption of MW energy by the media into heat occurs by two main mechanisms, dipolar polarization and ionic conduction. Dipolar polarization is the movement of molecular dipoles attempting to realign themselves along the alternating electric field, which causes friction and dielectric loss. Ionic conduction is the interaction of MWs on ions, making them oscillate and collide with neighboring molecules. Both of these phenomena generate energy loss in the form of heat. In such chemical synthesis, the loss tangent (tan δ) of the reaction solvent is the most important parameter to take into account. This physical property quantifies the ability to convert MW energy into heat, related to the solvent dielectric constant. Several tables describing tan δ values could be found in Refs. [4, 5]. However, as the molecular friction varies with the bulk viscosity, the dielectric properties of solvents strongly depend on temperature. Hence, several organic solvents lost their efficiency with increasing temperature.

Concerning MW-assisted synthesis of NPs a decrease in the average size and polydispersity compared to conventional heating is usually observed [26]. This is attributed to fast nucleation and growth rates when thermal gradients are minimized by the in-core volumetric heating. Moreover, rapid dissolution of the ionic precursor in nonpolar solvents contributes to acceleration of both nucleation and growth [27].

Specific effects of MW heating are described in the literature, in addition to the aforementioned thermal effects. Some studies have described *superheating* (liquid-phase temperature higher than the boiling point) by generation of hot spots, limiting the wall effect and promoting localized growth throughout the bulk solution [28]. Selective heating is another specific effect that concerns the ability of MW energy to be absorbed in preference by a highly polar substrate as metallic or metal oxide NPs, for example. Contrary to organic or polymer reactions, NP formation studies generally don't focus on reaction mechanisms in terms of electron trajectories. That is why most of MW nonthermal effects usually described for intermediates and transitions states can hardly be rationalized on nucleation and crystalline growth. Nevertheless, it has been demonstrated that MW frequency has an impact on NP synthesis in nonpolar solvents [29].

The use of MW energy for surface modification of nanomaterials involves two effects. First, MW energy interacts with the molecules in solution in a similar manner to what is described in organic chemistry. Second, MW energy also interacts with the material at the vicinity of the surface accounting for the interest in MW-assisted functionalization.

11.3 Microwave Surface Modification of Nanomaterials

MW surface engineering of nanomaterials is of course based, in large part, on the work initiated at the mesoscale. A large part of this work was conducted during the end of the 20th century by polymer and organic chemists for the MW-assisted functionalization of polymeric surfaces and polymer resins. The development of MWs in polymer chemistry was parallel to their study in organic chemistry, and the preparation of various polymers had already been extensively reviewed since the beginning of this century [30–34]. At the crossroads between the two centuries, both areas merged with the boom of solid-phase-supported organic chemistry with important applications in peptide synthesis, combinatorial chemistry, and preparation of microarrays. Nowadays MW solid-phase peptide synthesis (MW-SPPS) [13] and MW-assisted combinatorial chemistry [35–37] are well established and MW solid-phase synthesis has already been used for the building of small-molecule test libraries [38–42], peptide nucleic acids (PNAs), oligonucleotide microarrays [43–47], and the development of synthesis of multiple carbohydrate oligonucleotides on a solid support [8].

Most of these advances are now applied to modifications of nanomaterials of different natures. Nevertheless by doing a quick survey of the literature it rapidly appears that the use of MW irradiation for modification of the nanomaterial surface did not spread equally in all areas of nanoscience. Owing to their prior use for the modification of silica and silicon surfaces at the mesoscale MWs tend to be used more and more for the modification of nanoporous silica and silicon, and we will show some recent examples in this field. The second type of nanomaterial for which MWs are extensively used in surface modification is the family of carbon-based nanomaterials, especially carbon nanotubes (CNTs). Several descriptions are found of MW functionalization and modification of inorganic NPs' surfaces, and we herein classified them by function of the nature of the modification: silanization, formation of a metallic shell, polymerization, and modification of various small molecules at the NPs' surfaces). We will finally finish by introducing the first results on the use of MWs to functionalize plasmonic nanostructures.

11.3.1 *MW Modification of Porous Silicon-Based Nanomaterials*

Well-ordered nanoporous materials that have a pore size distribution from 2 to 50 nm have been extensively studied these past years for their applications in nanoscience. Advanced nanoporous silicon or silica-based materials have been designed with superior control in crystallite size, pore structure, pore size, surface area, and compositional flexibility and have achieved high activity, selectivity, and stability for a wide variety of applications, such as chemical separations, sensor development, biomedical applications, catalysis, and light-emitting diodes (LEDs) in optoelectronics [48–55]. Thus, the ability to decorate the silicon or silica framework and surface with specific ligands, organic molecules, or inorganic molecules creates powerful new capabilities for these materials.

MWs were already well described for the synthesis of mesoporous materials, and the subject was reviewed by Tompsett et al. [56] less than a decade ago. For the modification of the surface of silicon and silica-based nanoporous materials, no dedicated reviews could be found, in our knowledge, though several reports can be found in the literature. The use of MWs for the modification of such nanosurfaces can be divided into two different areas, depending on the nature of the modified surface: nonoxidized silicon surface (Si–H) requiring pretreatment of the surface (etching) for postfunctionalization and oxidized surfaces (SiO_x).

11.3.1.1 MW functionalization of porous silicon surfaces

Porous silicon is a nanostructured, highly porous material produced by the etching of a silicon wafer in a hydrofluoric acid solution. Materials with well-defined pore sizes at the nanoscale and extremely high surface area can be produced (Fig. 11.1). These porous silicon surfaces could be prepared in various forms, such as films, membranes, or particles, and the fine-tuning of their surfaces is needed for various biomedical, optical, and sensing applications. Since silicon can readily adsorb MW energy as an intrinsic property, heating in the near surface of the material can be observed. So the use of MW irradiation as a source of energy for silicon-based material modification could lead to an appreciable rate enhancement and drastically decreased reaction time.

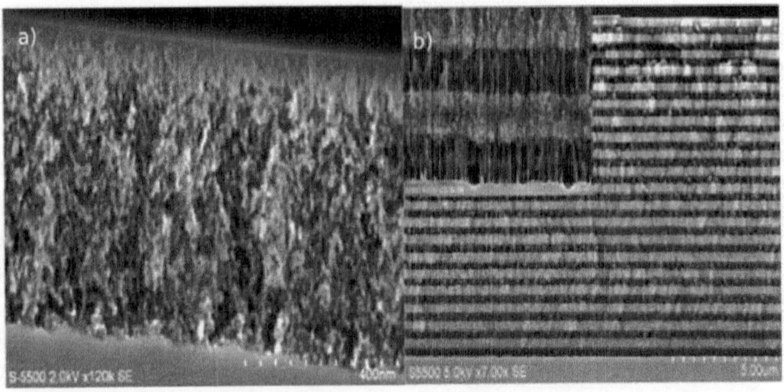

Figure 11.1 Cross-sectional SEM images of porous silicon nanostructures. A luminescent (a) monolayer and (b) a multilayer. These structures were prepared at the CIE-UNAM porous silicon laboratory. © 2013 de la Mora MB, Ocampo M, Doti R, Lugo JE, Faubert J. Published in Ref. [57] under CC BY 3.0 license. Available from: http://dx.doi.org/10.5772/52975.

Boukherroub et al. [58] were the first to describe the use of MWs for the modification of the surface of silicon-based material. They proposed an MW-assisted hydrosilylation reaction of hydrogen-terminated porous silicon surfaces with functional alkenes (Scheme 11.1). The reactions were performed in a commercial monomode reactor using an adapted open cylindrical Pyrex vessel. A remarkable increase in the rate of the hydrosilylation reaction and higher surface coverage were obtained. This technique yields highly stable organic monolayers and allows the introduction of different functional groups on the surface (acid and ester) required for immobilization of more complex structures on the surface. After its pioneering work the same team compared the reaction of hydrogen-terminated porous silicon with alkenes, aldehydes, and alkyl halides under MW irradiation [59] (Scheme 11.1). Once again the use of MW irradiation has a net effect on the hydrosilylation reaction rate. However, no

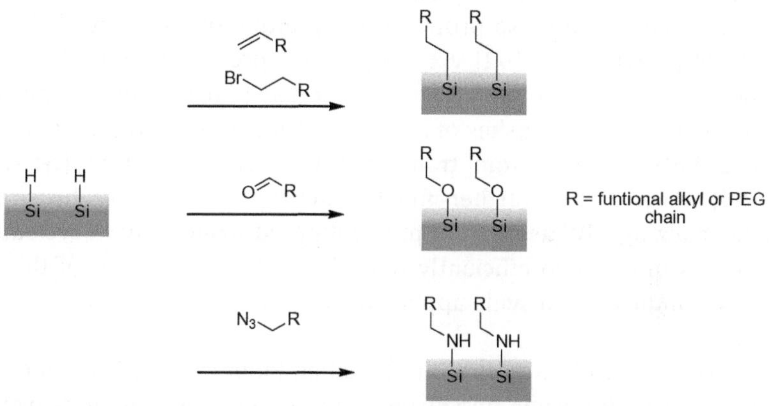

Scheme 11.1 Studied methodologies for MW modification of porous silicon surface.

specific MW effect was observed when the freshly prepared porous silicon surface was modified with an alkene bearing an aldehyde function. The reaction takes place at both the carbon–carbon and the carbon–oxygen double bonds.

In parallel to the work done by Boukherroub et al., Xiao's group proposed similar approaches for the addition of an organic layer on porous silicon materials, which can contribute to further functionalization of the surface by biomolecules [60]. They first proposed the formation of the robust Si–C bonds by MW-assisted radical reaction between alkyl bromides presenting carboxylic functions and silicon wafers, using a domestic MW oven in an open vessel [60] (Scheme 11.1). The power of the MW irradiation, the concentration of the reactant, and the presence of a radical initiator were investigated to optimize the reaction. The selected conditions were shown to unalter the nanostructure. Finally two proteins—bovine serum albumin (BSA) and lysozyme (Lys)—were immobilized on the surface via activation of the carboxylic function. A few years later Xiao's group proposed an alternative to the use of bromoalkanes, alkenes, or aldehydes for the grafting of alkyl or aryl substituent on porous silicon [61]. This new methodology utilized azido-functionalized molecules and created Si–N bonds at the surface. For this reaction MW irradiation was compared to photo-illumination and MWs were found to yield higher surface coverage in shorter time. Xiao's group also proposed the grafting of allyl polyethylene glycol (PEG) chains [62] via the protocol previously described by Boukherroub et al., and they further used this methodology to prepare polymer brushes of Si-g-P(poly(ethylene glycol) methacrylate PEGMA]) via atom transfer radical polymerization (ATRP) [63]. Finally, they further functionalized their polymer brushes by realizing MW-assisted copper-catalyzed azide–alkyne cycloadditions in order to efficiently tune the surface properties of these semiconductors for wide applications in electronics and biosensing [64].

More recently, Sun et al. [65] studied kinetics of hydrosilylation reaction with decene on silicon nanocrystals. Their findings were quite different from the results obtained by Boukherroub et al. and Xiao's groups since no evident acceleration effect on the hydrosilylation rate was observed using MW heating relative to conventional thermal heating. Sun et al. proposed that the results observed in prior reports could be explained by thermal effects, for example, shortened ramping time due to direct heating compared with long ramping times via heat conduction and convection (e.g., with oil bath or oven) [65].

In contradiction to these results van den Berg et al. [66] have recently shown that the rate of formation of covalently linked organic monolayers on HF-etched silicon carbide (SiC) surface is greatly increased by MW irradiation. Compared to typical thermal conditions (16 h, 130°C), densely packed, covalently attached monolayers on flat SiC surfaces were obtained upon MW treatment in 60 min at 100°C (60 W). This approach was extended to SiC microparticles with the same results. According to the authors MW assistance of monolayer formation using 1-alkenes on silicon carbide surfaces speeds up the reaction by 2 to 4 orders of magnitude. This methodology was also successfully extended to commercially available SiC microgrits and microspheres. Furthermore, by using 1,ω-dialkenes, the reaction permits easy access to "clickable" monolayers on SiC materials with complex structures that can be further functionalized by thiol-ene reaction.

11.3.1.2 MW functionalization of nanosilica surfaces

Mesoporous silica materials are new nanomaterials that stirred extensive research since their first development in the nineties by Mobil Oil Company. Among all the materials proposed since then, the two nanoporous materials, that is, Mobil composite material number 41 (MCM-41) and Santa Barbara amorphous type material No. 15 (SBA-15) are the most commonly used. These materials are characterized by large specific surface areas and pore sizes between 2 and 50 nm (Fig. 11.2), and this specific feature is at the base of their potential applications in various areas such as adsorption, chromatography, catalysis, sensor technology, and gas storage [53].

To further develop their functionalities modifications of the pore surface can be obtained through the coupling of inorganic and organic components by template synthesis. The use of MWs at diverse stages of the synthesis process of mesoporous silica (nucleation, elaboration of the nanosilica, digestion, etc.) has already been described and referenced in several reviews [53, 56, 68, 69], but to our knowledge the use of MWs in functionalized, synthesized mesoporous silica was not yet reviewed. Most of the reactions involved silanization of the silica surface using a functionalized silylating agent or postfunctionalization of the silica surface bearing reactive groups (Scheme 11.2).

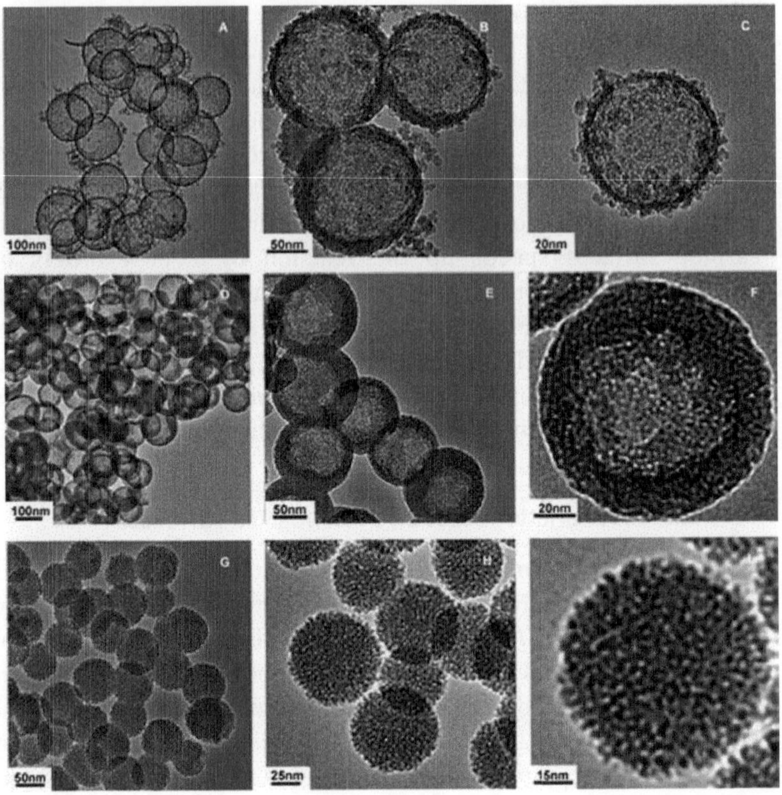

Figure 11.2 EM images of mesoporous silica nanospheres prepared with different molar ratios of CSS/NaOH. (A–C) CSS/NaOH = 0.2/1.5; (D–F) CSS/NaOH = 0.2/1.0; (G–I) CSS/NaOH = 0.2/0.5. Reproduced from Ref. [67], Copyright (2008), with permission of the Royal Society of Chemistry.

The first report on modification of mesoporous material under MW irradiation was described less than a decade ago by Pirouzmand et al. [70] and did not follow Scheme 11.2 as there was no covalent attachment onto the silica but rather immobilization in the pore. Pirouzmand et al. in fact proposed a novel ship-in-the-bottle strategy for the incorporation of metallophthalocyanine in MCM-41. The phthalocyanine complexes of cobalt, manganese, and iron were assembled and immobilized inside MCM-41 channels by MW irradiation of a mixture of ion-exchanged MCM-41 and 1,2-

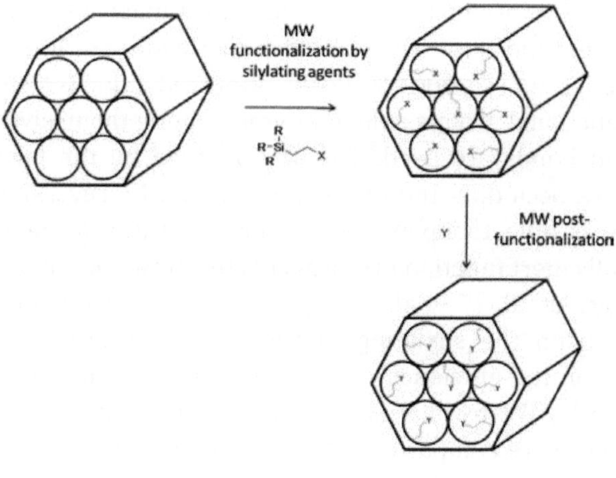

Scheme 11.2 MW modifications of mesoporous silica.

dicyanobenzene in a short period of time under solvent-free conditions. The authors describe this MW-assisted approach as efficient for the immobilization and stacking of metallophthalocyanine inside MCM-41 channels, together with preservation of MCM-41 textural properties.

A few years later Procopio et al. [71] reported an MW-assisted protocol for a rapid and efficient solvent-free functionalization of MCM-41 with trialkoxysilanes. Presence of solvent and MW exposition time were studied, as well as the effect of the leaving group on the silylating reagents. It was found that functionalization of mesoporous silica was much more efficient in neat conditions under MW irradiation (15 min) compared to classical functionalization, which was time and energy consuming (7 h in refluxing toluene). Each time yields were higher than those reported in the literature for classical conditions. Moreover in these improved conditions ethyl esters were more efficient than methyl esters for the silylating reagents. This methodology was then used by the same team for the preparation of a mesoporous silica-supported ErIII catalyst [72]. Finally, Proccopio's group realized a study of the MW-assisted grafting procedure of mesoporous silica by varying the functionalities of organosilane precursors [73]. The authors found that slight pore volume reduction is linearly correlated to the organic amount in the product and that the hybrid materials maintain the original mesoporous structure when the loading of linked organic groups does not exceed 10%. Pore volume reduction is also obtained when functional groups able to interact among themselves through hydrogen bonds are used. This is explained by the formation of a network occluding the pores but that can be reversible as acid washing is able to open the network. In the case of bulky but chemically inert functionalizing agents the network is not produced.

Fukaya et al. [74] also proposed an evaluation of the leaving group on the silylating reagent—(3-chloropropyl) dimethylsilanes—on the efficiency of MW functionalization of mesoporous silica. Both MW irradiation conditions and conventional heating conditions were compared, with heptane or toluene as the solvent. The authors showed that the effect of MW irradiation on the loading amount strongly depends on the leaving group as well as the solvent. The use of methoxysilane and *p*-anisylsilane in heptane caused a distinct acceleration. Considering the dielectric properties of the organosilanes it was suggested that strong MW-absorbing nature of the silylating agent is the key factor for the better results obtained. Also, using a silylating agent, Gonzalez et al. [75] proposed the MW-assisted sulfonic acid functionalization of SBA-15 and beta zeolite. The zeolite sample sulfonated with MW irradiations presented higher surface area, slightly higher pore volume, and mesoporosity, as well as a lower decrease in zeolite crystallinity than the conventional sulfonated sample. SBA-15 was sulfonated with MWs, achieving faster incorporation of a higher amount of sulfonic acid groups than allowed by conventional heating. The obtained sulfonated silica surfaces presented boosted selectivity toward etherification catalysis of glycerol. The same group extended its MW-assisted sulfonation methodology [76, 77] to the postsynthesis modification of silica aerogel and lyogel used as catalysts for the etherification of glycerol with isobutene. Once again the use of MWs allowed the loading of higher amounts of sulfonic acid groups in both silicas compared to conventional heating.

Lourenco et al. [78] recently reported the use of MW to postfunctionalize the pore surface of mesoporous material. They studied *N*-monoalkylation reactions of the amine-functionalized periodic mesoporous phenylene-silica (NH_2-PMO) under MW irradiation. A potassium-iodide-catalyzed method commonly used for the selective *N*-monoalkylation of aniline was adapted and optimized. Functionalization was achieved with about 87% of conversion in two reaction cycles with preservation of the ordered mesostructure and of the crystal-like molecular scale periodicity of the material.

The use of MW to postfunctionalize mesoporous silica was also largely used for the covalent immobilization of enzymes. Though proteins (including enzymes) are generally considered as too sensitive toward MW irradiation, several authors [79–83] have shown that short period of MW irradiation in ice water or using a cooling module could improve the grafting onto mesocellular siliceous foam and other supports. The specific activity of enzymes immobilized under MW irradiation was conserved, even increased in some of the cases, compared to a traditional method without MW irradiation. Moreover enzymatic deactivation under MW irradiation was very slight, and usually the enzymes retained their activities.

Recently, inorganic postfunctionalization of mesoporous silica has been also reported by several authors to prepare supported inorganic catalysts, which are widely recognized as an important class of industrial catalysts. Rodrigues et al. [84], for example, produce a cobalt catalyst supported in SBA-15 molecular sieves and evaluate its use in Fischer–Tropsch reaction. Two synthesis methods were evaluated, MW heating and conventional heating. The authors found that the selectivity was increased with the Co/SBA-15 prepared under MW heating and that the presence of Co did not modify the SBA-15 properties. The same year Yamashita and coworkers [85] immobilized silver NPs acting as catalysts in the pore of mesoporous silica by running MW in situ synthesis of the NPs. In this very nice work they moreover vary the size and morphology of the Ag NPs within the mesopore structure of SBA-15 by MW-assisted alcohol reduction. An enhancement of the catalytic activity using localized surface plasmon resonance (LSPR) properties was demonstrated. In 2014 the same group [86] proposed a similar preparation of supported Ag NPs with tunable particle sizes (1.9–3.9 nm) on CeO_2-SBA-15 supports through an efficient MW heating methodology. The authors showed that the strong charge polarization of CeO_2 clusters during the MW heating process contributed to the precise control of the Ag particle size relative to those obtained through oil-bath heating. The deposition of Ag NPs on CeO_2-SBA-15 through MW heating provided higher turnover frequency (TOF) numbers than those of Ag/SBA-15 without CeO_2 and Ag/CeO_2-SBA-15 prepared through conventional oil-bath heating, indicating the benefit of the MW methodology on efficiency of the obtained catalysts.

11.3.2 *MW Functionalization of Carbon-Based Nanomaterials*

As for the other types of nanomaterials, MW synthesis of carbon-based nanomaterials and especially single-walled and multiwalled carbon nanotubes (SWCNTs and MWCNTs) and carbon nanohorns (CNHs) has already been fairly well studied and described. The first review on interactions, responses, and applications of nanotubes with MWs was published in 2009 by Vazquez and Prato [87] and included a part on MW functionalization of nanotubes. This work was recently updated by the same group in a review describing nonconventional methods and media for the activation and manipulation of carbon nanoforms. In 2011 a review especially dedicated to the topic of MW functionalization of carbon-nanostructured materials was produced by Economopoulos et al. [88]. Therefore, in this part we will only shortly report the description of MW functionalization strategies of CNTs and present some of the more recent reports since Economopoulos et al.'s review. We will also focus on MW synthesis of CNT composites and organic/inorganic nanohybrids. Finally we will also present the reports of MW functionalization of carbon or diamond NPs.

11.3.2.1 MW functionalization of single-walled and multiwalled carbon nanotubes

Several strategies have been developed for the MW functionalization of CNTs (Scheme 11.3); the first one consists of creating by MW treatment new functionalities at the surface of the CNTs that can further react with molecules of interest.

One of the most popular modifications is the introduction of carboxyl groups, for example, by MW oxidation, as described by

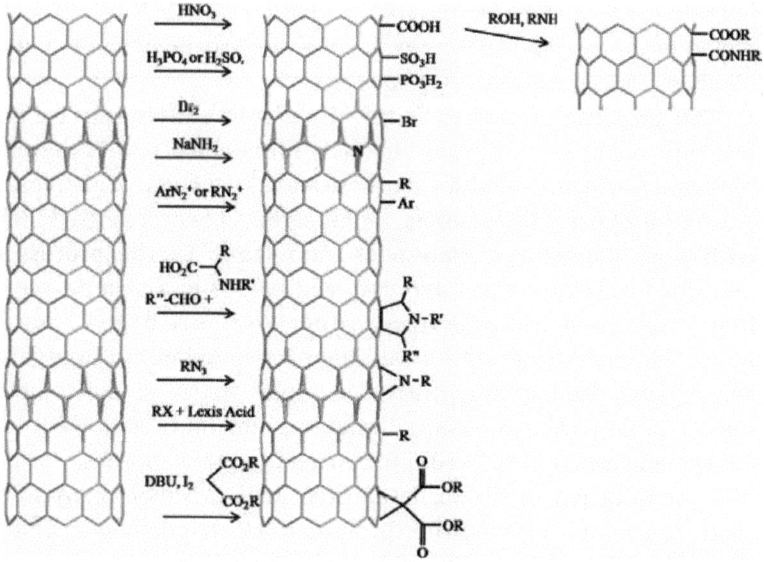

Scheme 11.3 MW functionalization methodologies employed to modify CNTs.

Kubota et al. [89], Wang et al. [90], and Yoshida et al. [91]. Sulfonic or phosphoric funtionalities can also be added onto the surface, such as those described by Kumar et al. [92], who proposed various single- and double-step chemical routes to covalently functionalize MWCNTs with different chemical groups (–COOH, $–SO_3H$, $–PO_3H_2$) using MWs. Brominated CNTs can also be obtained, as described by Colomer et al. [93], who reported for the first time a bromination reaction between double-walled carbon nanotubes (DWCNTs) and Br_2. Raman investigation showed a certain degree of functionalization of Br-DWCNTs, however, accounting for a mild alteration of the graphitic conjugated surface. The obtained Br-DWCNTs could then be further functionalized by cross-coupling reactions. More recently Lee et al. [94] realized nitrogen-doped graphene nanosheets using simple MW irradiation under the presence of sodium amide as a nitrogen source. Nitrogen content of the doped graphene could reach up to 8% within a minute. Binding configuration of nitrogen atoms on the graphitic layer consisted of various nitrogen-containing moieties such as pyridine-N, pyrrolic-N, and quaternary-N, depending on the irradiation power. The nitrogen-doped graphene showed an increase in electrical conductivity and highly enhanced capacitive performance.

The presence of carboxylic functions on the CNTs' surface has been utilized by several groups to further functionalize CNTs by simple amide or ester bond formations. Mallakpour and Zadehnazari [95–98] proposed, for example, modifications of carboxylated CNTs with *p*-aminophenol, amino acids, or vitamin C. The process is described as fast, one pot, and easy and results in a high degree of functionalization, while the structure of MWCNTs remains relatively intact. The modifications allowed better dispersion in the organic solvent and compatibility with a polymer matrix of poly(amide-imide) to prepare composites containing functionalized MWCNTs. Tahermansouri et al. [99–102] proposed the same approach for the MW modification of carboxylated CNTs with 2,3 diaminopyridine (and subsequent cyclization promoted by triphenylphosphite), creatinine, or thiourea followed by ninhydrine. The major advantage of this procedure was the decrease in the reaction time (down to 1 min) and of the number of steps in the procedure compared to conventional functionalization processes. A last example of this methodology is given by Davarpanah, Maghrebi, and Hosseinipour [103] for the MW-induced high surface functionalization of MWCNTs for long-term dispersion in water. The carboxylated MWCNTs were reacted with diversely *N*-alkylated ethanolamine and yielded highly functionalized material stable for a long-term period in polar solvents.

Another popular mode of CNT functionalization is the reaction with diazonium salts, initially described by Bahr and Tour [104], that has been popularized under MWs by several authors since then [105–107]. In 2010 Tagmatarchis and coworkers [108] functionalized the external surface of C60@SWCNT peapods by covalently grafting aryl units following an MW-assisted diazonium approach. In the modified peapods the fullerene remains encapsulated within SWCNTs regardless of the solubility achieved. This one-pot rapid MW-assisted methodology for MWCNT functionalization was used by Amiri et al. [109] in 2011 to add three different diamines, ethylenediamine (EDA), 1,6-hexamethylenediamine (HMDA), and 1,4-diaminobenzene (DAB), onto MWCNTs. Amiri et al.'s methodology did not include acid treatment, which was common in conventional methods. All the treated MWCNTs presented quite high extents of functionalization, which were distinguishably higher than what expected from sonication or other conventional methods. The same group [110–112] also reported the efficient functionalization of MWCNTs with two amino acids, arginine and lysine, under MW radiation via this simple and rapid procedure. Mallakpour et al. [113, 114] also described efficient amino acid grafting using the diazonium methodology to prepare their composites containing functionalized MWCNTs. Recently Mamane et al. [115] evaluated the effect of the MW irradiation time length on the functionalization of SWCNTs by the diazonium method. According to thermogravimetric analysis coupled with mass spectrometry and Raman spectroscopy, the optimal functionalization level was reached after 5 min of reaction. Prolonged reaction times can lead to undesired reactions such as defunctionalization, solvent addition, and polymerization of the grafted functions. Similar results were described by Liu et al. [116] for a SWNCT functionalization process that can be related to the diazonium methodology. The authors proposed functionalization through radical addition of 4-methoxyphenylhydrazine hydrochloride by a classical thermally activated procedure or via an MW-assisted method. The MW methodology was found to be more efficient in 5–15 min. But longer MW exposure times reduced the functionalization yield and led to a removal of groups, which were bonded in a previous stage, indicating the importance of the choice of optimal irradiation times.

Apart from the diazonium methodology the more commonly used methodology to chemically modify the surface of CNTs is, by far, the 1,3 dipolar cycloaddition route. Diels–Alder cycloaddition [117] or formation of a pyrrolidine cycle via reaction with azomethine ylides [105, 118–120] was already studied by several authors using MW. Very recently Gaudino et al. [121] showed interest in this reaction for the multifunctionalization of CNTs. They realized a one-shot multidecoration of SWCNTs with β-cyclodextrin and 1,4,7,10-tetraazacyclododecane-1,4,7,10-tetraacetic acid (DOTA) moieties under MW irradiation to realize a potential theranostic multicarrier nanoplatform. Bayazit and Coleman [122] conduced Raman spectroscopic investigations on the surface modification of SWCNTs through

1,3 dipolar cycloaddition of 3-phenyl phthalazinium-1-olate, which is a stable and reactive azomethine imine. For this reaction, MW heating was found to be more efficient than conventional and solvent-free heating. Moreover the authors concluded on the nondestructive covalent functionalization of the SWCNTs.

The cycloaddition was found to be a versatile way to functionalized CNTs, and other reactants than azomethine ylides can be used. For example, Bayazit et al. [123] recently explored the mechanism of MW-assisted 1,3-dipolar cycloaddition of pyridinium ylides to SWCNTs. In the reaction pyridinium ylides were generated in situ via the addition of triethylamine (NEt_3) to the Kröhnke salts and produced efficient covalent functionalization of the SWCNTs. Another example was given by Criado et al. [124], who studied the cycloaddition with arynes under MW irradiation. Experimental results pointed to the formation of SWCNTs with a high degree of functionalization, and theoretical calculations suggest that [4+2] cycloadducts are thermodynamically favored for aryne cycloadditions with zigzag SWCNTs, while [2+2] cycloadducts are favored for cycloaddition with armchair SWCNTs. A last example is the [2+1] nitrene cycloaddition proposed by Karousis et al. [108] for the functionalization of CNHs. Reaction of azides under MW irradiation led to the efficient covalent grafting of aziridino rings onto the graphitic network of CNHs.

Finally it is worth noting that other MW functionalization methodologies were also described for the CNT surface modification. The MW-assisted Bingel reaction [125, 126] was already previously reported in Economopoulos et al.'s review. But examples of electrophilic addition were also described in the literature. Xu et al. [127] were, in our knowledge, the first to report the MW-induced electrophilic addition of SWCNTs with alkylhalides using Lewis acid as a catalyst followed by hydrolysis. The reaction resulted in the attachment of alkyl and hydroxyl groups to the surface of the nanotubes. The MW-assisted reaction was found to be highly efficient, needing reaction times as low as several minutes, while the oil heating needed tens of hours. The authors also showed that iodoalkanes conducted to higher reaction efficiency with SWCNTs than chloro- and bromoalkanes. A year later electrophilic addition polymerization of 3-chloropropene under MW irradiation was reported by Tian et al. [128]. A Lewis acid was used as a catalyst, and under these conditions 3-chloropropene was polymerized. The hydrolysis in methanol resulted in the attachment of polymer chains and hydroxyl groups to the surface of the nanotubes. The results showed that 30 min of irradiation time was the optimal condition, as a longer time period did not improve the degree of functionalization.

11.3.2.2 MW synthesis of composites with CNT and formation of inorganic/organic nanohybrids

CNTs were successfully introduced into composites by MW irradiation methodologies. Chen and Gu [129] proposed the solid-state grafting of MWCNTs for the synthesis of a thermal polyurethane composite. Their one-step method reduced the reaction time to the order of minutes and allowed the polyurethane to be directly attached to the nanotubes, thus forming a highly solvent-dispersible composite. Nadagouda and Varma [130] accomplished the crosslinking reaction of polyvinyl alcohol with SWCNTs, MWCNTs, and C-60 using MW irradiation, and Huang et al. [131] grafted polyamide 6 (PA6) chains onto CNTs through a condensation reaction between the carboxylic groups of the MWCNTs and the terminal amine groups of PA6 using MW radiation heating. More recently a simple method for engineering chitosan-functionalized MWCNTs composites with a biomedically important polymer (hydroxyethyl methacrylate [HEMA]) was described by Mahmoodian et al. [132]. They realized a three-step functionalization, starting first by MW grafting of chitosan onto the surface and sidewalls of the CNTs. Second, HEMA monomers were grafted onto the polymeric matrix surface. The final step involved free-radical polymerization of HEMA monomers.

Organic/inorganic composites containing both CNTs and metallic or metal oxide NPs have stimulated extensive research as they combine both the extraordinary mechanical properties and the unique electrical properties of CNTs and the various assets brought by the nanoaparticles, depending on their constitution. MW NP synthesis is already well described and has so been extended to the synthesis of such composites. As we have already reviewed this part in our previous work [12] and because it has been extensively reviewed recently by Motshekga et al. [133] we invite readers to go back to these two articles.

11.3.2.3 MW functionalization of other carbon nanomaterials

There are fewer reports in the literature concerning the functionalization of carbon NPs or of nanodiamonds (NDs). The only report given on black carbon NPs was the work from Vukićević et al. [134] for the functionalization of carbon particle surfaces using iodine and azide-terminated poly(vinylidene fluoride) (PVDF). Azide-terminated PVDFs were more efficient than with iodine ones. Unfortunately, whatever the reactant was, implementation of MW irradiation did not result in any improvement of the functionalization process. The same year Chang et al. [135] proposed a simple and easy assisted functionalization of NDs with hydrophilic and hydrophobic moieties in two steps. The process involves surface graphitization of NDs, followed by radical-initiated surface grafting of oligomers with various functionalities in the presence of benzoyl peroxide in a domestic MW oven. Compared to another methodology for ND functionalization using ATRP, the MW-assisted surface modification process allowed more rapid functionalization (10 min rather than $>$20 h, in general), could be completed under ambient conditions, and did not require presurface functionalization of nanomaterials. More recently Desai and Mitra [136] proposed an MW methodology to functionalize NDs. The process is similar to the one utilized for the functionalization of CNTs. It consisted of treating NDs with HNO_3/H_2SO_4 in aqueous phase under MW radiation. The authors showed that the NDs' core was unaffected by MW treatment. The carboxylation led to altered colloidal behavior in terms of reduced agglomeration in both aqueous and polar organic solvents, which was accompanied by increased solubility. The same process was applied a year later by Galindo-Hernandez et al. [137] for the functionalization of carbon nanofibers. The reaction used only HNO_3 this time, and formation of carboxylic functions on the surface permitted the creation of extra spacing between contiguous nanofibers and an increase in intrawall volume. The enhanced functionalization allows for an increase in the hydrogen storage capacity.

11.3.3 *MW Functionalization of Inorganic Nanoparticles*

The use of MWs is well documented for the synthesis of inorganic nanocrystals and has been recently reviewed [21, 25]. MW technology for the synthesis of inorganic NPs has been receiving an exponential increase in acceptance, as evidenced by the rapidly increasing number of publications each year. The syntheses are described as more rapid and selective with, most of the time, both a shorter reaction time and an enhanced chemical reaction rate as compared to conventional syntheses. Strangely, by comparison to the enthusiasm for the use of MWs in the synthesis, the use of MWs for the surface modification of inorganic NPs has only been emerging for the past five years.

We decided to subdivide this part by differentiating the type of functionalization realized onto the NPs: (i) silanization and formation of TiO_2 shell, (ii) formation of a metallic shell, (iii) formation of an organic polymeric shell, and (iv) functionalization of small molecules at the NP surface.

11.3.3.1 MW formation of SiO_2 or TiO_2 shell on inorganic nanoparticles

The formation of a shell of silica or titania on the surface of NPs has been extensively used by several groups in order to increase the NPs solubility and protect them [138, 139]. Silica core–shell NPs are described to be nontoxic and easy to functionalize. However, silica encapsulation through conventional methods requires long reaction times ranging from hours to days, and it is still a challenge to achieve a uniform silica coating of monodisperse magnetic NPs with a fast encapsulation rate, so MWs could provide an interesting alternative for silica encapsulation.

The first example of MW-assisted silanization of inorganic NPs was provided by Feng et al. [140], who described a new MW methodology for the coating of europium oxide NPs. MW-assisted coating was done by treating directly the surface of commercially available Eu_2O_3 particles with (3-aminopropyl) trimethoxysilane (APTMS). This coating permitted to solubilize the NPs in water and protect them from degradation that could impair their optical properties. Moreover the coating provided an amine group for conjugation and use in an immunoassay model. With this model the authors proved that the fluorescence properties of europium NPs were preserved after the MW process. After Feng et al.'s work no other uses of MW were reported until five years later and the numerous studies of Suzuki's group. They first presented a silica coating of CeO_2 [141] or ZnO [142] NPs that was carried out by the MW irradiation method. NPs were condensed to a tetraethoxysilane (TEOS) precursor in an aqueous medium in the presence of NH_3 and yielded under MW irradiation the formation of a SiO_2 shell with a thickness of 5 nm on CeO_2 and 3 nm on ZnO. The surface characteristics of NPs were significantly modified after the silica coating and the NPs prepared by the MW irradiation method were suitable to apply as a UV shielding agent contained in sunscreens. The authors then applied their methodology to other types of inorganic NPs. The authors showed that the MW irradiation approach offers a unique, fast way to prepare silica-coated Al_2O_3 NPs [143] and provides by the conventional sol–gel approach a resolution for several important problems related to slow kinetics and the risk of particle aggregation. MW heating resulted in a thin but homogenous coverage of silica on the Al_2O_3 surface within a very short time, improving dispersibility. The same rapid, uniform hydrolysis and subsequent copolymerization of TEOS was realized on ZrO_2 NPs' surface [144]. The MW irradiation method provided a very short coating time and uniform heating in comparison to the conventional methodology, allowing the synthesis of ZrO_2 NPs uniformly, thinly coated with silica.

Suzuki's group also successfully extended the MW-assisted silanization to metallic NPs such as gold or silver NPs [145, 146]. Au@SiO_2 NPs were prepared by simply reacting the gold NPs with TEOS and ammonia under MW irradiation for 5 min. No precoating steps, as well as long reaction duration, were needed, as described for gold silanization in the literature. Au@SiO_2 NPs with a wide range of silica shell thicknesses (5–105 nm) were prepared within 5 min of MW irradiation by changing the concentration of TEOS only. The size uniformity and monodispersity were found to be better compared to the particles prepared by conventional methods. Preparation of Ag@SiO_2 was also carried out in a one-step manner

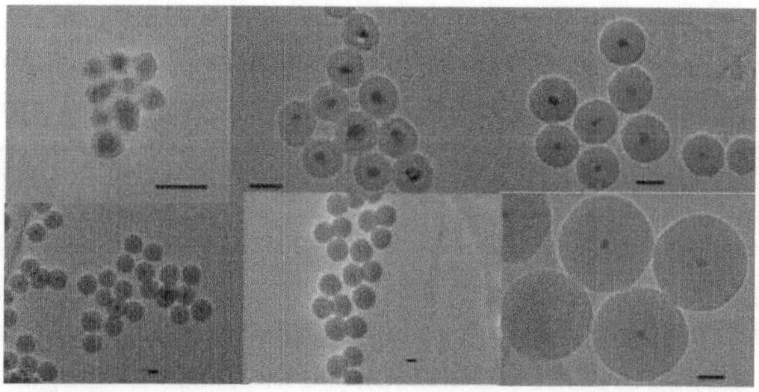

Figure 11.3 TEM micrographs of the $Ag@SiO_2$ NPs prepared by the MW technique at different TEOS concentrations. Reprinted from Ref. [145], Copyright (2011), with permission from Elsevier.

with TEOS using MWs and replacing ammonia for dimethylamine. Compared to the conventional techniques based on the Stöber method, which need 4–24 h for silica coating of Ag NPs, this new technique was capable of synthesizing monodispersed, uniform, and single-core-containing $Ag@SiO_2$ NPs within a very short reaction time (Fig. 11.3). In addition, this methodology allows for further straightforward surface functionalization of the prepared $Au@SiO_2$ or $Ag@SiO_2$ by posttreatment with (3-aminopropyl) triethoxysilane (APTES) and succinic anhydride.

MWs were also evaluated for the silanization of iron oxide NPs. The same year Stutz et al. [147] and Park et al. [148] proposed the use of the Stöber MW-assisted method developed by Suzuki's group for the formation of a silica shell around an NP. Stutz et al. reacted TEOS and ammonia with iron oxide NPs already obtained by the MW methodology. The surface functionalization of the $Fe_3O_4@SiO_2$ particle's surface with amino groups was realized in a secondary Stöber process with APTMS as a precursor. In contrast to many time-consuming procedures given in the literature, MW-assisted APTMS hydrolysis and condensation led in 3 min at 60°C to a functional coating on the surface of $Fe_3O_4@SiO_2$ particles. On the other hand, Park et al. proposed silica encapsulation based on a multistep protocol: (i) synthesis of NPs in an organic medium, (ii) followed by a phase transfer in water using tetramethylammonium hydroxide (TMAOH) as a transfer agent, and then (iii) MW-assisted silica encapsulation using TEOS as a silanization agent in the presence of ammonia. They demonstrated that uniform and spherical silica encapsulation of magnetic NPs was achieved within 10 min via MW irradiation after phase transfer. In addition, they showed that silica shell addition through MW synthesis was more effective than conventional heating methods.

More recently, the formation of a silica shell on other types of inorganic nanomaterials was described. In Posati et al.'s work [149], ZnAl-based hydrotalcite NPs (ZnAl-HTlc NPs) were covalently modified by an organic oligothiophene fluorescent compound (T4Si) by using direct MW-assisted silylation. Morphological and optical characterization proved that the MW-assisted method enables efficient grafting of the target fluorescent dye on the NPs' surface in a few minutes. The dye loading can be tuned by changing the MW irradiation time, suggesting the possibility of multifunctionalization. On the other side, Massaro et al. [150] proposed MW-assisted halloysite nanotube (HNT) external surface functionalization using 3-mercaptopropyltrimethoxysilane. Halloysite is a naturally occurring mineral nanoclay with a tubular shape that can be used in catalysis experiments. In their work Massaro et al. proposed the MW-assisted preparation of a halloysite-supported ionic-liquid-based Pd catalyst for Suzuki C–C coupling. The authors showed that they obtained better grafting yields in solvent-free MW conditions and that the tubular shape of the nanoclay was not lost upon grafting of the organosilane and the binding of the ionic liquid.

MW-assisted formation of a shell on an NP was also evaluated by several groups for the formation of a titania oxide shell. Once again Suzuki's group was the pioneer in this area, describing the synthesis of TiO_2-coated ZnO nanocomposite particles by the MW irradiation method [151]. The reaction using titanium tetra-*n*-butoxide as a chemical agent proceeded rapidly under MW irradiation compared to complicated and time-consuming conventional protocols. ZnO NPs were coated with 6–10 nm amorphous TiO_2 layers, and compared to uncoated ZnO NPs, they showed enhanced UV emission. Recently Kim et al. [152] proposed a similar protocol to prepare Au@TiO_2 core–shell NPs by MW-assisted hydrothermal synthesis using TiF_4. A TiO_2 shell layer with 60 ± 10 nm shell thickness was deposited on Au NPs by this methodology. Au NPs were red-shifted from 521 nm to 565 nm after TiO_2 shell formation.

11.3.3.2 MW formation of a metallic shell on inorganic nanoparticles

MW irradiation is also a convenient method for metallic core–shell NP synthesis. Many studies have been done in this field because of the increasing interest in these hybrid structures. These colloids are composed by the interaction of different crystalline materials, which generates collective effects. Their physical and chemical properties are modified, and the originality of these nanocrystals can be used as an enhancement for their application. MW heating allows for high precision and high efficiency, which helps control not only shell thickness but also shell crystallinity. Moreover the MW oven apparatus is suitable for hydrothermal synthesis, defined as a method of synthesis of single crystals that depends on the solubility of minerals in hot water under high pressure. Wu et al. proposed a relevant method for noble metal core–shell NPs by MW hydrothermal annealing [153]. These Au@Ag NPs were synthesized in water, and the Ag shell growth was carried out under drastic temperature and pressure conditions. It was showed that alloy interdiffusion occurred during the shell growth by UV-visible spectroscopy monitoring and discrete dipole approximation modeling. A further study [154] proposed a modified synthetic route using a $[AuBr_4]^-$ precursor instead of $[AuCl_4]^-$, allowing an accurate control of the shell thickness and alloy uniformity (Scheme 11.4).

The formation of a silver shell was also proposed by Zhi et al. in order to coat copper NPs [155] by an MW alcohol reduction process. These nanocrystals are known to be sensitive to oxidation under mild conditions; however, the silver shell improved the oxidation resistance of the material by physical isolation of oxygen and electronic interactions between Cu and Ag. Garcia et al. [156] have developed MW-assisted synthesis of Au@Rh NPs in ethylene glycol, which is the most efficient solvent for conversion of MW energy into heat. This process facilitates the isotropic overgrowth of Rh on Au seeds and allows a shell thickness control of about two or four monolayers. It has been demonstrated that the catalytic activity of this hybrid structure is enhanced compared to pure Rh NPs.

Liu et al. have elaborated a multilayer structure based on a magnetite core and a carbon–gold shell as a new platform for

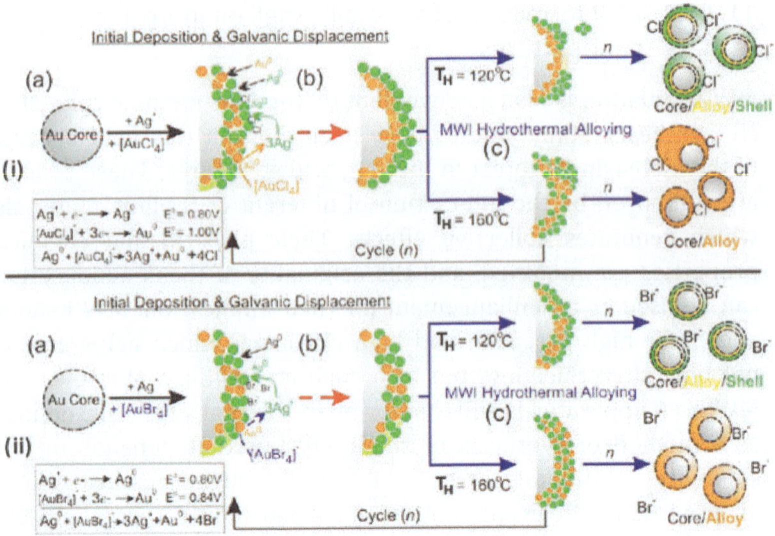

Scheme 11.4 Proposed core/alloy growth mechanisms using (i) $[AuCl_4]^-$ and (ii) $[AuBr_4]^-$ gold precursors to a presynthesized Au NP core. (a) The Ag+ and Au^{3+} precursors are added to a presynthesized Au NP core in the presence of a weak reducing agent. (b) In the moments before heating, reduction of the ions is possible, which will result in galvanic interactions occurring. Due to the standard reduction values (E_0) of the precursors, such redox reactions will be much more favorable using the $[AuCl_4]^-$ precursors ($\Delta E_0 = +0.20$ V) than the $[AuBr_4]^-$ ($\Delta E_0 = +0.04$ V). This results in a more segregated Au–Ag alloy shell using (i) $[AuCl_4]^-$, which leads to higher interactions with Cl^- ions in solution, and uncontrolled growth during heating. In contrast, when $[AuBr_4]^-$ is used, a more homogeneous Au–Ag alloy is deposited, which undergoes further annealing during heating. This improved alloy distribution and less interaction with Br^- ions lead to much improved alloy shell growth. Reprinted with permission from Ref. [154], Copyright 2013 American Chemical Society.

biosensing [157]. The deposition of carbon on magnetite was done under MW hydrothermal conditions with glucose as a precursor. Another original approach for magnetic core–shell NPs has been realized by Fashedemi et al. consisting of MW-assisted top-down nanosizing in ethylene glycol [158]. The supported Pd catalyst showed excellent electrochemical and electrocatalysis in an alkaline medium.

Figure 11.4 Typical SEM images of as-prepared (a) pure ZnO NRs and ZnO/ZnS core–shell NRs obtained via MW irradiation with different concentrations of TAA: (b) 0.0025 M, (c) 0.0075 M, (d) 0.0125 M, (e) 0.0175 M, and (f) 0.0225 M. Reproduced from Ref. [160], Copyright (2011), with permission of the Royal Society of Chemistry.

The synthesis of nanocomposites that are both luminescent and magnetic has been published by Zedan et al. [159]. They proved that the irradiation time has a considerable influence on the luminescent shell thickness. Purely luminescent ZnO@ZnS core–shell nanorods (NRs) were prepared with an MW refluxing system by Hu et al. [160]. The shell growth occurs with the surface sulfidation of ZnO NPs in aqueous media (Fig. 11.4). These hybrid nanomaterials showed enhanced photocatalytic activity concerning the decomposition of rhodamine B for a characteristic thickness of the ZnS shell. In the same field, Soltani et al. proposed the synthesis of ZnS@CdS and CdS@ZnS core–shell NPs [161]. Improvements in photocatalytic degradation of methylene were also observed.

11.3.3.3 MW formation of an organic polymer shell on inorganic nanoparticles

The functionalization of inorganic NPs with an organic polymeric shell is a promising approach for the design of composite materials with numerous properties, including thermal, mechanical, electronic, optical, and optoelectronic properties. Such organic–inorganic hybrid nanomaterials have been the subject of extensive study recently [162], especially because of the possibility to tune their properties according to the compositions of core or shell materials. MWs are already well described for the synthesis of polymers, and so several authors have studied their use in the direct polymerization at the surface of NPs with various polymer compositions and via direct or indirect polymer shell formation (Scheme 11.5).

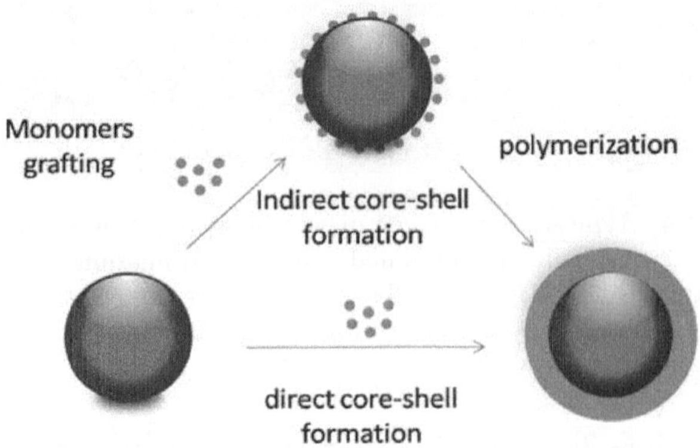

Scheme 11.5 Schematic of direct and indirect polymerization at the NP's surface.

Luo et al. [163] were the first to describe the formation of a polymeric shell on NPs under MW assistance. They developed a combined procedure of sol–gel and MW-assisted emulsion polymerization to prepare TiO_2/polystyrene core–shell nanospheres. The reaction consists in adding various amounts of a styrene monomer to the TiO_2 NPs in the presence of dodecyl sulfonate and ammonium persulfate. The obtained NPs have an effective core–shell structure and smooth surfaces, with the shell diameter depending on the initial monomer concentration. In 2009 Liebscher and coworkers [164] studied tin-catalyzed grafting polymerization of g-caprolactone onto prefunctional magnetite NPs under MW solvent-free conditions. Formation of a polycaprolactone shell around the magnetite core was observed with an increase in NP diameter (from 13–16 to 22–27 nm). The MW method provided much shorter reaction times (maximum 7 min instead of 5 h at 130°C as reported in the literature), and the obtained core–shell NPs retained their superparamagnetic behavior, still having relatively high-magnetization values. A few years later the same group applied the same grafting methodology to the preparation of magnetic core–shell NPs covered by polylactic acid [165]. This was obtained by solvent-free MW-assisted surface-initiated ring-opening polymerization of D,L-lactide. The magnetic NPs presented a high colloidal stability in water and retention of their magnetic properties. In 2009 Zhang et al. [166] described the preparation of atrazine (template molecule) magnetic molecularly imprinted polymer beads by MW-assisted suspension polymerization of PEG-coated NPs and methacrylic acid as a monomer. The time of the polymerization was dramatically shortened by using MW heating for the polymerization, which was less than 1/10th of that required by conventional heating. A few years later Chien et al. [167] utilized a similar methodology to prepare poly(acrylic)/SiO_2–TiO_2 core–shell NP hybrid thin films. They were successfully synthesized by MW-assisted grafting on the SiO_2–TiO_2 surface of a 3-(trimethoxysilyl) propylmethacrylate (MSMA) monomer and subsequent polymerization with various acrylic monomers. MW-assisted formation of core–shell NPs can also serve for the synthesis of microspheres from NPs. For example, You et al. [168] proposed an MW hydrothermal approach to synthesize magnetic Fe_3O_4/phenol–formaldehyde (PF) core–shell microspheres. They realized in situ polycondensation of phenol and formaldehyde with magnetic Fe_3O_4 clusters as the seeds in an aqueous solution without any surfactants. The amount of phenol and formaldehyde, the reaction temperature, and the MW heating time were studied and played a significant role in the shell thickness and size of the Fe_3O_4/PF microspheres. Fe_3O_4 clusters could be effectively coated within 10 min and the shell thickness finely controlled from 10 to 200 nm. Finally a last example of the formation of a polymer shell on NPs was reported recently by Wang et al. [169] for the formation of a dendrimer on magnetic NPs for the immobilization of bovine serum albumin (BSA) to develop a novel chiral magnetic nanoselector for direct separation of racemates. This nanomaterial was obtained by MW-assisted multistep synthesis. Starting from $Fe_3O_4@SiO_2$ core–shell NPs APTES was first grafted under MWs on the silica surface to create amine functionalities. Polyamidoamine (PAMAM) dendrimer-modified NPs were prepared by repeating Michael addition and amidation reactions under MW irradiation condition at the surface of the amino-functionalized $Fe_3O_4@SiO_2$ NPs. The effects of temperature and time, as well as the percentage of MNPs, on growing PAMAM onto the $Fe_3O_4@SiO_2$ NPs were investigated. The result of these experiments indicated that third-generation PAMAM-modified magnetite NPs could be obtained within 36 h using MW irradiation compared to the traditional heating methodology that has been reported to require as many as 18–30 days.

11.3.3.4 MW functionalization of small molecules at the inorganic nanoparticle surface

The first use of MWs for inorganic NP surface modifications was for the facile functionalization of gold NPs via MW-assisted 1,3 dipolar cycloaddition described by Sommer and Weck [170]. A library of alkyne substrates was successfully reacted with gold NPs bearing azide functions on their surfaces (Scheme 11.6). Unfortunately, the authors didn't give the results with non-MW-activated conditions

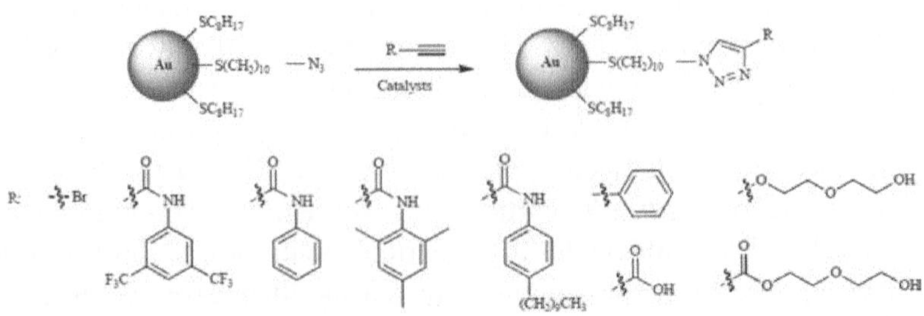

Scheme 11.6 Sommer's MW-assisted 1,3 dipolar cycloaddition [170] at the surface of gold NPs. Reprinted from Ref. [12] by permission of Eureka Science Ltd. Copyright © 2013, Eureka Science Ltd.

to compare with their MW results. In the study, the authors have shown that there were no noticeable changes on NP morphology but decomposition through particle aggregation occurred when the treatment exceeded a period of 15 min.

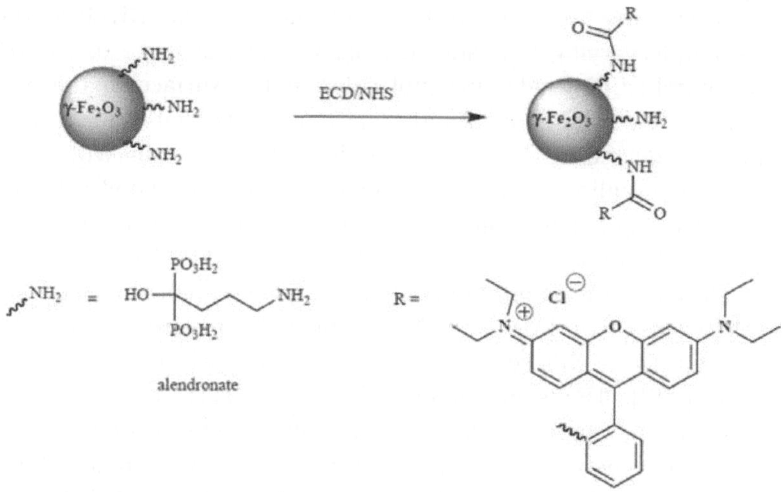

Scheme 11.7 Multimodal nanoplatform [170]. Reprinted from Ref. [12] by permission of Eureka Science Ltd. Copyright © 2013, Eureka Science Ltd.

More recently MW techniques were used to synthesize a multimodal nanoplatform with dual-imaging properties [171]. Maghemite (γFe_2O_3) nanocrystals, with an average diameter of 10 nm, were prefunctionalized with an antitumoral molecule having a bifunctional nature: alendronate or 4-amino-1-hydroxy-1-phosphonobutyl phosphonic acid (HMBP-$(CH_2)_3$–NH_3^+). Then, rhodamine B, a fluorophore that served as a model to study surface immobilization efficiencies, was used to study functionalization (Scheme 11.7). Nevertheless, the important aromatic structure of this fluorophore led to poor functionalization, yielding only around 10 molecules per NP in classical conditions [172]. To increase the conjugation of this molecule by carbodiimide coupling, the potential of the MW technique was studied and compared to grafting under the classical heating technique. With the MW methodology, when the rhodamine excess per amine function increased, the grafting rate gradually increased and reached a maximal rate around 600 molecules per NP (yield coupling around 50%), whereas without MW input, saturation was rapidly reached corresponding to an average of 12 molecules per nanocrystal, whatever the excess of rhodamine B used. Thus, the MW technology allowed increasing approximately by 50-fold the number of molecules per NP. Moreover, by simply varying the ratio of reagents, it is possible to control the number of rhodamine molecules at the surface of the NPs under MW irradiation. Another advantage is that the reaction time is decreased from 24 h to less than 20 min with MW-assisted coupling. In addition, similar morphology, good colloidal stability, and superparamagnetic behavior were conserved for the magnetofluorescent nanocrystals obtained with MW treatment.

Recently, the elaboration of new L-proline-functionalized superparamagnetic iron oxide NPs using MW functionalization was studied. These NPs could act as a new enantioselective organocatalyst [173]. The same nanoplatform was used to graft Boc-L-proline and Fmoc-L-proline followed by a deprotection step. A comparison between the grafting with and without the input of MWs was realized. In the case of Fmoc-protected proline, MW-assisted coupling was far more efficient for the grafting (a 30-fold increase) without tampering with the NP structure. This evaluation of the coupling efficiency was allowed due to the fluorenyl released during the deprotection reaction, which was quantified by UV spectroscopy. More recently MW-assisted carbodiimide coupling at the surface of NPs was extended to another nanoplatform consisting of iron oxide NPs, which were surface-functionalized with caffeic acid (CA), a catechol derivative [174]. This nanoplatform served to elaborate a dual-imaging magnetic resonance/fluorescence contrast agent targeting $\alpha v\beta 3$ integrins. The effect of the multivalency for the interactions between the cRGD ligand and the target $\alpha v\beta 3$ receptor was studied by means of surface plasmon resonance (SPR) and a solid-phase receptor-binding assay.

The potency of MW surface functionalization of iron oxide NPs using a clickable superparamagnetic nanoplatform was also demonstrated [175]. In this case, the nanoplatform was obtained by coating γFe_2O_3 NPs, with a new bifunctional molecule (1-hydroxy-1-phosphonopentynyl) phosphonic acid (HMBPyne) bearing an alkyne group. Coupling of several azide-functionalized molecules via 1,3-cycloaddition was evaluated with and without the input of MW energy (Scheme 11.8). Different techniques such as X-ray diffraction

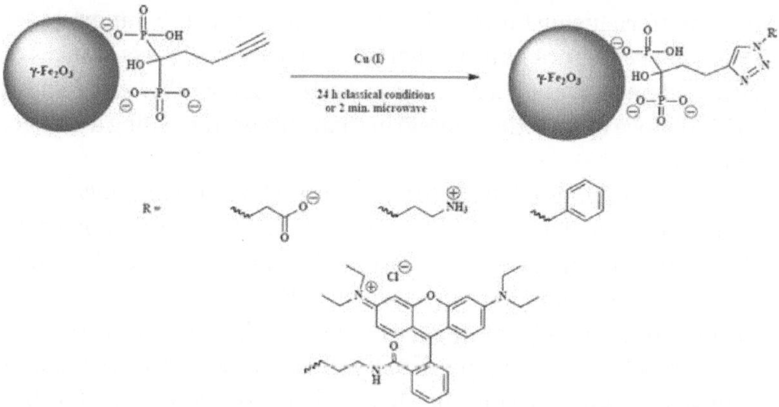

Scheme 11.8 Functionalization of iron oxide clickable nanoplatform [175]. Reprinted from Ref. [12] by permission of Eureka Science Ltd. Copyright © 2013, Eureka Science Ltd.

(XRD) spectra, transmission electron microscopy (TEM) diameter, surface charge (zeta potential), z-average hydrodynamic radius, superparamagnetic behavior, and infrared (IR) spectra were used before and after the MW treatment to check that MW conditions did not alter the NPs structure. No evidence of modification of the NPs' integrity could be found. Once again rhodamine B was used as a model molecule to prove that the coupling efficiency could be greatly enhanced using MW-assisted reaction with an increase of 40-fold of functionalization. The reaction was also extended to functional molecules, demonstrating the chemoselectivity of the process.

More recently another study has evaluated the use of MW irradiation for two classical coupling methods with NPs, click chemistry and carbodiimide chemistry [176]. The nanoplatforms were modified with rhodamine B (Rh-N3) when using the CuAAC methodology for click chemistry and 123-rhodamine (Rh-123) when using carbodiimide chemistry. Once again, this study has shown that using MWs the coupling is more efficient, with an increase of 4- to 10-fold in functionalization in comparison to the reaction without MWs. Then, to introduce $\alpha_v\beta_3$ targeting, the NPs were coupled to a cRGD motif. The peptide c(RGDfK) was chosen to be linked onto the carboxylated nanoplatform via the lysine residue, and an azido cRGD peptide was chosen to be linked onto the alkyne nanoplatform using CuAAC. Using a binding assay the importance of the enhancement of grafting density allowed by MW assistance was clearly demonstrated, as RGD-grafted NPs using MWs were shown to bind more efficiently to integrin than NPs grafted using a classical reaction.

A last example of the interest in the MW methodology can be found in the article by Toulemon et al. [177], who demonstrated very efficient and very fast assembling of iron oxide NPs on a self-assembled monolayer (SAM) by MWs. Irradiation with MWs promoted a significant improvement in the assembly reaction, which reached a maximum density of NPs after 1 h of reaction. In comparison with the conventional click chemistry reaction (without MWs), it would require up to 48 h to reach this performance. On the other hand, the MW-assisted click reaction has the advantage of preserving the specific reactions between alkyne groups (on the surface of SAMs) and azide groups (on the surface of NPs). Moreover this technique allows avoiding undesired reactions and side products, which limits purification processes.

11.3.4 *MW Functionalization of the Plasmonic Nanostructure*

Since a few years plasmonic applications have emerged as one of the most promising developments of nanotechnology. Plasmonic materials led to the production of several commercially viable technologies such as SPR, surface-enhanced Raman scattering (SERS), and surface plasmon fluorescence spectroscopy (SPFS), with increasing applications in biomedical and sensors domains. Hence the effective modification of these inorganic nanostructures is of crucial importance for their numerous applications, and so MW-assisted grafting of molecules onto these surfaces can be of great interest.

To our knowledge only three reports can be found in the literature related to MW surface modification of plasmonic nanostructures. The first one was described by Abel and Aslan [178] for the modification of silver NP films deposited onto glass slides, with oligonucleotides, using low-power MW heating. To assess the efficacy of the surface modification technique, identical assays were carried out with or without MW heating (conventional method). A close match in fluorescence emission measured from samples prepared using 10 s MW heating and the conventional method carried out for 24 h clearly indicates the benefit of the process. A year later Grell et al. studied the formation of SAMs of alkanethiols on gold thin films using MW heating [179]. SAMs of two model alkanethiols—11-mercaptoundecanoic acid (11-MUDA, to generate a hydrophilic surface) and undecanethiol (UDET, a hydrophobic surface)—were prepared in a semicontinuous or continuous fashion. Both SAMs were efficiently formed with a significant decrease in the amount of time for surface modification of gold thin films with SAMs from 24 h (conventional technique) to <10 min. Very recently, Vance and Sandros [180] described the preparation of an NP-amplified SPR imaging (SPRi) aptasensor for the detection of C-reactive protein (CRP) in serum. They proposed a 10 min MW-assisted methodology for the conventional immobilization procedures using cystamine/glutaraldehyde layers that normally require more than 3 h. MW-assisted surface functionalization not only decreased the functionalization procedure from several hours to minutes but also improved the detection of CRP and avidity using SPRi.

11.4 Conclusion

The use of MWs is becoming more and more widespread in all areas of chemistry. An MW apparatus can now be found in most modern laboratories, and so MWs represent one of the methodologies that have had the most important impact on chemical synthesis over the course of the past decades. Surface chemistry has also been impacted by the emergence of MW methodologies, but it has not been described evenly for all the different types of surfaces. In some area, for example, for silicon-based surfaces, MW irradiation was much more used than for other types of surfaces.

With the increasing importance of nanomaterials science and the great variety of possible targeted applications, there is a growing interest in modifying these surfaces. With the previous works done on organic and surface chemistry, MWs clearly appear as a method of choice for rapid and efficient functionalization of nanomaterial surfaces. Strangely there are really few reports on the use of MWs for the modification of nanomaterial surfaces before the past decades, but in the past couple of years there has been a growing interest in this field for various types of materials. In most of the cases the authors reported two major improvements brought by the use of MWs over classical conditions: a better functionalization yield and a lower reaction time. Moreover, in most of these reports under MW irradiation, authors indicated the nondegradability of the material as well as retention of its properties. Nevertheless, very few reports explained the obtaining of improved results. Better and direct *in-core* heating was often put forward by some of the authors, especially for conductive surfaces. But it is likely that the enhancements in grafting yields when employing MW effects could be mostly explained by thermal effects due to the rapid and direct heating compared to classical conditions. In addition when comparisons were given they are often made with literature methodologies, and so concentrations, vessel sizes, etc., could be different and led to irrelevant conclusions. So to properly compare the results extreme care must be taken to reproduce comparable conditions.

In conclusion MW functionalization of materials appears as an important improvement compared to classical methodologies with relatively few examples concluding on the absence of benefits from MWs. MWs are becoming an interesting tool for the precise functionalization and increasing development of nanomaterials. Moreover, they often permit us to come closer to greener functionalization solutions by reducing the solvent consumption or the use of aggressive reactants.

References

1. Gedye, R., Smith, F., Westaway, K., Ali, H., Baldisera, L., Laberge, L., and Rousell, J. (1986). The use of microwave ovens for rapid organic synthesis, *Tetrahedron Lett.*, **27**, pp. 279–282.

2. Caddick, S. (1995). Microwave assisted organic reactions, *Tetrahedron,* **51**, pp. 10403–10432.
3. Caddick, S., and Fitzmaurice, R. (2009). Microwave enhanced synthesis, *Tetrahedron,* **65**, pp. 3325–3355.
4. Kappe, C.O. (2004). Controlled microwave heating in modern organic synthesis, *Angew. Chem., Int. Ed.,* **43**, pp. 6250–6284.
5. Lidstrom, P., Tierney, J., Wathey, B., and Westman, J. (2001). Microwave assisted organic synthesis: a review, *Tetrahedron,* **57**, pp. 9225–9283.
6. Perreux, L., and Loupy, A. (2001). A tentative rationalization of microwave effects in organic synthesis according to the reaction medium, and mechanistic considerations, *Tetrahedron,* **57**, pp. 9199–9223.
7. Bardts, M., Gonsior, N., and Ritter, H. (2008). Polymer synthesis and modification by use of microwaves, *Macromol. Chem. Phys.,* **209**, pp. 25–31.
8. Bouillon, C., Meyer, A., Vidal, S., Jochum, A., Chevolot, Y., Cloarec, J.-P., Praly, J.-P., Vasseur, J.-J., and Morvan, F. (2006). Microwave assisted "click" chemistry for the synthesis of multiple labeled-carbohydrate oligonucleotides on solid support, *J. Org. Chem.,* **71**, pp. 4700–4702.
9. Colombo, M., and Peretto, I. (2008). Chemistry strategies in early drug discovery: an overview of recent trends, *Drug Discov. Today,* **13**, pp. 677–684.
10. Guénin, E., and Meziane, D. (2011). Microwave assisted phosphorus organic chemistry: a review, *Curr.Org. Chem.,* **15**, pp. 3465–3485.
11. Larhed, M., Moberg, C., and Hallberg, A. (2002). Microwave-accelerated homogeneous catalysis in organic chemistry, *Acc. Chem. Res.,* **35**, pp. 717-727.
12. Nehlig, E., Milosevic, I., Motte, L., and Guenin, E. (2013). Microwave assisted functionalization of materials: from the mesoscale to the nanoscale, *Curr. Org. Chem.,* **17**, pp. 528–544.
13. Pedersen, S. L., Tofteng, A. P., Malik, L., and Jensen, K. J. (2012). Microwave heating in solid-phase peptide synthesis, *Chem. Soc. Rev.,* **41**, pp. 1826–1844.
14. Al-Harahsheh, M., and Kingman, S. W. (2004). Microwave-assisted leaching: a review, *Hydrometallurgy,* **73**, pp. 189–203.
15. Jones, D. A., Lelyveld, T. P., Mavrofidis, S. D., Kingman, S. W., and Miles, N. J. (2002). Microwave heating applications in environmental engineering: a review, *Resour. Conserv. Recy.,* **34**, pp. 75–90.

16. Li, Y., and Yang, W. (2008). Microwave synthesis of zeolite membranes: a review, *J. Memb. Sci.*, **316**, pp. 3–17.
17. MacKenzie, K., Dunens, O., and Harris, A. T. (2009). A review of carbon nanotube purification by microwave assisted acid digestion, *Sep. Purif. Technol.*, **66**, pp. 209–222.
18. Roy, I., and Gupta, M. N. (2003). Applications of microwaves in biological sciences, *Curr. Sci.*, **85**, pp. 1685–1693.
19. Sun, W., Gao, S., Wang, L., Chen, Y., Wu, S., Wang, X., Zheng, D., and Gao, Y. (2006). Microwave-assisted protein preparation and enzymatic digestion in proteomics, *Mol. Cell. Proteomics*, **5**, pp. 769–776.
20. Morschhäuser, R., Krull, M., Kayser, C., Boberski, C., Bierbaum, R., Püschner Peter, A., Glasnov Toma, N., and Kappe, C. O. (2012). Microwave-assisted continuous flow synthesis on industrial scale, **1**, pp. 281–290.
21. Baghbanzadeh, M., Carbone, L., Cozzoli, P. D., and Kappe, C. O. (2011). Microwave-assisted synthesis of colloidal inorganic nanocrystals, *Angew. Chem., Int. Ed.*, **50**, pp. 11312–11359.
22. Bilecka, I., and Niederberger, M. (2010). Microwave chemistry for inorganic nanomaterials synthesis, *Nanoscale*, **2**, pp. 1358–1374.
23. Dahal, N., Garcia, S., Zhou, J., and Humphrey, S. M. (2012). Beneficial effects of microwave-assisted heating versus conventional heating in noble metal nanoparticle synthesis, *ACS Nano*, **6**, pp. 9433–9446.
24. Pan, L., Liu, X., Sun, Z., and Sun, C. Q. (2013). Nanophotocatalysts via microwave-assisted solution-phase synthesis for efficient photocatalysis, *J. Mater. Chem. A*, **1**, pp. 8299–8326.
25. Zhu, Y.-J., and Chen, F. (2014). Microwave-assisted preparation of inorganic nanostructures in liquid phase, *Chem. Rev.*, **114**, pp. 6462–6555.
26. Caponetti, E., Martino, D. C., Leone, M., Pedone, L., Saladino, M. L., and Vetri, V. (2006). Microwave-assisted synthesis of anhydrous CdS nanoparticles in a water/oil microemulsion, *J. Colloid Interface Sci.*, **304**, pp. 413–418.
27. Bilecka, I., Elser, P., and Niederberger, M. (2009). Kinetic and thermodynamic aspects in the microwave-assisted synthesis of ZnO nanoparticles in benzyl alcohol, *ACS Nano*, **3**, pp. 467–477.
28. Dallinger, D., Irfan, M., Suljanovic, A., and Kappe, C. O. (2010). An investigation of wall effects in microwave-assisted ring-closing metathesis and cyclotrimerization reactions, *J. Org. Chem.*, **75**, pp. 5278–5288.

29. Horikoshi, S., Abe, H., Sumi, T., Torigoe, K., Sakai, H., Serpone, N., and Abe, M. (2011). Microwave frequency effect in the formation of Au nanocolloids in polar and non-polar solvents, *Nanoscale,* **3**, pp. 1697–1702.
30. Bogdal, D., Penczek, P., Pielichowski, J., and Prociak, A. (2003). Microwave assisted synthesis, crosslinking, and processing of polymeric materials, *Adv. Polym. Sci.,* **163**, pp. 193–263.
31. Kempe, K., Becer, C. R., and Schubert, U. S. (2011). Microwave-assisted polymerizations: recent status and future perspectives, *Macromolecules,* **44**, pp. 5825–5842.
32. Sinnwell, S., and Ritter, H. (2007). Recent advances in microwave-assisted polymer synthesis, *Aust. J. Chem.,* **60**, pp. 729–743.
33. Sosnik, A., Gotelli, G., and Abraham, G. A. (2011). Microwave-assisted polymer synthesis (MAPS) as a tool in biomaterials science: how new and how powerful, *Prog. Polym. Sci.,* **36**, pp. 1050–1078.
34. Wiesbrock, F., Hoogenboom, R., and Schubert, U. S. (2004). Microwave-assisted polymer synthesis: state-of-the-art and future perspectives, *Macromol. Rapid Commun.,* **25**, pp. 1739–1764.
35. Blackwell, H. E. (2006). Hitting the SPOT: small-molecule macroarrays advance combinatorial synthesis, *Curr. Opin. Chem. Biol.,* **10**, pp. 203–212.
36. Kappe, C. O., and Stadler, A. (2006). Literature survey Part B: combinatorial chemistry and high-throughput organic synthesis. In *Microwaves in Organic and Medicinal Chemistry* (Wiley-VCH Verlag GmbH & Co. KGaA, Weinheim), pp. 291–391.
37. Lidstrom, P., Westman, J., and Lewis, A. (2002). Enhancement of combinatorial chemistry by microwave-assisted organic synthesis, *Comb. Chem. High T. Scr.,* **5**, pp. 441–458.
38. Bowman, M. D., Jeske, R. C., and Blackwell, H. E. (2004). Microwave-accelerated SPOT-synthesis on cellulose supports, *Org. Lett.,* **6**, pp. 2019–2022.
39. Dai, W.-M., Guo, D.-S., Sun, L.-P., and Huang, X.-H. (2003). Microwave-assisted solid-phase organic synthesis (MASPOS) as a key step for an indole library construction, *Org. Lett.,* **5**, pp. 2919–2922.
40. De Luca, L., Giacomelli, G., Porcheddu, A., Salaris, M., and Taddei, M. (2003). Cellulose beads: a new versatile solid support for microwave-assisted synthesis. Preparation of pyrazole and isoxazole libraries, *J. Comb. Chem.,* **5**, pp. 465–471.

41. Scharn, D., Wenschuh, H., Reineke, U., Schneider-Mergener, J., and Germeroth, L. (2000). Spatially addressed synthesis of amino- and amino-oxy-substituted 1,3,5-triazine arrays on polymeric membranes, *J. Comb. Chem.*, **2**, pp. 361–369.
42. Sun, L.-P., and Dai, W.-M. (2006). An Engineered linker capable of promoting on-resin reactions for microwave-assisted solid-phase organic synthesis, *Angew. Chem., Int. Ed.*, **45**, pp. 7255–7258.
43. Iyer, R. P., Coughlin, J. E., and Padmanabhan, S. (2005). Rapid functionalization and loadings of solid supports, *Org. Prep. Proced. Int.*, **37**, pp. 205–212.
44. Iyer, R. P., Padmanabhan, S., and Coughlin, J. E. (2006). Microwave-assisted functionalization of solid supports for rapid loading of nucleosides. In *Current Protocols in Nucleic Acid Chemistry* (John Wiley and Sons).
45. Lee, J. H., Hyun, H., Cross, C. J., Henary, M., Nasr, K. A., Oketokoun, R., Choi, H. S., and Frangioni, J. V. (2012). Rapid and facile microwave-assisted surface chemistry for functionalized microarray slides, *Adv. Funct. Mater.*, **22**, pp. 872–878.
46. Mahajan, S., Sethi, D., Seth, S., Kumar, A., Kumar, P., and Gupta, K. C. (2009). Construction of oligonucleotide microarrays (biochips) via thioether linkage for the detection of bacterial meningitis, *Bioconjugate Chem.*, **20**, pp. 1703–1710.
47. Padmanabhan, S., Coughlin, J. E., and Iyer, R. P. (2005). Microwave-assisted functionalization of solid supports: application in the rapid loading of nucleosides on controlled-pore-glass (CPG), *Tetrahedron Lett.*, **46**, pp. 343–347.
48. Anglin, E. J., Cheng, L., Freeman, W. R., and Sailor, M. J. (2008). Porous silicon in drug delivery devices and materials, *Adv. Drug Deliv. Rev.*, **60**, pp. 1266–1277.
49. Dhanekar, S., and Jain, S. (2013). Porous silicon biosensor: current status, *Biosens. Bioelectron.*, **41**, pp. 54–64.
50. Ge, M., Fang, X., Rong, J., and Zhou, C. (2013). Review of porous silicon preparation and its application for lithium-ion battery anodes, *Nanotechnology*, **24**, p. 422001.
51. Granitzer, P., and Rumpf, K. (2010). Porous silicon - a versatile host material, *Materials*, **3**, pp. 943–998.
52. Haidary, S. M., Corcoles, E. P., and Ali, N. K. (2012). Nanoporous silicon as drug delivery systems for cancer therapies, *J. Nanomater.*, **2012**, Article ID 830503 (15 pp).

53. Hoffmann, F., Cornelius, M., Morell, J., and Fröba, M. (2006). Silica-based mesoporous organic–inorganic hybrid materials, *Angew. Chem., Int. Ed.,* **45**, pp. 3216–3251.

54. Logar, N.Z., and Kaučič, V. (2006). Nanoporous materials: from catalysis and hydrogen storage to wastewater treatment, *Acta Chim. Slov.,* **53**, pp. 117–135.

55. Martin-Palma, R. J., Manso-Silvan, M., and Torres-Costa, V. (2010). Biomedical applications of nanostructured porous silicon: a review, *J. Nanophoton.,* **4**, pp. 042502-042502-20.

56. Tompsett, G. A., Conner, W. C., and Yngvesson, K. S. (2006). Microwave synthesis of nanoporous materials. *ChemPhysChem,* **7**, pp. 296–319.

57. de la Mora, M. B., Ocampo, M., Doti, R., Lugo, J. E., and Faubert, J. (2013). Porous silicon biosensors. In *State of the Art in Biosensors: General Aspects,* http://www.intechopen.com/books/state-of-the-art-in-biosensors-general-aspects/porous-silicon-biosensors.

58. Boukherroub, R., Petit, A., Loupy, A., Chazalviel, J.-N., and Ozanam, F. (2003). Microwave-assisted chemical functionalization of hydrogen-terminated porous silicon surfaces, *J. Phys. Chem. B,* **107**, pp. 13459–13462.

59. Petit, A., Delmotte, M., Loupy, A., Chazalviel, J.-N., Ozanam, F., and Boukherroub, R. (2008). Microwave effects on chemical functionalization of hydrogen-terminated porous silicon nanostructures, *J. Phys. Chem. C,* **112**, pp. 16622–16628.

60. Guo, D.-J., Xiao, S.-J., Xia, B., Wei, S., Pei, J., Pan, Y., You, X.-Z., Gu, Z.-Z., and Lu, Z. (2005). Reaction of porous silicon with both end-functionalized organic compounds bearing a-bromo and w-carboxy groups for immobilization of biomolecules, *J. Phys. Chem. B,* **109**, pp. 20620–20628.

61. Wang, J., Guo, D.-J., Xia, B., Chao, J., and Xiao, S.-J. (2007). Preparation of organic monolayers with azide on porous silicon via S-N bonds, *Colloids Surf., A,* **305**, pp. 66–75.

62. Guo, Y., Zhou, J., Song, Y., and Zhang, L. (2009). An efficient and environmentally friendly method for the synthesis of cellulose carbamate by microwave heating, *Macromol. Rapid Commun.,* **30**, pp. 1504–1508.

63. Liu, X., Liu, H.-B., Guo, P.-F., and Xiao, S.-J. (2011). Construction of multiple generation nitriloacetates from poly(PEGMA) brushes on planar silicon surface for enhancement of protein loading, *Phys. Status Solidi A,* **208**, pp. 1462–1470.

64. Liu, X., Zheng, H.-N., Ma, Y.-Z., Yan, Q., and Xiao, S.-J. (2011). Microwave irradiated click reactions on silicon surfaces via derivertization of covalently grafted poly(PEGMA) brushes, *J. Colloid Interface Sci.*, **358**, pp. 116–122.

65. Sun, W., Qian, C., Mastronardi, M. L., Wei, M., and Ozin, G. A. (2013). Hydrosilylation kinetics of silicon nanocrystals, *Chem. Commun.*, **49**, pp. 11361–11363.

66. van den Berg, S. A., Alonso, J. M., Wadhwa, K., Franssen, M. C. R., Wennekes, T., and Zuilhof, H. (2014). Microwave-assisted formation of organic monolayers from 1-alkenes on silicon carbide, *Langmuir*, **30**, pp. 10562–10565.

67. Feng, Z., Li, Y., Niu, D., Li, L., Zhao, W., Chen, H., Li, L., Gao, J., Ruan, M., and Shi, J. (2008). A facile route to hollow nanospheres of mesoporous silica with tunable size, *Chem. Commun.*, pp. 2629–2631.

68. Cao, Y., Wei, H.-J., and Xia, Z.-N. (2009). Advances in microwave assisted synthesis of ordered mesoporous materials, *Trans. Nonferrous Met. Soc. China*, **19**, pp. s656–s664.

69. Tian, B., Liu, X., Yang, H., Xie, S., Yu, C., Tu, B., and Zhao, D. (2003). General synthesis of ordered crystallized metal oxide nanoarrays replicated by microwave-digested mesoporous silica, *Adv. Mater.*, **15**, pp. 1370–1374.

70. Pirouzmand, M., Amini, M. M., and Safari, N. (2007). Microwave-assisted immobilization of metallophthalocyanines in mesoporous channels of MCM-41, *J. Porphyrins Phthalocyanines*, **11**, pp. 118–124.

71. Procopio, A., De Luca, G., Nardi, M., Oliverio, M., and Paonessa, R. (2009). General MW-assisted grafting of MCM-41: study of the dependence on time dielectric heating and solvent, *Green Chem.*, **11**, pp. 770–773.

72. Oliverio, M., Procopio, A., Glasnov, T. N., Goessler, W., and Kappe, C. O. (2011). Microwave-assisted grafting to MCM-41 silica and its application as catalyst in flow chemistry, *Aust. J. Chem.*, **64**, pp. 1522–1529.

73. Pasqua, L., Procopio, A., Oliverio, M., Paonessa, R., Prete, R., Nardi, M., Casula, M., Testa, F., and Nagy, J. (2013). Hybrid MCM-41 grafted by a general microwave-assisted procedure: a characterization study, *J. Porous Mat.*, **20**, pp. 865–873.

74. Fukaya, N., Yamashita, H., Haga, H., Tsuchimoto, T., Onozawa, S.-Y., Sakakura, T., and Yasuda, H. (2010). Microwave-assisted organic functionalization of silica surfaces: effect of selectively heating silylating agents, *J. Organomet. Chem.*, **696**, pp. 825–828.

75. Gonzalez, M. D., Cesteros, Y., Llorca, J., and Salagre, P. (2012). Boosted selectivity toward high glycerol tertiary butyl ethers by microwave-assisted sulfonic acid-functionalization of SBA-15 and beta zeolite, *J. Catal.*, **290**, pp. 202–209.

76. Gonzalez, M. D., Salagre, P., Taboada, E., Llorca, J., and Cesteros, Y. (2013). Microwave-assisted synthesis of sulfonic acid-functionalized microporous materials for the catalytic etherification of glycerol with isobutene, *Green Chem.*, **15**, pp. 2230–2239.

77. Gonzàlez, M. D., Salagre, P., Taboada, E., Llorca, J., Molins, E., and Cesteros, Y. (2013). Sulfonic acid-functionalized aerogels as high resistant to deactivation catalysts for the etherification of glycerol with isobutene, *Appl. Catal. B*, **136–137**, pp. 287–293.

78. Lourenco, M. A. O., Siegel, R., Mafra, L., and Ferreira, P. (2013). Microwave assisted N-alkylation of amine functionalized crystal-like mesoporous phenylene-silica, *Dalton Trans.*, **42**, pp. 5631–5634.

79. Chen, F., Zhang, F., Du, F., Wang, A., Gao, W., Wang, Q., Yin, X., and Xie, T. (2012). A novel and efficient method for the immobilization of thermolysin using sodium chloride salting-in and consecutive microwave irradiation, *Bioresour. Technol.*, **115**, pp. 158–163.

80. Du, Z., Wang, A., Zhou, C., Zhu, S., and Shen, S. (2009). Microwave-assisted lipase covalent immobilization and improved its thermal stability using alcohols to quench excessive activated groups on support surface, *J. Chem. Eng. Jpn.*, **42**, pp. 441–446.

81. Plagemann, R., von Langermann, J., and Kragl, U. (2014). Microwave-assisted covalent immobilization of enzymes on inorganic surfaces, *Eng. Life Sci.*, **14**, pp. 493–499.

82. Wang, A., Wang, M., Wang, Q., Chen, F., Zhang, F., Li, H., Zeng, Z., and Xie, T. (2010). Stable and efficient immobilization technique of aldolase under consecutive microwave irradiation at low temperature, *Bioresour. Technol.*, **102**, pp. 469–474.

83. Wang, A., Zhou, C., Du, Z., Liu, M., Zhu, S., Shen, S., and Ouyang, P. (2009). Enhancement of microwave-assisted covalent immobilization of penicillin acylase using macromolecular crowding and glycine quenching, *J. Biosci. Bioeng.*, **107**, pp. 219–224.

84. Rodrigues, J. J., Fernandes, F. A. N., and Rodrigues, M. G. F. (2013). Study of Co/SBA-15 catalysts prepared by microwave and conventional heating methods and application in Fischer Tropsch synthesis, *Appl. Catal. A.*, **468**, pp. 32–37.

85. Fuku, K., Hayashi, R., Takakura, S., Kamegawa, T., Mori, K., and Yamashita, H. (2013). The synthesis of size- and color-controlled

silver nanoparticles by using microwave heating and their enhanced catalytic activity by localized surface plasmon resonance, *Angew. Chem., Int. Ed.*, **125**, pp. 7594–7598.

86. Qian, X., Kuwahara, Y., Mori, K., and Yamashita, H. (2014). Silver nanoparticles supported on CeO_2-SBA-15 by microwave irradiation possess metal–support interactions and enhanced catalytic activity, *Chem. Eur. J.*, **20**, pp. 15746–15752.
87. Vazquez, E., and Prato, M. (2009). Carbon nanotubes and microwaves: interactions, responses, and applications, *ACS Nano*, **3**, pp. 3819–3824.
88. Economopoulos, S. P., Karousis, N., Rotas, G., Pagona, G., and Tagmatarchis, N. (2011). Microwave-assisted functionalization of carbon nanostructured materials, *Curr. Org. Chem.*, **15**, pp. 1121–1132.
89. Kubota, K., Sano, M., and Masuko, T. (2005). Microwave irradiation for chemical modification of carbon nanotubes for better dispersion, *Jpn. J. Appl. Phys.*, **44**, pp. 465–468.
90. Wang, Y., Iqbal, Z., and Mitra, S. (2005). Microwave-induced rapid chemical functionalization of single-walled carbon nanotubes, *Carbon*, **43**, pp. 1015–1020.
91. Yoshida, S., and Sano, M. (2006). Microwave-assisted chemical modification of carbon nanohorns: oxidation and Pt deposition, *Chem. Phys. Lett.*, **433**, pp. 97–100.
92. Kumar, P., Park, J.-S., Randhawa, P., Sharma, S., Shin, M.-S., and Sekhon, S. (2011). A study on the effect of different chemical routes on functionalization of MWCNTs by various groups (–COOH, $-SO_3H$, $-PO_3H_2$), *Nanoscale Res. Lett.*, **6**, p. 583.
93. Colomer, J. F., Marega, R., Traboulsi, H., Meneghetti, M., Van Tendeloo, G., and Bonifazi, D. (2009). Microwave-assisted bromination of double-walled carbon nanotubes, *Chem. Mater.*, **21**, pp. 4747–4749.
94. Lee, K. H., Oh, J., Son, J. G., Kim, H., and Lee, S.-S. (2014). Nitrogen-doped graphene nanosheets from bulk graphite using microwave irradiation, *ACS Appl. Mater. Interfaces*, **6**, pp. 6361–6368.
95. Mallakpour, S., and Zadehnazari, A. (2013). Effect of amino acid-functionalization on the interfacial adhesion and behavior of multi-walled carbon nanotubes/poly(amide-imide) nanocomposites containing thiazole side unit, *J. Polymer Res.*, **20**, pp. 1–12.
96. Mallakpour, S., and Zadehnazari, A. (2014a). A facile, efficient, and rapid covalent functionalization of multi-walled carbon nanotubes with natural amino acids under microwave irradiation, *Progr. Org. Coat.*, **77**, pp. 679–684.

97. Mallakpour, S., and Zadehnazari, A. (2014b). A convenient strategy to functionalize carbon nanotubes with ascorbic acid and its effect on the physical and thermomechanical properties of poly(amide-imide) composites, *J. Solid State Chem.*, **211**, pp. 136–145.

98. Mallakpour, S., and Zadehnazari, A. (2014c). Efficient functionalization of multi-walled carbon nanotubes with p-aminophenol and their application in the fabrication of poly(amide-imide)-matrix composites, *Polym. Interfaces*, **63**, pp. 1203–1211.

99. Azizian, J., Tahermansouri, H., Chobfrosh Khoei, D., Yadollahzadeh, K., and Delbari, A.S. (2012). Microwave-induced chemical functionalization of carboxylated multi-walled nanotubes with 2,3-diaminopyridine, *Fuller., Nanotub., Carbon Nanostruct.*, **20**, pp. 183–190.

100. Tahermansouri, H., and Abedi, E. (2014). One-pot functionalization of short carboxyl multi-walled carbon nanotubes with ninhydrin and thiourea via microwave and thermal methods and their effect on MKN-45 and MCF7 cancer cells, *Fuller., Nanotub., Carbon Nanostruct.*, **22**, pp. 834–844.

101. Tahermansouri, H., Azadfar, M., and Keshel, S. H. (2013). Functionalization and toxicity effect of multi-walled carbon nanotubes with urea derivatives via microwave irradiation, *Fuller., Nanotub., Carbon Nanostruct.*, **21**, pp. 568–578.

102. Tahermansouri, H., and Ghobadinejad, H. (2013). Functionalization of short multi-walled carbon nanotubes with creatinine and aromatic aldehydes via microwave and thermal methods and their influence on the MKN45 and MCF7 cancer cells, *C. R. Chim.*, **16**, pp. 838–844.

103. Davarpanah, M., Maghrebi, M., and Hosseinipour, E. (2014). Microwave-induced high surface functionalization of multi-walled carbon nanotubes for long-term dispersion in water, *Appl. Phys. A*, **115**, pp. 167–175.

104. Bahr, J. L., and Tour, J. M. (2001). Highly functionalized carbon nanotubes using in situ generated diazonium compounds, *Chem. Mater.*, **13**, pp. 3823–3824.

105. Brunetti, F. G., Herrero, M. A., Munoz, J. D. M., Giordani, S., Diaz-Ortiz, A., Filippone, S., Ruaro, G., Meneghetti, M., Prato, M., and Vazquez, E. (2007). Reversible microwave-assisted cycloaddition of aziridines to carbon nanotubes, *J. Am. Chem. Soc.*, **129**, pp. 14580–14581.

106. Dyke, C. A., and Tour, J. M. (2003). Solvent-free functionalization of carbon nanotubes, *J. Am. Chem. Soc.*, **125**, pp. 1156–1157.

107. Rubio, N., Herrero, M. A., de la Hoz, A., Meneghetti, M., Prato, M., and Vazquez, E. (2010). Versatile microwave-induced reactions for the multiple functionalization of carbon nanotubes, *Org. Biomol. Chem.*, **8**, pp. 1936–1942.

108. Karousis, N., Ichihashi, T., Yudasaka, M., Iijima, S., and Tagmatarchis, N. (2010). Microwave-assisted functionalization of carbon nanohorns via [2+1] nitrenes cycloaddition, *Chem. Commun.*, **47**, pp. 1604–1606.

109. Amiri, A., Maghrebi, M., Baniadam, M., and Zeinali Heris, S. (2011). One-pot, efficient functionalization of multi-walled carbon nanotubes with diamines by microwave method, *Appl. Surf. Sci.*, **257**, pp. 10261–10266.

110. Amiri, A., Zardini, H. Z., Shanbedi, M., Maghrebi, M., Baniadam, M., and Tolueinia, B. (2012). Efficient method for functionalization of carbon nanotubes by lysine and improved antimicrobial activity and water-dispersion, *Mater. Lett.*, **72**, pp. 153–156.

111. Ghiadi, B., Baniadam, M., Maghrebi, M., and Amiri, A. (2013). Rapid, one-pot synthesis of highly-soluble carbon nanotubes functionalized by L-arginine, *Russ. J. Phys. Chem. A*, **87**, pp. 649–653.

112. Zardini, H. Z., Amiri, A., Shanbedi, M., Maghrebi, M., and Baniadam, M. (2012). Enhanced antibacterial activity of amino acids-functionalized multi walled carbon nanotubes by a simple method, *Colloid Surf., B*, **92**, pp. 196–202.

113. Mallakpour, S., Abdolmaleki, A., and Rostami, M. (2014). Hybrid S-valine functionalized multi-walled carbon nanotubes/poly(amid-imide) nanocomposites containing trimellitimidobenzene and 4-hydroxyphenyl benzamide moieties: preparation, processing, and thermal properties, *J. Mater. Sci.*, **49**, pp. 7445–7453.

114. Mallakpour, S., and Zadehnazari, A. (2013). Functionalization of multi-wall carbon nanotubes with amino acid and its influence on the properties of thiadiazol bearing poly(amide-thioester-imide) composites, *Synth. Met.*, **169**, pp. 1–11.

115. Mamane, V., Mercier, G., Abdul Shukor, J., Gleize, J., Azizan, A., Fort, Y., and Vigolo, B. (2014). Chemi- vs physisorption in the radical functionalization of single-walled carbon nanotubes under microwaves, *Beilstein J. Nanotechnol.*, **5**, pp. 537–545.

116. Liu, J., Zubiri, M. R. I., Vigolo, B., Dossot, M., Fort, Y., Ehrhardt, J.-J., and McRae, E. (2007). Efficient microwave-assisted radical functionalization of single-wall carbon nanotubes, *Carbon*, **45**, pp. 885–891.

117. Delgado, J. L., de la Cruz, P., Langa, F., Urbina, A., Casado, J., and Lopez Navarrete, J. T. (2004). Microwave-assisted sidewall functionalization of single-wall carbon nanotubes by Diels-Alder cycloaddition, *Chem. Commun.*, pp. 1734–1735.
118. Georgakilas, V., Kordatos, K., Prato, M., Guldi, D. M., Holzinger, M., and Hirsch, A. (2002). Organic functionalization of carbon nanotubes, *J. Am. Chem. Soc.*, **124**, pp. 760–761.
119. Li, J., and Grennberg, H. (2006). Microwave-assisted covalent sidewall functionalization of multiwalled carbon nanotubes, *Chem. Eur. J.*, **12**, pp. 3869–3875.
120. Wang, Y., Iqbal, Z., and Mitra, S. (2005). Rapidly functionalized, water-dispersed carbon nanotubes at high concentration, *J. Am. Chem. Soc.*, **128**, pp. 95–99.
121. Calcio Gaudino, E., Tagliapietra, S., Martina, K., Barge, A., Lolli, M., Terreno, E., Lembo, D., and Cravotto, G. (2014). A novel SWCNT platform bearing DOTA and [small beta]-cyclodextrin units. "One shot" multidecoration under microwave irradiation, *Org. Biomol. Chem.*, **12**, pp. 4708–4715.
122. Bayazit, M. K., and Coleman, K. S. (2012). Probing the selectivity of azomethine imine cycloaddition to single-walled carbon nanotubes by resonance raman spectroscopy, *Chem. Asian J.*, **7**, pp. 2925–2930.
123. Bayazit, M. K., Celebi, N., and Coleman, K. S. (2014). A theoretical and experimental exploration of the mechanism of Â microwave assisted 1,3-dipolar cycloaddition of pyridinium ylides to single walled carbon nanotubes, *Mater. Chem. Phys.*, **145**, pp. 99–107.
124. Criado, A., Vizuete, M., Gomez-Escalonilla, M. J., Garcia-Rodriguez, S., Fierro, J. L. G., Cobas, A., Pena, D., Guitian, E., and Langa, F. (2013). Efficient cycloaddition of arynes to carbon nanotubes under microwave irradiation, *Carbon*, **63**, pp. 140–148.
125. Economopoulos, S. P., Pagona, G., Yudasaka, M., Iijima, S., and Tagmatarchis, N. (2009). Solvent-free microwave-assisted Bingel reaction in carbon nanohorns, *J. Mater. Chem.*, **19**, pp. 7326–7331.
126. Umeyama, T., Tezuka, N., Fujita, M., Matano, Y., Takeda, N., Murakoshi, K., Yoshida, K., Isoda, S., and Imahori, H. (2007). Retention of intrinsic electronic properties of soluble single-walled carbon nanotubes after a significant degree of sidewall functionalization by the Bingel reaction, *J. Phys. Chem. C*, **111**, pp. 9734–9741.
127. Xu, Y., Wang, X., Tian, R., Li, S., Wan, L., Li, M., You, H., Li, Q., and Wang, S. (2008). Microwave-induced electrophilic addition of single-walled

carbon nanotubes with alkylhalides, *Appl. Surf. Sci.,* **254**, pp. 2431–2435.

128. Tian, R., Wang, X., Xu, Y., Li, S., Wan, L., Li, M., and Cheng, J. (2009). Microwave-assisted functionalization of single-walled carbon nanotubes with 3-chloropropene, *J. Nanopart. Res.,* **11**, pp. 1201–1208.
129. Chen, Y., and Gu, H. (2009). Microwave assisted fast fabrication of Fe_3O_4-MWCNTs nanocomposites and their application as MRI contrast agents, *Mater. Lett.,* **67**, pp. 49–51.
130. Nadagouda, M. N., and Varma, R. S. (2007). Microwave-assisted synthesis of crosslinked poly(vinyl alcohol) nanocomposites comprising single-walled carbon nanotubes, multi-walled carbon nanotubes, and buckminsterfullerene, *Macromol. Rapid Commun.,* **28**, pp. 842–847.
131. Huang, S., Liu, T., Zhang, W.-D., Tjiu, W.C., He, C., and Lu, X. (2010). Grafting polyamide 6 onto multi-walled carbon nanotubes using microwave irradiation, *Polym. Interfaces,* **59**, pp. 1346–1349.
132. Mahmoodian, H., Moradi, O., and Shariatzadeh, B. (2014). Grafting chitosan and polyHEMA on carbon nanotubes surfaces: "grafting to" and "grafting from" methods, *Int. J. Biol. Macromol.,* **63**, pp. 92–97.
133. Motshekga, S. C., Pillai, S. K., Sinha Ray, S., Jalama, K., and Krause, R. W. M. (2012). Recent trends in the microwave-assisted synthesis of metal oxide nanoparticles supported on carbon nanotubes and their applications, *J. Nanomater.,* **2012**, Article ID 691503 (15 pp).
134. Vukićević, R., Hierzenberger, P., Hild, S., and Beuermann, S. (2010). Functionalization of carbon black nanoparticles with poly(vinylidene fluoride), *J. Polym. Sci. Pol. Chem.,* **48**, pp. 4847–4854.
135. Chang, I. P., Hwang, K. C., Ho, J.-A. A., Lin, C.-C., Hwu, R. J. R., and Horng, J.-C. (2009). Facile surface functionalization of nanodiamonds, *Langmuir,* **26**, pp. 3685–3689.
136. Desai, C., and Mitra, S. (2013). Microwave induced carboxylation of nanodiamonds, *Diamond Relat. Mater.,* **34**, pp. 65–69.
137. Galindo-Hernandez, F., Portales, B., Dominguez, J. M., and Angeles-Beltran, D. (2014). Porosity and fractal study of functionalized carbon nanofibers: effects of the functionalization degree on hydrogen storage capacity, *J. Power Sources,* **269**, pp. 69–80.
138. Pujari, S. P., Scheres, L., Marcelis, A. T. M., and Zuilhof, H. (2014). Covalent surface modification of oxide surfaces, *Angew. Chem., Int. Ed.,* **53**, pp. 6322–6356.
139. Turcheniuk, K., Tarasevych, A. V., Kukhar, V. P., Boukherroub, R., and Szunerits, S. (2013). Recent advances in surface chemistry

strategies for the fabrication of functional iron oxide based magnetic nanoparticles, *Nanoscale,* **5**, pp. 10729–10752.

140. Feng, J., Shan, G., Maquieira, A., Koivunen, M. E., Guo, B., Hammock, B. D., and Kennedy, I. M. (2003). Functionalized europium oxide nanoparticles used as a fluorescent label in an immunoassay for atrazine, *Anal. Chem.,* **75**, pp. 5282–5286.

141. Siddiquey, I. A., Furusawa, T., Hoshi, Y.-I., Ukaji, E., Kurayama, F., Sato, M., and Suzuki, N. (2008). Silica coating of CeO_2 nanoparticles by a fast microwave irradiation method, *Appl. Surf. Sci.,* **255**, pp. 2419–2424.

142. Siddiquey, I. A., Furusawa, T., Sato, M., and Suzuki, N. (2008). Microwave-assisted silica coating and photocatalytic activities of ZnO nanoparticles, *Mater. Res. Bull.,* **43**, pp. 3416–3424.

143. Siddiquey, I. A., Furusawa, T., Sato, M., Bahadur, N. M., and Suzuki, N. (2011). Fabrication of silica coated Al_2O_3 nanoparticles via a fast and facile route utilizing microwave irradiation, *Mater. Chem. Phys.,* **130**, pp. 583–586.

144. Siddiquey, I. A., Furusawa, T., Sato, M., Bahadur, N. M., Uddin, M. N., and Suzuki, N. (2011). A rapid method for the preparation of silica-coated ZrO_2 nanoparticles by microwave irradiation, *Ceram. Int.,* **37**, pp. 1755–1760.

145. Bahadur, N. M., Furusawa, T., Sato, M., Kurayama, F., Siddiquey, I. A., and Suzuki, N. (2011). Fast and facile synthesis of silica coated silver nanoparticles by microwave irradiation, *J. Colloid Interface Sci.,* **355**, pp. 312–320.

146. Bahadur, N. M., Watanabe, S., Furusawa, T., Sato, M., Kurayama, F., Siddiquey, I. A., Kobayashi, Y., and Suzuki, N. (2011). Rapid one-step synthesis, characterization and functionalization of silica coated gold nanoparticles, *Colloids Surf., A,* **392**, pp. 137–144.

147. Stutz, C., Bilecka, I., Thunemann, A. F., Niederberger, M., and Borner, H. G. (2012). Superparamagnetic core-shell nanoparticles as solid supports for peptide synthesis, *Chem. Commun.,* **48**, pp. 7176–7178.

148. Park, J. C., Gilbert, D. A., Liu, K., and Louie, A. Y. (2012). Microwave enhanced silica encapsulation of magnetic nanoparticles, *J. Mater. Chem.,* **22**, pp. 8449–8454.

149. Posati, T., Melucci, M., Benfenati, V., Durso, M., Nocchetti, M., Cavallini, S., Toffanin, S., Sagnella, A., Pistone, A., Muccini, M., Ruani, G., Zamboni, R. (2014). Selective MW-assisted surface chemical tailoring of hydrotalcites for fluorescent and biocompatible nanocomposites, *RSC Adv.,* **4**, pp. 11840–11847.

150. Massaro, M., Riela, S., Cavallaro, G., Gruttadauria, M., Milioto, S., Noto, R., and Lazzara, G. (2014). Eco-friendly functionalization of natural halloysite clay nanotube with ionic liquids by microwave irradiation for Suzuki coupling reaction, *J. Organomet. Chem.*, **749**, pp. 410–415.

151. Bahadur, N. M., Furusawa, T., Sato, M., Kurayama, F., and Suzuki, N. (2010). Rapid synthesis, characterization and optical properties of TiO2 coated ZnO nanocomposite particles by a novel microwave irradiation method, *Mater. Res. Bull.*, **45**, pp. 1383–1388.

152. Kim, Y.-S., Rai, P., and Yu, Y.-T. (2013). Microwave assisted hydrothermal synthesis of Au@TiO2 core-shell nanoparticles for high temperature CO sensing applications, *Sens. Actuators B-Chem.*, **186**, pp. 633–639.

153. Wu, W., Njoki, P. N., Han, H., Zhao, H., Schiff, E. A., Lutz, P. S., Solomon, L., Matthews, S., and Maye, M. M. (2011). Processing core/alloy/shell nanoparticles: tunable optical properties and evidence for self-limiting alloy growth, *J. Phys. Chem. C*, **115**, pp. 9933–9942.

154. Njoki, P. N., Wu, W., Lutz, P., and Maye, M. M. (2013). Growth characteristics and optical properties of core/alloy nanoparticles fabricated via the layer-by-layer hydrothermal route, *Chem. Mater.*, **25**, pp. 3105–3113.

155. Zhi, C., Dai, M., Masato, M. M., and Yuji, W. (2013). Facile synthesis of bimetallic Cu/Ag nanoparticles under microwave irradiation and their oxidation resistance, *Nanotechnology*, **24**, p. 265602.

156. Garcia, S., Anderson, R. M., Celio, H., Dahal, N., Dolocan, A., Zhou, J., and Humphrey, S. M. (2013). Microwave synthesis of Au-Rh core-shell nanoparticles and implications of the shell thickness in hydrogenation catalysis, *Chem. Commun.*, **49**, pp. 4241–4243.

157. Liu, S., Lu, F., Liu, Y., Jiang, L.-P., and Zhu, J.-J. (2013). Synthesis, characterization, and electrochemical applications of multifunctional Fe_3O_4@C–Au nanocomposites, *J. Nanopart. Res.*, **15**, p. 1331.

158. Fashedemi, O. O., Julies, B., and Ozoemena, K. I. (2013). Synthesis of Pd-coated FeCo@Fe/C core-shell nanoparticles: microwave-induced 'top-down' nanostructuring and decoration, *Chem. Commun.*, **49**, pp. 2034–2036.

159. Zedan, A., Abdelsayed, V., Mohamed, M., and El-Shall, M. S. (2013). Rapid synthesis of magnetic/luminescent (Fe3O4/CdSe) nanocomposites by microwave irradiation, *J. Nanopart. Res.*, **15**, pp. 1–8.

160. Hu, Y., Qian, H., Liu, Y., Du, G., Zhang, F., Wang, L., and Hu, X. (2011). A microwave-assisted rapid route to synthesize ZnO/ZnS core-shell

nanostructures via controllable surface sulfidation of ZnO nanorods, *CrystEngComm,* **13**, pp. 3438–3443.

161. Soltani, N., Saion, E., Yunus, W. M. M., Erfani, M., Navasery, M., Bahmanrokh, G., and Rezaee, K. (2014). Enhancement of visible light photocatalytic activity of ZnS and CdS nanoparticles based on organic and inorganic coating, *Appl. Surf. Sci.,* **290**, pp. 440–447.

162. Kao, J., Thorkelsson, K., Bai, P., Rancatore, B. J., and Xu, T. (2013). Toward functional nanocomposites: taking the best of nanoparticles, polymers, and small molecules, *Chem. Soc. Rev.*, **42**, p. 2654.

163. Luo, H. L., Sheng, J., and Wan, Y. Z. (2008). Preparation and characterization of TiO2/polystyrene core-shell nanospheres via microwave-assisted emulsion polymerization, *Mater. Lett.,* **62**, pp. 37–40.

164. Nan, A., Turcu, R., Craciunescu, I., Pana, O., Scharf, H., and Liebscher, J. (2009). Microwave-assisted graft polymerization of ε-caprolactone onto magnetite, *J. Polym. Sci. Pol. Chem.,* **47**, pp. 5397–5404.

165. Nan, A., Turcu, R., and Liebscher, J. (2012). Magnetite-polylactic acid core–shell nanoparticles by ring-opening polymerization under microwave irradiation, *J. Polym. Sci. Pol. Chem.,* **50**, pp. 1485–1490.

166. Zhang, Y., Liu, R., Hu, Y., and Li, G. (2009). Microwave heating in preparation of magnetic molecularly imprinted polymer beads for trace triazines analysis in complicated samples, *Anal. Chem.,* **81**, pp. 967–976.

167. Chien, W.-C., Yu, Y.-Y., Chen, P.-K., and Yu, H.-H. (2011). Microwave-assisted synthesis and characterization of poly(acrylic)/*SiO2–TiO2 core–shell* nanoparticle hybrid thin films, *Thin Solid Films,* **519**, pp. 5274–5279.

168. You, L.-J., Xu, S., Ma, W.-F., Li, D., Zhang, Y.-T., Guo, J., Hu, J. J., and Wang, C.-C. (2012). Ultrafast hydrothermal synthesis of high quality magnetic core phenol-formaldehyde shell composite microspheres using the microwave method, *Langmuir,* **28**, pp. 10565–10572.

169. Wang, Y., Su, P., Wang, S., Wu, J., Huang, J., and Yang, Y. (2013). Dendrimer modified magnetic nanoparticles for immobilized BSA: a novel chiral magnetic nano-selector for direct separation of racemates, *J. Mater. Chem. B,* **1**, pp. 5028–5035.

170. Sommer, W. J., and Weck, M. (2007). Facile functionalization of gold nanoparticles via microwave-assisted 1,3 dipolar cycloaddition, *Langmuir,* **23**, pp. 11991–11995.

171. Benyettou, F., Guénin, E., Lalatonne, L., and Motte, L. (2011). Microwave assisted nanoparticle surface functionalization, *Nanotechnology,* **22**, p. 055102.

172. Bertorelle, F., Wilhelm, C., Roger, J., Gazeau, F., Ménager, C., and Cabuil, V. R. (2006). Fluorescence-modified superparamagnetic nanoparticles: intracellular uptake and use in cellular imaging, *Langmuir,* **22**, pp. 5385–5391.

173. Nehlig, E., Motte, L., and Guénin, E. (2013). Nano-organocatalysts synthesis: Boc vs Fmoc protection, *Catal. Today,* **208**, pp. 90–96.

174. Bolley, J., Lalatonne, Y., Haddad, O., Letourneur, D., Soussan, M., Perard-Viret, J., and Motte, L. (2013). Optimized multimodal nanoplatforms for targeting avb_3 integrins, *Nanoscale,* **5**, pp. 11478–11489.

175. Guénin, E., Hardouin, J., Lalatonne, Y., and Motte, L. (2012). Bivalent alkyne-bisphosphonate as clickable and solid anchor to elaborate multifunctional iron oxide nanoparticles with microwave enhancement, *J. Nanopart. Res.,* **14**, pp. 1–10.

176. Bolley, J., Guenin, E., Lievre, N., Lecouvey, M., Soussan, M., Lalatonne, Y., and Motte, L. (2013). Carbodiimide versus click chemistry for nanoparticle surface functionalization: a comparative study for the elaboration of multimodal superparamagnetic nanoparticles targeting avb^3 integrins, *Langmuir,* **29**, pp. 14639–14647.

177. Toulemon, D., Pichon, B. P., Leuvrey, C., Zafeiratos, S., Papaefthimiou, V., Cattoen, X., and Bégin-Colin, S. (2013). Fast assembling of magnetic iron oxide nanoparticles by microwave-assisted copper(I) catalyzed alkyne-azide cycloaddition (CuAAC), *Chem. Mater.,* **25**, pp. 2849–2854.

178. Abel, B., and Aslan, K. (2012). Surface modification of plasmonic nanostructured materials with thiolated oligonucleotides in 10 seconds using selective microwave heating, *Ann. Phys.,* **524**, pp. 741–750.

179. Grell, T. A. J., Alabanza, A. M., Gaskell, K., and Aslan, K. (2013). Microwave-accelerated surface modification of plasmonic gold thin films with self-assembled monolayers of alkanethiols, *Langmuir,* **29**, pp. 13209–13216.

180. Vance, S. A., and Sandros, M. G. (2014). Zeptomole detection of C-reactive protein in serum by a nanoparticle amplified surface plasmon resonance imaging aptasensor, *Sci. Rep.,* **4**, p. 5129.

Index

2-ethyl-2-oxazoline 91–95, 97–98, 105
  polymerizations of 95
2-oxazine 91–92
2-oxazoline 74, 84, 91–98, 106

ability 4, 7–8, 104, 121, 127, 132, 203–4, 266, 270, 277, 292, 370–73
absence 59, 76–77, 89–90, 143, 204, 244, 276, 310, 317, 332–33, 404
accessibility 181–83, 267
acetonitrile 3–4, 93, 95–96, 98, 105
acid, glycolic 84, 88–89
acid functionalization, microwave-assisted postsynthesis 186
acidity 177–78, 180, 184, 186
acid sites 178, 182, 184, 314
acid treatment 179, 181–82, 385
activation energy 10, 12, 46–47, 55, 243, 271
adsorption 167, 175, 233, 237–38, 311, 377
agents, sulfonating 167, 183, 186
agglomeration 277, 285, 287
aging temperature 174–75
alcohols, vinyl 87, 90
alkenes 375–76
alloy MNPs 278, 281, 285
alloys 147, 152, 154, 156, 282
alumina 6–7, 61, 119, 122–23, 133–34, 244
  sample of 6–7
amide-imide 99–100, 384
ammonia 238, 390–92
applications
  catalytic 237, 285
  commercial 135, 230
  single-mode 121
aqueous solution 166, 176, 273–74, 276, 311, 315, 325, 333, 335, 339, 397
arc discharge method 196, 200–201
Arrhenius plot 46–47, 63
as-synthesized materials 333
atoms 57, 59, 61, 264, 279–80, 282
atom transfer radical polymerization (ATRP) 75, 78, 376, 388
ATRP *see* atom transfer radical polymerization
autoclaving 168–69, 171, 174, 179–80
average diameter, 277, 283, 334–35, 399
average velocity, 63–65
azides, 102–3, 386

band gap 308, 311, 317, 319–21, 327, 329, 331, 335, 340

benzene 3, 86, 202, 216, 326, 329–31, 336
BET *see* Brunauer–Emmett–Teller
beta zeolite 180–81, 380
bimetallic MNPs 278, 282, 287, 290
bovine serum albumin (BSA) 104, 376, 398
Brønsted acid sites 165, 178, 180–81, 183
Brunauer–Emmett–Teller (BET) 169–70, 172, 210, 333, 335
BSA *see* bovine serum albumin
bulk metals 37, 151, 264
bulk temperature 9, 11, 15, 17, 19–20

capping agents 199, 264, 273–75, 277, 282, 292
carbon 122, 153, 202, 204, 212, 217, 289–90, 328, 382, 394
  amorphous 204, 207, 209
carbon-based materials 204, 370
carbon-based nanomaterials 373, 382
carbon dot 199
carbon flakes 207–8
carbon nanohorns 212, 382
carbon nanomaterials 388
carbon nanotubes 145, 198, 200, 202, 204, 206, 208, 210, 212, 214, 216, 218, 373
  multiwalled, 200, 286, 382
catalysis 90, 167, 175, 230, 236, 251, 283, 313, 316, 373, 377
catalysts, metal 201–2, 207
catalytic activity 83, 90, 185, 285, 381, 393
catalytic properties 166, 278, 282, 287–88
cation exchange 163, 165–67
cationic 91, 94, 167, 186, 276
cationic exchange capacity (CEC) 165, 168–69, 172–73
cationic ring-opening polymerization (CROP) 91, 95
cavity 41, 119, 214, 232, 268
  single-mode 39, 42–43, 66
CB *see* conduction band
CBD *see* chemical bath deposition
CEC *see* cationic exchange capacity
cells 95, 339–41, 344, 346
CEM discover bench-top microwave reactor 87, 90
CEM discover microwave reactor 76, 78, 82, 84
ceramic materials 117, 126, 130, 133–35
ceramic products 115–16, 130
ceramics 61, 115–22, 124–28, 130–32, 134, 164
  microwave processing of 123, 135
ceramics processing 117, 123, 135
change, microwave electromagnetic field energies 47
channels 56, 181, 232–33
character, living 94, 96–98
characterization 100, 168, 175, 181, 184, 242
charged particles 39, 55, 57, 64, 70, 270
charges 33–34, 36
chemical bath deposition, microwave-assisted 313
chemical bath deposition (CBD) 338
chemical bonds 240, 267
chemical processes 2, 6, 9
chemical properties 165, 263, 307, 316, 393
chemical reactions
  microwave-enhanced 72
  rate of 57, 62–63
chemical research 197

chemical synthesis 124, 239, 241, 243, 249, 371, 403
  microwave-assisted 277
chemical synthesis purposes 197
chemical transformations 288
chemistry
  carbodiimide 401
  combinatorial 372
  microwave-assisted 286, 292
cladding, composite 146
clay products, structural 116
clays 116, 133, 164–65, 167, 169, 171, 173, 186
click chemistry 74, 102–3, 401
closed-vessel microwave chemistry workstation 321
CNT bundles 207–8
CNT fibers, purified 207–9
CNTs
  carboxylated 384
  dispersion of 205
  solubility of 196
  synthesis of 200–201
coating 146, 149–50, 389–90, 400
  microwave-processed 149–50
  processed 145, 149–50
colors 21, 48, 53
compacts, microwave-sintered 153–54
complex conjugates 34–36
composite materials 166, 287, 396
composites 11, 116, 142–50, 152, 154, 156, 384–85, 387
  microwave-cured 148
  microwave-processed 143
compositions 119–20, 126, 129–30, 146, 151, 165, 174, 280, 283–84, 396
conditions 22–23, 35, 128, 179, 247, 251, 271, 275, 393, 404
  conventional 9–11, 13–14, 19–21, 23–25, 73, 91, 241
conduction band (CB) 181, 308, 337
contributions 39, 105–7, 180, 370
controlled radical polymerizations (CRPs) 74–75, 106, 403
control VACNTs 206
convective heating 265, 271–72, 286, 288
conventional sintering 149, 151–53, 156
copolymerizations 94–95, 98, 100
  microwave-supported 79, 101
copolymers 88, 92, 95, 98, 102–3
copper 38, 151, 153, 204
copper powders 37, 48–49, 52–54
copper powder sample 50–54
core–shell MNPs 278–80
covalent functionalization 196, 210, 212
CROP
  see cationic ring-opening polymerization
  microwave-supported, 93, 95–96
CRPs *see* controlled radical polymerizations
crystalline materials 169, 254, 256, 393
crystallinity 172, 179, 185, 238, 250, 252, 318
crystallization 116, 127–29, 143, 150, 244, 248, 318
crystal structures 56, 243, 279, 282, 321–22
CS-beta 184–85
CuO, conventional heating of 46
cycloaddition 211, 386
  1,3-dipolar 210–11

data, experimental 75–77
decrystallization 39, 69, 71–72
deformation 155
degradation rate 310–11, 317, 324, 334
delamination 168, 170, 173, 186

densification 130–31, 134, 145, 153
deoxidation 51, 54–55
deposition, chemical vapor 196, 202
dielectric constants 3–4, 55, 118, 266, 271
dielectric heating 73–74, 83, 86, 99–100, 182, 242, 255, 265, 270, 275, 280, 282–83, 287, 290
dielectric heating mechanism 2–3, 10, 16
dielectric loss 37, 134, 266, 270, 371
dielectric loss factor 118–19, 121, 129
dielectric materials 33–34, 37, 55, 59
dielectric properties 3, 5, 118–20, 135, 142, 246–47, 347, 371, 380
diffusion 151–52, 154, 177, 251, 284
dipoles 2, 12, 56, 203, 240–42, 266, 270, 306
direction 58, 60, 64, 330
direct microwave-assisted synthesis of unsupported MNPs 273–291
dispersion relation 34–35, 57, 60
displacement 33–34, 59
disturbances 62, 68, 70
DM values 77, 79, 83–84, 86–87, 98, 105
drug adsorption/delivery 236, 238–39, 251
DSSCs *see* dye-sensitized solar cells
dye-sensitized solar cells (DSSCs) 325, 337, 342, 344–46

ε caprolactone 81 87, 101
  ring-opening polymerization of 86, 104
economy, global 115–16, 229, 231
EDX data 51, 54, 148
E-field heating 45–46
electrical field 240
electric field 4, 32–33, 35, 37, 56, 59–60, 66, 120, 132, 151, 203–4, 248, 256, 270
electrodes 201, 323, 345
electromagnetic field 38, 48, 54–55, 57, 118, 142, 156, 164, 240
  alternating 33, 56–57, 59
electromagnetic radiation 2, 4, 239–40, 270
electromagnetic waves 5, 10, 32, 37, 48, 55, 57, 61, 207
electrons 55, 58, 60–61, 204, 270, 308, 337–38
emulsion polymerizations 75, 79–80
energy, kinetic 55, 61, 64–66, 68
energy consumption 230, 256–57
energy dissipation 34–35, 119
energy requirements 257
energy savings 130–31
energy transfer mechanism on microwave processing 55–69
energy transfers 1, 35–36, 242, 271
enyl-2-oxazoline 92–94
enzymes 381
ethylene glycol 87–88, 103, 272, 280, 288–89, 291–93, 318, 376, 393–94
exchanged samples 177–78
excitation 10, 58, 240–41
exposure 53, 252, 255, 275–76
exposure of metal powders to microwave fields results 272
extrinsic 122–23, 130

facets 267, 315–16
firing 115, 130–32
formaldehyde 217, 320, 397
frameworks 106, 181, 229–31, 233, 238, 248
free energy, standard 44–45
free energy of formation of oxides 44–45
fullerenes 196, 199, 384
function, carboxylic 376, 384, 388
functional groups 210, 212, 271, 274–75, 292, 321, 375, 380

gases 10, 49, 130, 152, 233, 237–38, 285, 328, 335
gas heating 115–16, 130
gas sensors 313, 324–26, 329
geometries 119, 131–32
glass-ceramic glazes 150
  microwave-processed 151
glass-ceramics 116, 127–29
glasses
  lithium disilicate 119
  quartz 3, 6, 8
glass transition temperatures 98, 127
glycerol 183, 185, 280, 329, 380
grafting 376, 381, 392, 399–400, 404
graphene 199, 213
graphitic materials 218
green chemistry 74, 164
  microwave-assisted 98
ground state (GS) 10, 12–13
groups of hypotheses 9
GS *see* ground state
guest molecules 236, 251

hardness, higher 149, 151, 153
heating
  internal 38, 98
  rapid 38, 124, 126, 128, 134, 288, 306, 347
heating rate 3, 6, 10–11, 14, 121, 132
heating temperature 150, 199, 216
heat treatments 127–28, 130, 202, 284
hectorites 163–65, 168–70
  synthesis of 165, 168–69
H-field 46
H-field heating 42–43, 45–47
highlight 98–99, 101, 103, 267, 307
high-temperature applications 130–33, 156
high-temperature microwave processing 130
hybrid heating 122–23
hydrocalumites 163, 165, 174–75
  synthesis of 166, 175
hydrogels 93–94
hydrolysis 149, 176, 178, 386–87
hydrotalcites 163–65, 174–75
hydrotalcite synthesis 165
hydrothermal method 320, 322, 326
  conventional 320, 324, 331
  facile microwave 328
  microwave-assisted 335
hydrothermal synthesis 320, 393
hydrothermal treatment 168–69, 175, 328
  microwave-assisted 314–15

illumination, sun 339–41
imaginary parts 4, 32, 34, 36–37
immobilization 285, 375, 378–79, 398
increasing sintering temperature 154, 156
inorganic nanoparticles 370, 389, 393, 396

intense microwave irradiations 55, 64, 70
intense microwaves 63, 65
interactions
  dipole 240
  electronic 279, 282, 393
  microwave-material 119, 124, 156
$InVO_4$ nanocrystal photocatalysts 322–23
ion exchange 166–67, 175–76
ionic conduction 164, 203, 240, 266, 268, 270, 272, 288, 371
ionic liquids 74, 98–100, 243, 270, 342, 392
  room-temperature 292–93
iron oxide NPs 391, 400, 402
irradiation times, prolonged 87–89
isobutene 183, 185, 380

Joule heating 37, 203–4

kinetic investigation 89, 94–97
kinetics
  chemical 249
  first-order 89, 94–98

lattices 56–57, 60, 65, 67, 150, 196
layered double hydroxides (LDHs) 165
layers 5, 125, 164–65, 173, 231–32, 239, 316, 330, 332
  outer 5–6, 13
LDHs *see* layered double hydroxides
Lewis acidity 165–66, 182
L-lactic acid 89–90
L-lactide 90, 397
localized surface plasmon resonance (LSPR) 381
loss factor 4–8
loss tangents 4, 266, 270–71, 371
low-temperature microwave applications 124–26
low-temperature microwave-assisted hydrothermal treatment 313
low-temperature microwave heat 126
low temperatures 123, 309
LSPR *see* localized surface plasmon resonance

MACBD *see* microwave-assisted chemical bath deposition
MACBD method 339–40
macroinitiators 82, 92
magnetic fields 32, 37, 39, 41, 56–57, 69, 236, 242, 306
magnetite 40–42, 394
magnetite samples 40
magnetrons 48, 239, 267–68
Magtrieve™ 16–19, 28
  temperature of 16–17
MARS *see* microwave accelerated reactor system
MASPOS *see* microwave-assisted solid-phase organic synthesis
material chemistry 203, 249, 370
materials science 115, 232, 305, 370–71
material synthesis 239
matrix composites, microwave-sintered copper metal 147
maximum surface temperature 19–20
maximum temperatures 6, 124, 126
Maxwell's equations 32–34, 36–37, 55, 59–60
MB *see* methyl blue
MC *see* microsized composite

MCM-41 288, 377–79
mechanical properties 95, 125–26, 134, 143, 151, 153–54, 156, 196, 202
medium, acidic 177, 180–81, 183–85
mesoporous materials 164, 175, 236, 378, 380
mesoporous silica 370, 377, 379–81
metal ions 233–34, 246, 266, 272, 275, 280–82, 292, 308
metallophthalocyanines 378–79
metal–organic frameworks 230–31, 233, 235
metal–organic frameworks (MOFs) 229–34, 236–38, 244–46, 249, 251
metal precursors 265–67, 271, 273–75, 277, 281, 287
metals, combinations of 278–79
metal tungstate 316
microwave-assisted methodologies 279, 282–83
methyl blue (MB) 310, 313, 317–18
methyl methacrylate 75, 78–81, 99
methyl tosylate 92–95, 98, 105
microsized composite (MC) 144
microspheres 90, 332, 334, 377, 397
  nanostructured 334
  surface of 332
microstructures, uniform 134, 153
microwave absorbers 18, 98
  weak 13, 17
microwave absorption 13, 106, 121, 146, 151, 328
microwave absorption capabilities 154, 266
microwave accelerated reactor system (MARS) 345
microwave activation 9, 13, 29
microwave applications 3, 9, 124
microwave applicators 119
microwave-assisted chemical bath deposition (MACBD) 313
microwave-assisted heating
  application of 73–74
  temperature-controlled 89
microwave-assisted heating in wet MNP synthesis 265
microwave-assisted hydrothermal reaction 345–46
microwave-assisted method 287, 325, 328, 337, 347
microwave-assisted modification of zeolites 175–185
microwave-assisted organic chemistry 27
microwave-assisted oxidation 28
microwave-assisted polyol methods 283–84, 292
microwave-assisted preparation 273, 276, 280, 283
microwave-assisted protocols 76, 283
microwave-assisted reactions 74, 277, 310–11, 343
microwave-assisted solvothermal process 318
microwave-assisted synthesis, green 274
microwave cavity 14, 119, 156, 268
  single-mode 142
microwave conditions 1, 9–11, 13–15, 25–26, 28
microwave dielectric heating 26–27, 182, 267, 270
microwave dielectric heating effects 28
microwave drying cycles 126
microwave effects 9, 27–28, 39, 56, 69, 81, 153
  non-thermal 71
microwave energy, applying 31

microwave equipment 124, 131–32
microwave experiments 11, 17, 21–23
microwave exposure 146, 286, 324
microwave field 53–54, 67, 70, 119, 121, 128, 132
microwave frequencies 37–38, 56, 58, 60, 118, 182
microwave furnace 87, 153–54, 167
microwave-heated transformations 28
microwave heating
  direct 53, 122
  hybrid 122
microwave heating method 144
microwave heating technique 142, 145, 307
microwave H-field heating 39
microwave hydrothermal method 317, 321, 331
microwave-induced plasma (MIP) 204
microwave interactions 48, 118
microwave irradiation methods 275, 284
microwave irradiation power 310–11, 333, 339–41
microwave magnetic fields 34, 69, 71–72
microwave method 288, 345
microwave ovens 4, 39, 168, 171, 269, 306, 314, 317, 320, 323, 346, 404
  domestic 2, 90, 100, 146, 268, 306, 313, 324
microwave perturbations 66
microwave power 16, 40, 49, 51, 66, 68, 80, 101, 126, 146, 268, 320
microwave-processed material 142
microwave processes 18, 135
microwave processing, energy flow diagram of 66–67
microwave processing of materials 120, 156, 186
microwave radiation 2–3, 10, 99, 104, 146–47, 154–55, 319, 331
microwave reactions 164, 315–16
microwave reactors 6, 16, 73, 86, 92–93, 99–100, 104–6, 268–69, 277, 322
  bench-top 95, 97
  single-mode 79, 94, 97
microwaves effect 169, 172
microwave sintering 31, 37, 48, 54, 131, 134–35, 143–45, 147, 149, 152–56, 346
microwave solvothermal synthesis 334
microwave support 75–76, 79, 83, 85, 90, 93–94
microwave susceptors 143, 147
microwave synthesis 16, 272, 285, 347–48
microwave-synthesized orthovandates 324
microwave synthesizer 318, 331, 335
  commercial multimode 318
microwave system 83, 310–11, 329, 332, 339–41, 344, 346
microwave technique 1
microwave technology 7, 135, 148, 164
microwave temperatures 317
microwave treatment 32, 148–49
MIL *see* Materials of Institute Lavoisier
MIL-53 237–38, 246, 250–51, 255–56
  crystallized 251–53
MIL-100 238–39, 248

MIP *see* microwave-induced plasma
MNP syntheses 265, 268–70, 272, 286
  microwave-assisted 265, 267
Mn values 85–87, 90, 105
moderate-temperature applications 127–129
modification, chemical 290
MOF materials 237, 245, 257
MOFs
  see metal–organic frameworks
  carboxylate 234–35
molar ratios 168–69, 171, 378
molecular masses 77, 83, 171, 173
molecular weights 76, 79, 82, 84, 86–87, 89, 93, 98, 103
monolayers 279, 374, 393
monomer conversions 76–77, 83–84, 96–97, 99
monometallic, microwave-synthesized 285
mordenite 163, 177, 180–82
MW and UTS irradiation 251, 254–55, 257
MW-assisted synthesis 197, 199, 204, 245
MW chemistry 214, 218
MW energy 197, 239–40, 244, 252, 369, 371–72, 374, 393, 401
MW energy for surface modification of nanomaterials 372
MW field 204, 240, 242, 248, 255
MW formation 389, 393, 396
MW functionalization 380, 382, 388, 400, 402
MW heating, heating curves of 197–98
MW heating mechanisms 197–98
MW irradiation method 390, 392
MW methodologies 370, 382, 385, 388, 391, 399, 402–3
MW radiation 205, 211, 241, 243, 385, 388
MWs in surface chemistry 370
MW synthesis 239, 382, 392
MW temperature 252–54

NaCl and low-temperature microwave-assisted hydrothermal treatment 313
nanocomposites 8, 165, 289, 395
nanomaterials
  inorganic 196, 392
nanoparticles, magnetic iron core 88–89
nanoparticle synthesis 306
nanoporous materials 236, 377
nanorods 306–7, 309, 325, 343–46
nanosheets 307, 315, 325, 332, 334
nanosized composite (NC) 144
nanotechnology 196, 370–71, 402
nanotubes 207, 307, 341, 344, 382, 386–87
nanowires 306–7, 309, 341, 346
NC *see* nanosized composite
neat reagents, reaction of 15–16, 18
nitroxide-mediated polymerization (NMP) 75, 82
NMP *see* nitroxide-mediated polymerization
nonthermal effect 43, 56, 81, 241, 243, 347
NPs
  inorganic 271, 370, 373, 389–90, 396
  surface of 277, 389, 396, 400, 402
  synthesis of 270, 372, 391

nucleation 116, 127, 243–44, 254–58, 266, 271, 280–81, 288, 339, 371–72, 377
  homogeneous 246, 248, 266

oleic acid 274, 327, 332–33
optical image 207–9
ordered motions 55, 57–58, 62, 65, 69
organic chemistry 73, 369, 372
organic linkers 230, 233–34, 236–37, 339
organic synthesis 74, 81, 102, 104, 165, 239–40
orthovanadates, microwave-synthesized 324
oscillating 4, 66, 240
overheating 1, 14–15
oxidants 16, 18, 205, 216
oxide layers 50–51, 54
oxygen 45, 47, 49–51, 54, 202, 279, 308, 346, 393

particles, powder 38, 61
partition functions 64, 66, 70
PDC *see* polymer-derived ceramic
penetration depth 5–6, 13, 37, 134
peptide synthesis 372
  solid-phase 372
permeability 33, 36–39
permittivity 32–33, 35, 37, 39, 120, 216
perturbations 60, 62–63, 70
phases, metastable 309–10
phase transfer catalysis (PTC) 15, 28
phase transformations 154–55, 322
phase velocity 35, 55, 57, 60, 62
photoanodes 317, 337, 340–42, 344
photocatalytic activities
  high 311, 319
  higher 310, 318, 321, 333, 335
  highest 315–16, 318, 320, 325
  visible light 322
photocatalytic performance 308, 311, 316–17
photocatalytic properties 328–29, 335
polarization, dipolar 203, 270, 371
polyesters 74, 84, 106
  biodegradable 84
polylactide 84
polymer brushes 376
polymer chemistry 7, 73, 105–6, 372
polymer classes 73–74
polymer-derived ceramic (PDC) 215
polymeric materials 8
polymerization rates 76–77, 92, 99
polymerizations
  microwave-supported 75–76, 85, 99, 101, 105
polymer networks 94
polymer syntheses 90, 103, 396
pores, 144, 167, 183, 233–34, 236–38, 251, 378, 380–81
porosity 119, 146–47, 232–33, 236
porous materials 230, 232–33, 374
porous silicon 374, 376
porous silicon surfaces 374–75
potassium persulfate 80–81, 101
powders, composite 144, 147
power levels 86–87, 99, 255, 324
precursors 152, 166, 205, 271, 314, 320–21, 326, 328, 331, 335, 345, 390–91, 393–94
precursor solution 243, 310–11, 341

pressure 14, 69, 196–97, 201, 236, 242, 245–46, 249–50, 306, 346
processes
  low-temperature 124, 127
  polyol 272, 287
processing
  chemical 240
  composite 7, 142
processing temperatures, lower 128, 131
product crystallinity 247, 252–54, 257–58
properties, acidic 163, 167, 175–76, 179
PTC *see* phase transfer catalysis
purification 207, 229, 231, 237
purification of CNTs 205, 217–18

QDs *see* quantum dots
quantum dots (QDs) 311, 337–38, 340–41, 347
quartz tube 201, 207–8

RAFT polymerization 76–77
Raman spectra 206, 209, 211–12
rapid synthesis 244, 246, 250, 333, 345
ratio of oxygen 50–51, 54
raw materials 317, 320–22, 326
reaction conditions 73, 93–94, 244, 246–48, 306
reaction medium 25, 100, 248, 256, 265, 267, 270, 272, 277, 284, 288
reaction mixture, nonstirred 19–20
reaction rates 9, 13–14, 25, 65, 68, 74, 76, 99, 241, 248, 257, 271, 282, 321
reaction schemes 75, 79, 81, 85, 91, 103
reaction temperatures 15, 21, 76, 80, 82, 88, 92, 249, 268, 274, 277, 306, 318, 345, 397
reaction vessel 14, 16–19, 22–23, 104, 204, 253–54, 265, 272, 371
reactors 7, 23, 83–84, 95, 102, 105, 174–75, 239, 241, 268
reagents 9, 12–14, 168, 182, 202, 204, 207, 400
reconstruction 203, 207–9
reducing agents 264, 274, 277
reduction
  chemical 264, 273
  rate of 45–47, 280
$ReFeO_3$, synthesize perovskite-type samples 328
refluxing 174–75, 179–80, 196, 206
reinforcement 143, 147, 150
reproducibility 105, 264–65, 268, 306
RhB, degradation of 317–18, 335
rhodamine 309, 395, 399–401
room temperature 48, 118, 122, 124, 129, 215, 247, 285, 288
rotation 2, 10, 22–23, 25, 56, 62, 203
route
  energy-saving microwave heating 333
  microwave-assisted hydrothermal 315, 345

SAED *see* selected-area electron diffraction
sample holder 123
sample materials 119, 134
samples
  as-synthesized $ZnGa_2O_4$ 335
  ceramic 122
  crystallized as-synthesized 251
  dealuminated 180–82

delaminated 169–70
microwave-assisted
hydrothermal 321
microwaved 177, 181
microwave-dried 126
microwaved ZSM-5 182
microwave-exchanged 178
microwave-irradiated 314
microwave-processed 142
microwave-sintered 151–53
perovskite-type 325
premixed microwave 153
processed 142
sintered 40, 151–52
sulfonated beta 184–85
synthesized 185, 251, 309
sample surface 5, 50–52
sample temperature 40, 123
SAMs *see* self-assembled monolayer
saponites, synthesis of 165, 172
SBUs *see* secondary building units
secondary building units (SBUs) 233–35
selected-area electron diffraction (SAED) 148, 329–30
selective heating 39, 124, 142, 182, 198–99, 214, 255, 271, 372
self-assembled monolayer (SAMs) 338, 402–3
semiconductor nanomaterials 305–48
sensitized solar cells (SSCs) 325, 337, 339, 341–43, 345, 347
sensitizers 337, 339–40, 347
separation 229–30, 237, 307, 309
set temperature 213
shell thickness 279, 392–93, 397
silica 244, 370, 373–74, 378, 380, 389–90
silica shell 389, 391–92
silica surface 377, 398
single-walled carbon nanotubes 200, 288
sintering 50, 52, 66, 116, 130–35, 143, 145–46, 151, 285
sintering behavior of metal powders involving microwave-enhanced chemical reaction 48, 72
sintering temperatures 40, 54, 130, 132, 152, 156
sintering time 143, 151, 156
S-lactide 88–90
SnO_2 nanocrystals 342–43
solar cells 305–7, 313, 329, 337–38
solar energy 308, 336
solid copper tubing 49, 52–54
solvent-free polymerizations 85, 98
solvents 3, 9, 14–15, 86, 98, 196–97, 204, 210, 230, 247, 265–66, 270–73, 277, 371
solvothermal reactions, microwave-assisted 327–28, 342
solvothermal synthesis 306
SPFS *see* surface plasmon fluorescence spectroscopy
SPR *see* surface plasmon resonance
SPRi *see* SPR imaging
SPR imaging (SPRi) 403
SSCs *see* sensitized solar cells
stannous octanoate 85, 87–88, 90, 104
starting materials 94, 213, 218, 271, 309–10, 318, 324–25, 327, 329, 333
step-growth polymerizations 74, 81, 83
Stille coupling, microwave-supported 83–84
stirring 15, 19, 23, 25, 243, 276
storage 229–30, 345

stress 57, 59–60, 62, 64, 147
structures, core–shell 279–82
sulfonation 11, 183, 185
sulfonic acid groups 167, 182–83, 380
superstructures 329–30
support 21, 214, 288, 290, 292–93
solid 17–18, 372
support materials 287–88, 292
support media 285–87
surface, nanomaterial 373, 404
surface area
high 147, 217, 237, 246, 250–51, 322, 334, 342, 374
higher 179, 181, 184, 331, 380
larger 249, 335
surface chemistry 370, 403–4
surface coverage, higher 375–76
surface defects 339–40
surface functionalization 237, 391
surface modification 370, 372–73, 385, 389, 403
surface plasmon fluorescence spectroscopy (SPFS) 402
surface plasmon resonance (SPR) 400, 402
surfactants 99, 196, 317, 346, 397
susceptors 48, 52, 123, 204, 243
SWCNTs, synthesis of 201–2
synthesis
controlled 102, 273
conventional 242, 389
direct 167, 273, 291
efficient 73
facile 288
heated 74, 76, 97
large-scale 105, 197
microwave-heated 88
microwave-supported 79, 82, 86, 97, 104
nanomaterial 370
one-pot 290
solid-phase 372
synthesis and characterization 100
synthesis methods 251–54, 317, 324–25, 381
conventional 248, 257, 274
microwave-assised hydrothermal 320
microwave plasma 152
synthesis parameters 186, 245
synthesis process 201, 377
synthesis temperature 314
synthesis time 163–64, 168, 242, 245
synthesize perovskite-type samples 325
system, microwave digestion 315, 327

techniques 1, 106, 145, 149, 176, 205, 245–46, 339–40, 375, 401–2
hybrid synthesis 246, 250
microwave-assisted hydrothermal 327
Teflon 3–4, 6
telecommunication 2, 197
temperature
approximate 116–17
critical 121–22, 134
elevated 116, 166, 247, 272, 339
highest 17–19, 23, 147
inverse 46
lower 16, 21, 24, 47, 66, 124–25, 128, 145, 197, 202, 277
process 40
temperature acquisition 86, 88, 97, 106
temperature dependence 45–46
temperature distribution 15, 251

temperature gradient 15, 17, 272
temperature homogeneity 13, 25
temperature measurements, proper 14–15, 25
temperature pattern 10–11
temperature profile 16–17, 19, 26
temperature regimes 14, 124, 182
tetrabutylammonium bromide 21, 99–100
thermal effects 9, 14–15, 78, 243, 371, 376, 404
thermal energy 10, 38, 45, 65–66, 125, 142, 239–41, 243, 266
thermal gradients 9, 11, 15, 22–23, 142, 371
thermal microwave effects 39
thermal treatment, intermediate microwave 178
thermovision camera 15–16, 18–21, 23, 25
thiol-ene reaction 92, 94, 102, 377
TiO_2 surface 338–39
toluene 17–18, 86, 215–16, 329, 336, 380
transfer agents, chain 75–77
transition state (TS) 10, 12–13, 57
TS *see* transition state

ultrasound 169, 171–72, 230, 249, 257
UTS conditions 250–51, 254, 257–58
UTS irradiation 246, 250–51, 253–55, 257
UTS power 254, 257
UTS time 254, 257
UV irradiation 315, 318, 324, 326–27
UV lamps 318, 320, 326, 331, 335
UV light irradiation 311, 314–16, 318, 322–23, 329–32, 334, 336

vacuum 3, 33, 55, 57–58, 61, 154, 335
variable-frequency microwave (VFM) 127–29, 206–7
velocity space 62–63, 66
VFM *see* variable-frequency microwave
visible light 284, 308, 314–15, 324, 326–27
visible light irradiation 309–12, 316–17, 319, 321, 323–27, 332–33, 335–36
volumetric heating 1, 5, 10–11, 38, 125, 132–34, 198

water splitting 307, 321–22, 324, 327
wavelengths 1–2, 5, 55, 57, 61, 197, 239, 249, 268, 318, 327–28, 331, 335, 340
waves 32–36, 48, 57, 59, 61, 64, 268
  longitudinal 60, 66, 68
weight loss temperatures, significant 207
whitewares 115–16, 124–25

X-ray diffraction (XRD) 51, 69, 146, 148–50, 155, 168–69, 172, 176, 179, 181, 185, 213, 401
XRD *see* X-ray diffraction
XRD patterns 54, 172, 175, 213, 291

Young's modulus 150–51

zeolite dealumination 166, 183
zeolites
  modification of 186
  production of 244
zeolite structure 180, 182–83
zeolitic materials 242, 244
ZnO nanorods 313–15
$ZnWO_4$, synthesize nanocrystalline 317
$ZnWO_4$ nanoparticles, synthesized 317–18
ZSM-5 163, 180–82, 243

An environmentally friendly book printed and bound in England by www.printondemand-worldwide.com

PEFC Certified

This product is from sustainably managed forests and controlled sources

www.pefc.org

PEFC/16-33-415

This book is made of chain-of-custody materials; FSC materials for the cover and PEFC materials for the text pages.